Sparsity and Its Applications

Sparsity and Its Applications

Edited by
DAVID J. EVANS
Professor of Computing, University of Technology, Loughborough

The right of the University of Cambridge to print and sell all manner of books was granted by Henry VIII in 1534. The University has printed and published continuously since 1584.

CAMBRIDGE UNIVERSITY PRESS
Cambridge
London New York New Rochelle
Melbourne Sydney

Published by the Press Syndicate of the University of Cambridge
The Pitt Building, Trumpington Street, Cambridge CB2 1RP
32 East 57th Street, New York, NY 10022, USA
10 Stamford Road, Oakleigh, Melbourne 3166, Australia

First published 1985

Printed in Great Britain at the University Press, Cambridge

Library of Congress catalogue card number: 84–45714

British Library cataloguing in publication data

Sparsity and its application.
1. Sparse matrices
I. Evans, D. J. (David John) 1928–
512.9'434 QA188

ISBN 0 521 26272 0

CONTENTS

PREFACE

This volume is based on a series of lectures planned and prepared by a group of invited internationally known experts at a Meeting on the Advanced research topic 'Sparsity and its Applications' at the University of Technology, Loughborough, England in April 1983.

Sparse matrix research is an increasingly important subject which plays a central role in large scale scientific computing. The contributions enclosed deal specifically with the design, analysis, implementation and applications of computer algorithms in order for effective mathematical software in Sparse Matrix Technology to be used as tools in science and engineering. In addition, an introductory section is included which briefly surveys the field of sparsity and summarises the major ideas in each contribution.

I wish to thank the contributors for the care they took in preparing their contributions, D. Newham for the organisation of the meeting and to my Secretary Judith Briers for carrying out all the numerous tasks involved in the completion of the manuscript. Finally, support from the Science and Engineering Research Council under the auspices of the Informatics Training Group of the European Economic Communities' Scientific and Technical Research Committee (CREST) is gratefully acknowledged.

July, 1984. DAVID J. EVANS

CONTRIBUTORS

A. BRAMELLER, Power Systems Laboratory, University of Manchester Institute of Science and Technology, P.O. Box 88, Manchester M60 1QD

A. BRANDT, Weizmann Institute of Science, Israel.

IAIN S. DUFF, Computer Science and Systems Division A.E.R.E. Harwell, Oxon OX11 ORA.

DAVID J. EVANS, Department of Computer Studies, University of Technology, Loughborough, Leics. LE11 3TU

A. JENNINGS, Civil Engineering Department, Queens University of Belfast, BT7 1NN, N. Ireland.

S. MCCORMICK, Institute for Computational Studies, Colorado State University, P.O. Box 1852, Fort Collins, Colorado 80522, U.S.A.

N.M. MISSIRLIS, Department of Applied Mathematics, University of Athens, Panepistimiopolis 621, Athens, Greece.

FRANS J. PETERS, Eindhoven University of Technology, 5600 MB Eindhoven, The Netherlands.

J. RUGE, Institute for Computational Studies, Colorado State University, P.O. Box 1852, Fort Collins, Colorado 80522, U.S.A.

FRANZ STEIDLER, Wild HeerBrugg Ltd., CH 9442, Heerbrugg, Switzerland.

R.P. TEWARSON, Department of Applied Mathematics and Statistics State University of New York at Stony Brook, Long Island, New York, 11794, U.S.A.

INTRODUCTION

Sparse Matrix research has been an active area of study for some 15 years or so by groups of applied numerical analysts and computer specialists and scientists and engineers oriented towards computer applications in chemical engineering, structural analysis, geodesy, physics, image processing, etc. The combination of these two dedicated and vigorous groups has greatly contributed to the progress achieved to date and to the many new ideas that have emerged.

The many themes of the subject matter of this topic are still graphy theory to model sparse matrices and their operations, optimal ordering strategies to minimise the storage and solution times for diverse applications, while the competition between direct and iterative methods of solution continues with the introduction of new hybrid methods i.e. semi-direct methods and improved acceleration strategies i.e. preconditioning methods. The consideration of efficient data structures has recently become prominent and the influence of parallelism in algorithm design and their implementation on new machine architectures the next main target to surmount.

For the individual papers Duff [1] presents a good introduction to the data structures required for the storage and manipulation of sparse matrices appropriate to direct methods of solution. Then general techniques for the solution of general sparse unsymmetric equations and their use in mathematical software packages discussed. Brammeller [2] also discusses direct methods, i.e. Gaussian elimination and matrix factorisation in addition to optimum ordering strategies to attain maximum sparsity for problems involving physical networks in electricity and gas supply, traffic flows, etc.

Evans [3] briefly surveys graph theory, sparse elimination methods for symmetric equations and the sparse eigenvalue problem with the main emphasis on iterative methods for the sparse linear systems derived from the discretisation of elliptic partial differential equations. New block S.O.R. methods are introduced and the powerfully convergent preconditioned conjugate gradient methods discussed and compared. Missirlis [4] continues with the preconditioning approach for first order iterative methods and analyses a new extrapolation procedure for successive overrelaxation i.e. ESOR.

The sparse non-linear systems which arise in modelling fluid flow problems, etc. are discussed by Tewarson [5] where by renumbering the variables, algorithms for non-linear Gauss Seidel and Newton's method are derived which minimise run time and storage. In Jennings [6], the transformation, binary search and vector iterative methods for the sparse eigenvalue problem are discussed for the symmetric problems which arise in vibration frequency analyses. Pivotal strategies which are relevant for Gaussian elimination methods applicable to mathematical models of nuclear spectroscoptic resonance theory leading to systems of linear and non-linear ordinary differential equations are discussed by Zlatev [7].

In [8], Brameller discusses the aspect of sparsity in transportation problems where a method to solve the linear programming problem is developed which is more effective than the simplex method while in [9] the effect of sparsity on network changes using diakoptical considerations are considered. An updating strategy where the inverse matrix is not explicitly available is used to derive the new solution and also shown to be applicable for non-linear optimisation problems. In Brandt et al [10], the multigrid method is introduced where the basic concepts, foundations and current strategies are discussed for discrete systems which occur in geodetic problems and also for systems derived from finite element discretisation on irregular grids.

Then in Peters [11] the new concepts of parallel computer architectures are discussed in relation to sparse matrix computations. Two topics, the parallelism which arises from the sparseness and that from the vector inner products are explored. Finally, in [12] and [13] sparsity applications in Geodesy and Photogrammetry are discussed by Steidler. The various sparse matrices which arise from the data handling techniques such as Least Squares and geodetic network adjustments, finite element approximation and photogrammetric bundle adjustment are discussed and suitable solution algorithms such as conjugate gradient, preconditioned methods, Gaussian elimination, banded and nested dissection methods are compared.

David J. Evans
Loughborough, England.

1 DATA STRUCTURES, ALGORITHMS AND SOFTWARE FOR SPARSE MATRICES

Iain S. Duff
A.E.R.E., Harwell, Oxon.

Abstract. We discuss data structures for storing and manipulating sparse matrices, stressing techniques which are appropriate when solving sets of linear equations by direct methods. We consider general techniques for the solution of sparse unsymmetric equations and examine the embodiment of these methods in mathematical software packages.

1. Introduction

Most of the dramtic advances in the manipulation of sparse matrices and the direct solution of sparse linear equations would not have been possible without the development of sophisticated data structures for storing and handling sparse matrices. We discuss some of the principal structures in section 2 indicating why they are appropriate for the solution of equations. Since it is often helpful to think of the matrix pattern as a graph (and indeed the associated data structures are equivalent) we introduce graph theory in section 3.

Many techniques for the solution of general sets of sparse linear equations are now well-established. We describe these in section 4 emphasising methods which have been employed by software packages. Recent developments, including hybrid iterative-direct schemes and the separation of the analysis of structure from numerical pivoting, are considered in section 5.

Finally, the packaging of the previously discussed algorithms is of vital importance if our sparse codes are to have wide acceptance. We discuss such mathematical software considerations in section 6.

Throughout we illustrate our discussion through runs on test matrices. These matrices have occurred in practical problems and are included in the test bed described by Duff et al (1982). In each case we identify the matrix by its order (n) and number of non-zeros (τ).

2. Data structures

One of the main objectives of a sparse matrix data structure is to economize on storage while at the same time facilitating subsequent operations on the matrix. Thus the ideal scheme will be dependent on the particular operations involved. We will concentrate on structures appropriate to the solution of sparse sets of equations

$$A\underline{x} = \underline{b} \tag{2.1}$$

by direct methods based on Gaussian elimination. Here a factorization

$$PAQ = LU, \tag{2.2}$$

where P and Q are permutation matrices and L and U are (unit) lower and upper triangular matrices respectively, precedes the solution of (2.1) by the forward elimination

$$L\underline{y} = P\underline{b}$$

followed by the back substitution

$$U\underline{z} = \underline{y}\,. \tag{2.3}$$

Finally a permutation

$$\underline{x} = Q\underline{z}$$

yields the desired solution $\underline{x}$.

We can divide the storage schemes into static schemes where the structure does not change during the operations on the matrix and dynamic schemes where it does.

The storage of A as a two-dimensional array has the merit of being static and simple but is quite infeasible for matrices of large dimension. It is worth remarking that the meaning of "large" changes with advances in hardware. There are instances where systems of order 500 can be usefully held as a two-dimensional array. The main problem with such a scheme lies as much in its manipulation as in its storage since

$$2n^3/_3 + 0\ (n^2)$$

arithmetic operations are required by the factorization (2.2), and even if tests for zero entries help avoid some real arithmetic the time for such tests will be significant.

At the other extreme, we can choose to store only the entries which are non-zero. This necessitates holding integer indexing information to identify the position of each non-zero entry in the coefficient matrix. This indexing information and indeed any storage over and above that for the nonzeros themselves is termed overhead storage. If the entries can be in arbitrary order, a simple scheme might be to hold the row and column indices of entry A(k) in IRN(k) and ICN(k) respectively, $k = 1, 2 \ldots \tau$ where τ is the number of nonzeros in the matrix A. The benefit of this scheme is its simplicity and friendly interface. The defects are that it is difficult to perform any operations on a matrix held in this format since a scan of all the data ($O(\tau)$) is required to locate the existence (and value) of any entry and the overhead storage is high. For this reason it is not uncommon to permit input to sparse routines using this form but to use a sorted form when performing operations on the data. We revisit this in section 6.

When considering data structures appropriate for the solution of sparse linear systems, it is important to realize that access is not random. Matrix entries are usually required by rows or columns (or both). Additionally, since such access is needed at each step of the elimination, it is important that access is rapid and easy. For example, if a scan of a one-dimensional array of length n was required at each stage, the total work merely to access these entries would be

$$n^2$$

which would, in most instances, dominate the arithmetic required by the factorization.

If we wish access by rows, for example, an obvious scheme might be to hold all the entries for each row contiguously (in, say, A(k), $k1 \le k \le k2$) and to hold the column index in a corresponding integer array (in, say, ICN(k), $k1 \le k \le k2$). All that is then needed for access to any row is a pointer to the start and finish of the row in arrays A and ICN (viz. k1 and k2). If the rows are held in order, that is with entries in row i preceding those in row i + 1 (i = 1, 2 .. n-1), and there is no wasted space, a single pointer array (IPTR, say) can provide both row starts and row ends. The entries in row i are

$$\left.\begin{matrix} & A(jj) \\ \text{in columns} & \\ & ICN(jj)\end{matrix}\right\} \quad IPTR(i) \le jj \le IPTR(i+1)-1$$

This row pointer/column index scheme is widely used by sparse codes. For the matrix in Figure 2.1 the appropriate arrays are given in Figure 2.2. Notice that an extra entry (the n+1 st) has been added to

Figure 2.1 An example matrix

$$\begin{pmatrix} 1.0 & 0 & 0 & 2.0 \\ 0 & 7.0 & 0 & 6.0 \\ 0 & 3.0 & 5.0 & 4.0 \\ 8.0 & 0 & 0 & 0 \end{pmatrix}$$

Figure 2.2 Matrix of Fig. 2.1 stored using row pointer/column index scheme

A:	1.0	2.0	6.0	7.0	3.0	4.0	5.0	8.0
ICN:	1	4	4	2	2	4	3	1
IPTR:	1	3	5	8	9			

the IPTR array so that the end of row n can be identified. Also note that we have not imposed any ordering within the rows.

If the rows are not in sequence, another integer array of length n is required either to delineate the last entry of each row or, more commonly, to record the number of entries in each row. Even with this more general scheme the overhead storage is only

$$\tau + 2n \qquad \text{integers}$$

which is normally substantially less than if both row and column indices are stored.

As a static scheme, this technique often requires the least storage. It is used by Duff et al (1982) to hold the Harwell-Boeing test problems and is also employed, in some form or other, in most sparse matrix packages. At virtually no cost in ease of access, it is possible to economise even further on storage. It was observed by Sherman (1975) that some structures (particularly the factors of matrices from the discretization of partial differential equations) have many rows whose leading non-zeros are in the same columns as the trailing non-zeros of the previous row. Thus a single column index (entry in ICN) can refer to more than one matrix entry (entry in A). All that is required to implement this scheme is another integer array pointing to the start of each row in ICN. With this scheme the entries in row i are

$$A(k) \qquad IPTR(i) \leq k \leq IPTR(i+1) - 1$$

in columns

$$ICN(k) \qquad IPICN(i) \leq k \leq IPICN(i) + \ell - 1$$

where

$$\ell = IPTR(i+1) - IPTR(i) \quad \text{which}$$

may be held in a row length array, LENR(i) say. We give examples of possible savings of this scheme over the row pointer/column index scheme in Table 2.1.

If this method is used as a dynamic storage scheme then one must either leave gaps in the structure to accommodate additions to the data structure (caused by, for example, fill-ins during matrix factorization) or must be prepared to copy a row to the end of the structure and add the required entries. The old copy of the row is no longer required and is now redundant. Fortunately, the worst excesses of data fragmentation can be avoided by a very simple garbage collection mechanism whereby all the active data is moved to one end of the arrays so that no gaps are left between rows. This storage scheme is used in a dynamic fashion

Order	900	130	445
Non-zeros	4322	713	3809
Non-zeros in factors	17798	717	20609
Number of entries in ICN	5580	71	7611

Table 2.1 The use of compressed index scheme

by the Harwell package MA28 (Duff, 1977) inter alia and, in Table 2.2, the overhead due to garbage collection can be seen to be low, particularly if a little extra space ('elbow room') is allocated above that needed to store the factors themselves.

Some matrix algorithms (for example, MA28) require access to the structure both by rows and columns and clearly the row pointer/column index scheme is not well-suited for that purpose. One solution is to hold a completely separate data structure where only the pattern is held by columns. We show in Figure 2.3 the contents of the column pointer/row index (IPC/IRN) arrays appropriate for the matrix of Figure 2.1

Figure 2.3 Column oriented storage for matrix of Fig. 2.1

IRN:	1	4	2	3	3	1	2	3
IPC:	1	3	5	6	9			

An alternative data structure which gives equivalent access by rows and columns and requires no elbow room is a linked-list scheme. Here each entry is stored as a 3-tuple : the real value and pointers to the position in the data structure of the next entry in the row and the next entry in the column. If the matrix is stored in A, the row pointers in NEXTR and the column pointers in NEXTC then a linked list scheme for the Figure 2.1 matrix is shown in Figure 2.4 where arrays IPTR and IPC are needed to identify the first entry in each row and column respectively.

Order	199	292	822
Non-zeros	701	2208	4841
Minimum space (max. no. collections)	.38	1.54	2.17
Min + 2n	.26	1.04	1.79
Much space (very few collections)	.24	.83	1.76

Table 2.2 Times in IBM 370/168 seconds illustrating overhead of garbage collections

Figure 2.4 Linked list scheme for matrix of Fig. 2.1

A	1.0	2.0	3.0	4.0	5.0	6.0	7.0	8.0
NEXTR	2	-1	5	-3	4	-2	6	-4
NEXTC	8	6	-2	-4	-3	4	3	-1
IPTR	1	7	3	8				
IPC	1	7	5	2				

The last entry of each row (column) is flagged negative and we have chosen to hold the negated row (column) index in that link. It is very easy to add or delete entries when using this scheme since, when adding say, the entry can be placed anywhere in storage and only two pointers need be adjusted to incorporate the entry into the data structure. The symmetry of access to rows and columns is aesthetically pleasing and can be useful when coding.

There are two main problems with the linked list data schemes. The first concerns the amount of indirect addressing and concomitant lack of locality of reference. The indirect addressing for every entry will be inefficient on conventional machines and will degrade vectorization on vector processors. The lack of locality of reference can be disastrous on machines with paged virtual memory. The more serious problem involves access to row and column indices, necessary, for example, when adding a multiple of one row to another. These could, of course, be stored along with the links and real values but the overhead storage of $4\tau + 2n$ integers is not acceptable. The MA18 code (Curtis and Reid, 1971) used negative links at the end of each row and column (as we did in Figure 2.4) to identify the row and column indices but this led to an appreciable amount of scanning of links as is illustrated in Table 2.3. Of more minor

Order	199	130	822	900
Non-zeros	701	1282	4841	4380
Number of multiplications (in 1000s)	2.5	2.6	6.3	536
Number of array look-ups to find row and column indices (in 1000s)	68	274	587	61939

Table 2.3 Overhead of scanning links for indices

concern, when using this form of linked list on a machine with short word integers (for example, INTEGER*2 on the IBM), is that the link pointers would have to be full integers unless we wish to restrict the number of nonzeros to the short integer limit (32767 on the IBM). For the row pointer/column index scheme, the use of short integers only imposes this restriction on the order of the matrix.

Link lists are, however, very useful in other contexts. Examples are: a linked list of all the columns in a particular row (perhaps generated temporarily at each pivot step), a list linking the columns of the first nonzero entry in each row and a linked list of all rows or columns with the same number of nonzeros. All these uses of linked lists are employed in sparse codes. The deletion of entries from the middle of such a list is facilitated by holding backward pointers to the previous entry in the chain (a two-way linked list).

Bit maps have been suggested as an ingredient of a sparse storage scheme. Here all or part of the pattern is held as a string of bits, a one indicating the presence of an entry; a zero its absence. A bit map scheme for the matrix of Figure 2.1 is shown in Figure 2.5 where the bit map is held by rows.

Figure 2.5 Bit map storage for matrix of Fig. 2.1

A:	1.0	2.0	7.0	6.0	3.0	5.0	4.0	8.0
B:	1001010101111000							

The bit map scheme requires

$$n^2 \quad \text{bits}$$

of overhead storage which can be high even if we pack many bits to a word. Other defects lie in portability and the difficulties in adding or deleting entries.

A popular scheme for the storage of sparse data in other disciplines (for example, computer science) is that of hash coding. Here a map is set up from the domain of interest, in our case $i \times j, 1 \leqslant i, j \leqslant n$ to the array in which the data is held, A(k), k=1... IA, say Since this is a many-one map some mechanism must be included to handle collisions. Again, the regular way in which we wish to access the data (unlike many computer science applications) militates against such a scheme. There are, however, instances in working with sparse matrices when some form of hash coding can be useful.

We end this section with a discussion of a particular manipulation on sparse data structures which occurs commonly in the factorization (2.2). We consider the addition of a multiple of one row (the pivot row) of the matrix to another row (the non-pivot row) where the matrix is stored in the row pointer/column index scheme. We assume that an integer array, IQ say, containing all positive entries is available and that there is sufficient space to hold temporarily two copies of the pivot row. Assume the second copy has been made. Then a possible scheme is:

(i) Scan the pivot row and for each entry determine its column from the ICN value. Set the corresponding entry in IQ to the negation of the position of that entry within the compact form of the pivot row. Hold the original IQ entry in the appropriate position in ICN.

For each non-pivot row do:

(ii) Scan the non-pivot row. For each column check the corresponding entry in IQ. If it is negative, set it positive and update value of entry in non-pivot row. If it is positive, continue to next entry in non-pivot row.

(iii) Scan pivot row. If the corresponding IQ value is positive, set it negative. If it is negative, then there is fill-in to the non-pivot row. The new entry is added and we continue to the next entry in the pivot row.

(iv) Finally reset IQ using information from both copies of the pivot row.

We illustrate the situation before and after each step in Figure 2.6. *In this figure the upper row for IQ gives the contents of the locations indicated in the lower row. Notice that it is not necessary to keep each row in column order when using this scheme. Conceptually simpler sparse vector additions can result when they are kept in order but the scanning of sparse arrays in parallel and the extra work to keep the columns in order can lead to inefficiencies which we observe in Table 2.4.

Figure 2.6 Addition of two sparse vectors

Before (i)

IQ	i_1	i_2	i_3	i_4	i_5
	j_2	j_1	j_4	j_3	j_5

Pivot row	A	α_1 α_2 α_3	+ second copy.
	ICN	j_1 j_2 j_3	

Non-pivot row	A	β_1 β_2 β_3 β_4
	ICN	j_2 j_3 j_4 j_5

Before (ii)

IQ	-2	-1	i_3	-3	i_5
	j_2	j_1	j_4	j_3	j_5

Pivot row	A	α_1 α_2 α_3	+	α_1 α_2 α_3
	ICN	i_2 i_1 i_4		j_1 j_2 j_3

No change to non-pivot row.

Before (iii)

IQ	2	-1	i_3	3	i_5
	j_2	j_1	j_4	j_3	j_5

Pivot row unchanged

Non-pivot row	A	$\beta_1 + \zeta\,\alpha_2$	$\beta_2 + \zeta\,\alpha_3$	β_3	β_4
	ICN	j_2	j_3	j_4	j_5

Before (iv)

IQ	-2	-1	i_3	-3	i_5
	j_2	j_1	j_4	j_3	j_5

Pivot row unchanged

Non-pivot row	A	$\beta_1 + \zeta\,\alpha_2$	$\beta_2 + \zeta\,\alpha_3$	β_3	β_4	$\zeta\,\alpha_1$
	ICN	j_2	j_3	j_4	j_5	j_1

Order	199	292	822
Non-zeros	701	2208	4841
Number of multiplications (in 1000s)	2.5	28	6.3
Number of comparisons to keep rows/cols in order (in 1000s)	9.6	72	80
Number of compares when adding sparse vectors	3.8	39	16

Table 2.4 Cost of keeping rows (columns) in column (row) order

3. Graph theory

A graph G (V, E) consists of a set of vertices, V and edges E where the members e of E are ordered pairs of vertices (v_i, v_j) from V. If $(v_i, v_j) \varepsilon$ E we say there is an edge in the graph from v_i to v_j. If we identify the vertices with the rows and columns of the matrix and the edges with the non-zero entries, we see that the same data structure can be used for the graph and its associated sparse matrix.

This is a useful equivalence since some things are better visualized in terms of the graph rather than the matrix. This is particularly true for symmetric systems where the corresponding graph does not have directions associated with its edges (e.g. $(v_i, v_j) \equiv (v_j, v_i)$). This undirected graph has three important properties:

(i) It is invariant under symmetric permutations of its associated matrix (only the labelling of the vertices is changed).

(ii) In grid-based problems (for example, discretizations of pdes) there is a correspondence between the underlying grid structure and the corresponding graph (these can be equivalent) so when working with the graph one is in a sense working closer to the underlying problem.

(iii) It is natural to represent cliques (complete subgraphs) in a graph by an index list of the vertices in the clique rather than by storing all the interconnecting edges.

The development of (iii) has led to very efficient techniques for selecting pivots when solving sparse positive definite symmetric systems but the details are beyond the scope of this paper. We do, however, consider such techniques briefly in section 5.

4. Basic techniques

In this section, we will study some basic techniques used in the solution of sparse linear equations when the coefficient matrix is unsymmetric and has no special structure.

For full systems, pivoting (i.e. choice of P and Q in (2.2)) is used to guard against numerical instability in the factorization. The most common form of pivoting is partial pivoting where the largest entry in the column of the reduced matrix is used as pivot. This strategy does not guarantee stability but has been found to work well in practice. The problem with using partial pivoting on sparse systems is that the entries chosen as pivot may be very unsuitable on sparsity grounds, for example, the largest entry in the column may lie in a full row of the reduced matrix. Thus, for sparse systems, it is common to relax the partial pivoting condition and allow entry a_{ij} to be acceptable as a pivot on numerical grounds if

$$|a_{ij}| \geqslant u.\max_k |a_{kj}| \qquad 0 < u \leqslant 1 \qquad (4.1)$$

This is termed threshold pivoting and, if the value of u (often set by the user) is chosen close to zero, there will usually be several entries in column j which are adjudged numerically suitable for use as pivots. This added freedom is used to maintain sparsity in the factors.

The stability of the factorization is determined by the growth in the numerical size of entries of the reduced matrices. Equation (4.1) bounds the growth in the largest matrix element to the factor $(1 + {}^1/u)$ at each step. This factor may seem high when u is small, but it is only a bound and most entries are affected by only very few pivot steps. We show, in Table 4.1, the effect of altering u both on sparsity and stability. From these results and many other experiments, we believe a value of u around 0.1 is a good compromise between stability and sparsity.

Clearly, the amount of fill-in which can be generated at a single pivot step by entry a_{ij} is limited to

$$(r_i - 1) \times (c_j - 1) \qquad (4.2)$$

where r_i and c_j are the number of non-zero entries in row i (the row count) and column j (the column count) of the reduced matrix respectively. Expression (4.2) is often called the Markowitz count since Markowitz (1957) recommended choosing as pivot the non-zero entry which minimized (4.2)

Order	199	541	1454
Non-zeros	701	4285	3377
Non-zeros in factors			
Threshold 0.5	1633	15105	10405
0.25	1539	14754	9690
10^{-1}	1556	14685	9106
10^{-2}	1418	15068	8623
10^{-4}	1391	16848	8568
10^{-6}	1391	18627	8596
10^{-10}	1391	20436	8596
Maximum entry in reduced matrix			
Threshold 0.5	154	7.9(5)	68
0.25	154	7.9(5)	119
10^{-1}	161	7.9(5)	126
10^{-2}	7.3(4)	6.9(6)	2.4(5)
10^{-4}	4.1(6)	3.0(18)	1.3(6)
10^{-6}	2.9(8)	8.4(20)	2.9(9)
10^{-10}	2.9(8)	1.4(34)	2.9(9)
Scaled residual			
Threshold 0.5	6.0(-16)	1.1(-16)	6.6(-16)
0.25	1.9(-15)	1.1(-16)	2.7(-15)
10^{-1}	1.5(-14)	1.1(-16)	7.9(-15)
10^{-2}	1.5(-11)	1.2(-14)	6.9(-12)
10^{-4}	6.9(-10)	2.3(-5)	1.1(-10)
10^{-6}	1.8(-8)	1.1(-5)	3.6(-8)
10^{-10}	1.8(-8)	1.6(-2)	3.6(-8)

Table 4.1 Effect of varying threshold parameter. The notation a(b) is used to denote the number $a \times 10^b$

over the entire reduced matrix. In spite of many attempts to develop better (and more complicated!) ordering strategies, the Markowitz strategy is still the preferred sparsity pivoting technique in use today for general systems of equations.

Our final strategy (for example in MA28) might thus be to search through the non-zero entries in order of increasing Markowitz count until one is found satisfying the threshold condition (4.1). This is effected by searching the rows and columns in ascending order of row and column count since entries with low Markowitz count are likely to come early in this sequence. Indeed, in some applications (for example, matrices arising from discretizations of pdes) it pays to restrict the searching to only the first few rows and/or columns. This was recommended by Zlatev et al (1978) and is now incorporated in the MA28 code some runs of which we show in Table 4.2. The accuracy of solution was in all cases similar and very satisfactory.

Order	199	541	900	1454
Non-zeros	701	4285	4322	3377
Non-zeros in factors				
Markowitz	1556	12571	34998	9106
Search restricted				
1 row	1903	12952	34514	10049
3 rows	1709	12641	34514	9372
10 rows	1616	12595	34514	9427
Time				
Markowitz	.078	.802	3.28	.626
Search restricted				
1 row	.072	.707	3.11	.410
3 rows	.067	.675	3.10	.380
10 rows	.069	.699	3.10	.464

Table 4.2 Effect of restricting pivot search in MA28. Time in seconds on an IBM 3081K. The matrix of order 541 has different entries from that used in Table 4.1

When the matrix is symmetric or structured (for example, banded) then a whole host of highly efficient algorithms are available. These are discussed at length by other contributions to this volume and so we will not consider them in this paper. One of the principal differences is that the analysis of the sparsity pattern can be decoupled from the numerical factorization and can be implemented very efficiently. Often it is possible to use a static storage scheme for the numerical factorization. Among the recent advances in the direct solution of sparse linear systems has been the extension of this decoupling to more general matrices. We discuss such advances in the next section.

5. Recent developments

In the last year or two there have been several attempts to use some of the highly efficient techniques for symmetric systems on more general matrices. We discuss some of these in this section and consider approaches which combine aspects of direct and iterative methods. Additionally, we reflect on the performance of recent methods on vector processors.

A natural way to view a sparse matrix, A, which arises in a finite element discretization is as a sum of element matrices, $B^{(\ell)}$,

$$A = \sum_{\ell} B^{(\ell)} \tag{5.1}$$

where each $B^{(\ell)}$ has nonzeros only in a few rows and columns corresponding to variables present in element ℓ. It is, however, possible to view any matrix in this way, the $B^{(\ell)}$ may be artificial "elements", low rank matrices, or single rows of A, for example. The important thing about the formulation (5.1) is that, if the element-wise summation

$$a_{ij} = a_{ij} + b_{ij}^{(\ell)} \tag{5.2}$$

is compared with the basic Gaussian elimination operation

$$a_{ij} = a_{ij} - a_{ik}\left[a_{kk}\right]^{-1} a_{kj}, \tag{5.3}$$

then the elimination (5.3) can proceed before all of the summations (5.2) so long as all summations have been performed for the terms in the triple product in (5.3). Thus assembly (5.2) and elimination (5.3) can be performed in tandem without the matrix, A, ever being formed explicitly as in (5.1).

This observation is the basis of frontal schemes originally developed (Irons, 1970) for the solution of structural analysis problems where the assembled matrix A is positive definite. Here, rows and columns become pivotal as soon as they are fully summed. The method can be extended to unsymmetric element problems (Hood, 1976) or indeed to any general unsymmetric matrix (Duff, 1981) by employing numerical pivoting, if necessary performing more summations so that numerically suitable fully summed variables are available as pivots. In the MA32 code (Duff, 1981) threshold pivoting of the form (4.1) is used although, in the equation entry (when each $B^{(\ell)}$ is a row of the matrix), ordinary partial pivoting can be employed without loss of efficiency. A significant feature of frontal methods is that the "front" (corresponding to variables assembled but not yet eliminated) is held in a full matrix (the frontal matrix) so that the innermost elimination loop can be written similarly to code for full matrices rather than requiring the sparse data structures of section 2. Typically, the inner loops might have the form shown in Figure 5.1. Clearly loop 5.1(a) should be more efficient than loop 5.1(b) since

Figure 5.1 Innermost loops in typical full and sparse codes

```
      DO 10 I = 1, N
      A(I,J) = A(I,J) + AMULT * A(I,K)
10    CONTINUE
```

(a) Full matrix code (direct addressing)

```
      DO 10 II = I1, I2
      I = IRN (II)
      W(I) = W(I) + AMULT * A(II)
10    CONTINUE
```

(b) Sparse matrix code (indirect addressing)

access to data is by fixed increment. The difference in efficiency is even more marked on vector machines where the full code vectorizes but the sparse code does not. The effect of vectorization is apparent in the result in Table 5.1 where we show times for runs of MA32 on the same problems

Dimensions of grid of elements or equations	64 x 64	16 x 16
Order of matrix	4096	5445
IBM 3081K	20.	109
CRAY-1S	2.0	5.8

Table 5.1 Times (in seconds) for matrix factorization using the unsymmetric frontal code MA32 on the CRAY-1S and the IBM 3081K

on the CRAY-1 and the IBM 3081K at Harwell. The scalar speed differential between the two machines is about 2 but a factor of over 10 is recorded in Table 5.1.

If the summation (5.1) is written in the form

$$A = ((\ldots((B^{(1)} + B^{(2)}) + B^{(3)}) + B^{(4)}) + B^{(5)}) + \ldots$$

then the frontal schemes we have just described correspond to summing the "elements" in the order given by the brackets and, if numerical considerations permit, performing eliminations after each summation. A different grouping of the summation, for example,

$$A = ((\ldots(((B^{(1)} + B^{(2)}) + (B^{(3)} + B^{(4)})) + ((B^{(5)} + B^{(6)}) + (B^{(7)} + B^{(8)}))) + \ldots$$

corresponds to a different strategy since, after the summation of $B^{(1)}$ and $B^{(2)}$ and subsequent eliminations, the remaining part of the frontal matrix (which has itself the same form as the contribution from an original element) is temporarily set aside while elements $B^{(3)}$ and $B^{(4)}$ are assembled. General groupings of this sort thus give rise to more than one frontal matrix and such methods are sometimes termed multifrontal. Now there is more freedom to determine the assembly and elimination order and so we might expect that multifrontal schemes can be designed to keep the fill-in and number of arithmetic operations low. Duff and Reid (1983b) have implemented the highly popular and successful minimum degree algorithm within a multifrontal scheme. The assembly order and a tentative pivot order are first determined using only the sparsity structure of the matrix. Thereafter, numerical pivoting is performed within the frontal matrices so perturbing the tentative pivot order. Harwell Subroutine Library codes MA27 (Duff and Reid, 1982) and MA37 (Duff and Reid, 1983a) have been written using this scheme. They solve equations (2.1) where the coefficient matrix is indefinite symmetric and unsymmetric respectively. In the first case, 2 x 2 pivoting (Bunch and Parlett, 1971) is used and in the second threshold pivoting similar to equation (4.1) is employed. Although the front is again held in a full matrix, it is likely to be of lower order than in the uni-front case (reflecting the lower operation count), and the overhead involved in administering the multiple fronts is likely to be significant. Thus we would not expect to find such a marked improvement due to vectorization of the inner loop. This is apparent from the results in Table 5.2 where we also observe that the total separation of the analysis from the factorization results in a fast analysis time. Additionally, this separation enables the analysis to be performed in a predictable and low amount of storage and the results of the analysis will give a fairly accurate estimate of the time and space required for subsequent factorization. Additionally, if no pivoting is required a static storage scheme could be used for the factorization and, even with pivoting, only a limited form of dynamic scheme is required.

The separation of analysis and factorization for general systems is not new. It was used by Sherman (1978) in the NSPIV (subsequently NSPFAC) code. A row ordering was determined a priori and, during numerical factorization, the largest entry in each row (later, in NSPFAC, modified to a threshold test) was used as pivot. Although this code had the merit of being simple, partial pivoting gave too much fill-in to be generally acceptable and even with threshold pivoting the a priori choice of a good row ordering was difficult and the performance could be very sensitive to that ordering.

Order	147	1561	900
Non-zeros	2449	10681	4322
Analysis			
IBM 3081K	.015	.11	.13
CRAY 1S	.013	.14	.08
Numerical factorization			
IBM 3081K	.065	1.30	.65
CRAY 1S	.024	.34	.18
Solution using factors			
IBM 3081K	.006	.080	.044
CRAY 1S	.003	.029	.017

Table 5.2 Times (in seconds) for the multifrontal code MA37 on the IBM 3081K and the CRAY 1S

More recently George and Heath (1980) and George and Ng (1983) used a symmetric analysis on the structure of A^TA to obtain a column ordering for A and from the structure of the resulting Cholesky factor have devised static data structures and algorithms for the subsequent numerical factorization.

The first approach (George and Heath, 1980) was designed for the solution of least squares problems but can be used to solve unsymmetric systems. It uses a decomposition of the form

$$A = QU \tag{5.2}$$

where Q is orthogonal and U upper triangular utilizing the fact that U has the same form as the Cholesky factor of the normal equations. Although the orthogonal decomposition (5.2) yields a stable algorithm, the method is slower than traditional techniques and is inconvenient for the solution of subsequent systems since the Q matrix (which will be quite full) is not kept. As we see, however, in Table 5.3 (from Heath, private communication) the technique can be competitive in terms of storage.

A recent observation of George and Ng (1983) that the pattern of the Cholesky factor of A^TA includes the pattern of the LU factors of any factorization of A where pivoting is performed on the rows of A, has lead them to another scheme which separates analysis and factorization.

Order	113	199	541
Non-zeros	655	701	4285
Total storage			
LU	3897	6493	39846
QU	5494	5167	31798
Time in IBM 3081 seconds			
LU	.07	.11	1.22
QU	.30	.25	5.31

Table 5.3 Factorization statistics comparing a QU with an LU (using an old version of MA28) decomposition. Results of Mike Heath

They order the columns of A according to the minimum degree ordering on A^TA and then perform threshold pivoting within each column in turn. They report good results on nearly symmetric matrices but since the static structure for L and U includes that for any partial pivoting the storage requirements and time can be poor for more general systems. We reproduce some of the results from George and Ng (1983) in Table 5.4.

While the methods we have discussed for separating analysis and factorization all work well on nearly symmetric matrices it is true to say that similar techniques for truly unsymmetric systems have not yet appeared. Another range of techniques which are proving difficult to extend to unsymmetric problems are iterative methods based on Krylov sequences. Substantial advances have been made in the solution of symmetric systems using preconditioned conjugate gradients but attempts to generate similar methods for general systems have not been completely successful.

A common form of preconditioning is to use a partial factorization of the coefficient matrix. Thus, for example, if A is symmetric and positive definite, a suitable preconditioning matrix might be $\hat{L}$ where

$$A = \hat{L}\hat{L}^T + E$$

Order	541	991	778
Non-zeros	4282	6027	5272
Total storage (in thousands of words)			
LU	40	179	93
George + Ng	45	279	85
Total time for one-off problem (in seconds on an IBM 4341)			
LU	5.0	147	36.3
George + Ng	24.3	322	33.9

Table 5.4 A comparison of the George and Ng (1983) method with an LU factorization using an old version of MA28. From results in their paper

with E not at rounding level. Then if the matrix

$$\hat{L}^{-1} A \hat{L}^{-T}$$

is used in a conjugate gradient iteration, the number of iterations can be substantially cut if $\hat{L}$ is chosen suitably. Another way of viewing preconditioning is to look upon it as using some iterative method to compensate for the error obtained when using $\hat{L}$ as a factor in a direct method. Thus, if

$$A = L\,U + E \tag{5.3}$$

where permutations have been omitted to simplify the discussion, we solve using the forward elimination and back substitution of (2.3) to obtain a solution $\underline{x}^{(1)}$ to

$$L\,U\,\underline{x} = \underline{b}.$$

A possible correction to $\underline{x}^{(1)}$ so that the resulting $\underline{x}^{(2)}$ better satisfies

$$A\,\underline{x} = \underline{b}$$

can be obtained through solving

$$A\,\underline{\delta x} = \underline{r}^{(1)} \tag{5.4}$$

where $\underline{r}^{(1)} = \underline{b} - A\,\underline{x}^{(1)}$
and updating $\underline{x}^{(1)}$ by the solution $\underline{\delta x}^{(1)}$, obtained from (5.4) by using the partial factors L and U. This is just simple iterative refinement and can be shown to converge when the largest eigenvalue of

$$(I - (LU)^{-1} A)$$

is less than 1.0 so that E in (5.3) need not be negligible. One way of obtaining L and U (and E) is to employ a drop tolerance, that is to drop all entries from the structure less than a certain value. This idea, first suggested by Zlatev and Nielsen (1977), is used by the sparse code Y12M (Zlatev et al, 1981) and in a new version of MA28. We show results from using the MA28 code in Table 5.5. A significant feature of such an approach is that the storage requirements can be substantially reduced so larger problems can be handled. Iterative refinement is not a very rapidly convergent iterative method and so methods of accelerating the convergence further are under investigation.

Drop Tolerance	0.	10^{-4}	10^{-2}	.05
Non-zeros in "factors"	21027	18350	16455	11162
Factorization time	2.18	1.72	1.41	.53
Number of iterations	1	2	3	8
Time for solution	.066	.094	.121	.235

Table 5.5 Use of drop tolerances by MA28 in solution of five-diagonal system of order 1000 with semi-bandwidth 100. Times in seconds on IBM 3081K

6. Mathematical Software

A recent survey of software implementing sparse matrix algorithms has been given by Duff (1982). In this section, we are not so much concerned with the availability of codes but rather with the general design philosophy behind mathematical software for solving sparse problems.

Perhaps more than in any other area of mathematical software, the packaging is all important. The complexities of sparse data structures were described in section 2 and it is evident that very few users of library codes would be happy to work with such structures. Thus it is highly desirable that the packaging presents a clean user interface and screens the user from the internal data structures. This may require some very sophisticated sorting of the users' input and, in our experience, the code for doing the sorting can be more complicated to write than the code which operates on the internal data structures. As we can see from Table 6.1, the price for performing such a sort may be high and, while the cost may be difficult to justify in other areas of mathematical software, experience with a wide user community has taught us that even such a high overhead is quickly repaid by the savings in overall costs to the user (not to mention the lengthening of his life!). Indeed, much of the popularity of direct methods over iterative for the solution of linear equations stems from the much better packaging of the former.

For sparse linear equations, the row pointer/column index scheme is fairly easily understood and is used by the YSMP code, inter alia (Eisenstat et al, 1977). The main problem with this input format lies in its inflexibility when the matrix does not naturally occur in the correct orientation. The input demanded by MA28 (Duff, 1977) and Y12M

Order	147	1561	1072	900
Non-zeros	2449	10681	12444	4322
Time for sort	.018	.08	.09	.06
Time for analysis	.015	.11	.21	.13

Table 6.1 Illustration of the overhead involved in sorting data on input to MA37. Times in seconds on an IBM 3081K

(Zlatev et al, 1981) is less severe requiring only that the user supply row and column indices for each non-zero which can be in any order. SPARSPAK (George et al, 1980) permits the user to input the matrix in several ways where the input is performed by using subroutine calls rather than by passing appropriately set arrays. Indeed the SPARSPAK package is more like a programming environment than a set of subroutines.

Such is the complexity of sparse matrix software that current trends are towards programs and whole packages rather than just a few subroutines. For example, some statistics for the MA32 and MA37 codes which we have referred to earlier are given in Table 6.2 together with those for the SPARSPAK package, which of course will normally be accessed as a library so all 132 K bytes need not be included in the user's program. Of course, sparse techniques are also embedded in many very large scale application codes in oil reservoir modelling, structural analysis, nuclear safety codes etc.

An important mode of communication with sparse packages is that of reverse communication. Here, whenever some data is required by the program, for example the values of a Jacobian, the product of a matrix with a vector or the solution of a set of linear equations, control is returned to the user rather than through calling a lower level subroutine. This is very useful in sparse applications since there are so many different ways of storing (or generating) the matrix and of forming its product with a vector, for example, that a fixed calling sequence is

	MA32	MA37	SPARSPAK
Number of "cards"	3409	3266	12026
Number of Fortran statements	2295	2121	5527
Number of subroutines	22	18	129
Length compiled code in K bytes IBM H ext OPT=3	40	38	132

Table 6.2 Statistics on lengths of three sparse matrix packages

inadequate. Thus, the structure from the user's viewpoint might be

```
      DO 10 IT = 1, MAXIT
      Call subr(..., iflag)
      IF (IFLAG.EQ.0) terminate
      IF (IFLAG.LT.0) fail
      Return the product of A and x in v
10    CONTINUE
```

We have found the flexibility of this approach to be very popular with users of the Harwell Subroutine Library.

7. Conclusions

We have tried in this chapter to introduce the reader to many of the basic ideas used in the construction of sparse matrix algorithms and codes. We have in particular studied the manipulation of sparse data structures upon which the efficiency of the code is crucially dependent. We have also given an indication of active areas in sparse matrix research and have considered some recent developments in algorithms and software.

References

Bunch, J. R. & Parlett, B. N. (1971). Direct methods for solving symmetric indefinite systems of linear equations. Siam. J. Numer. Anal 8 pp. 639-655

Curtis, A. R. & Reid, J. K. (1971). Fortran subroutines for the solution of sparse sets of linear equations. AERE Report R 6844. HMSO, London

Duff, I. S. (1977). MA28 - a set of Fortran subroutines for sparse unsymmetric linear equations. AERE Report R 8730, HMSO, London

Duff, I. S. (1981). MA32 - A package for solving sparse unsymmetric systems using the frontal method. Harwell Report AERE R 10079 HMSO, London

Duff, I. S. (1982). A survey of sparse matrix software. Harwell report AERE R 10512. To appear in Sources and Development of Mathematical Software. W. R. Cowell (Ed). Prentice-Hall

Duff, I. S., Grimes, R. G., Lewis, J. G. & Poole, W. C. Jr. (1982). Sparse matrix test problems. SIGNUM Newsletter 17 (2)

Duff, I. S. & Reid, J. K. (1982). MA27 - A set of Fortran subroutines for solving sparse symmetric sets of linear equations. AERE Harwell Report R 10533, HMSO, London

Duff, I. S. & Reid, J. K. (1983a). The multifrontal solution of unsymmetric sets of linear equations. Harwell Report CSS 133. To appear in Siam. J. Sci. Stat. Comput.

Duff, I. S. & REID, J. K. (1983b). The multifrontal solution of indefinite sparse symmetric linear systems. ACM Trans. Math Softw. 9 pp. 302-325

Eisenstat, S. C., Gursky, M. C., Schultz, M. H. & Sherman, A. H. Yale sparse matrix package. I. The Symmetric Codes. Report 112 Department of Computer Science, Yale Univ.

George, A. & Heath, M. T. (1980). Solution of sparse linear least squares problems using Givens rotations. Lin. Alg and its Applics. 34 pp. 69-83

George, A., Liu, J. & Ng, E. (1980). User Guide for SPARSPAK: Waterloo Sparse Linear Equations Package. Report CS-78-30 (Revised Jan. 1980). Department of Computer Science. Univ. of Waterloo

George, A. & Ng, E. (1983). A partial pivoting implementation of Gaussian elimination for sparse systems. Report CS-83-15, Department of Computer Science, Univ. Waterloo, Canada

Hood, P. (1976). Frontal solution program for unsymmetric matrices. Int. J. Numer. Meth. Engng. 10 pp. 379-399

Irons, B. M. (1970). A frontal solution program for finite element analysis. Int. J. Numer. Meth. Engng. 2 pp. 5-32

Markowitz, H. M. (1957). The elimination form of the inverse and its application to linear programming. Management Science 3 pp. 255-269

Sherman, A. H. (1975). On the efficient solution of sparse systems of linear and non-linear equations. Report 46. Ph. D. Thesis. Dept. Comput. Sci. Yale Univ.

Sherman, A. H. (1978). Algorithms for sparse Gaussian elimination with partial pivoting. ACM Trans. Math. Softw. 4 pp. 330-338 and pp. 391-398

Zlatev, Z., Barker, V. A. & Thomsen, P. G. (1978). SSLEST: A Fortran IV subroutine for solving sparse systems of linear equations. User's Guide. Report NI-78-01, Numerisk Institut, Lyngby, Denmark

Zlatev, Z & Nielsen, H. B. (1977). Preservation of sparsity in connection with iterative refinement. Tech. Report NI-77-12. Numerisk Institut, Lyngby, Denmark

Zlatev, Z., Wasniewski, J. & Schaumburg, K. (1981). Y12M Solution of Large and Sparse Systems of Linear Algebraic Equations. Lecture Notes in Computer Science 121. Springer-Verlag.

2 DIRECT METHODS FOR SPARSE LINEAR EQUATIONS

A. Brameller
Power System Laboratory, University of Manchester Institute of Science & Technology, PO Box 88, Manchester M60 1QD, England

SPARSITY

Introduction

Much of modern engineering and non-engineering practice requires the analysis of large and complex problems which are defined by a set of linear equations, perhaps having several thousand variables. These problems may relate to a physical network such as electricity supply, gas distribution, communication systems, mechanical and civil engineering structures or to a physical problem such as traffic flow, cash flow, generic problems, behaviour and social science problems. The range of problems is large and diverse. Most problems can be mathematically modelled as a network and solved by the network approach.

Many of these problems are characterised by sparsity, that is the associated coefficient matrix contains a large proportion of zero elements. Physically this means that only a small number of branches of the network representation (graph) is connected to each node or junction. In an electrical power system, for example, the ratio between the number of branches and the number of nodes is about 1.5 or less, which means that for a network of 1,500 nodes the associated coefficient matrix contains only 0.1% of non-zero elements.

There are many methods which can be used to solve a set of simultaneous equations. In choosing a suitable method, consideration should be given to the computational efficiency, that is the simplicity of the method, the number of arithmetical operations and the computer storage requirements. One important aspect affecting the computational efficiency is the degree of sparsity exhibited by the equations. By exploiting sparsity and only processing and storing the non-zero elements, a considerable improvement in efficiency can be made.

Gauss elimination

The set of simultaneous equations to be solved may be written in matrix form as:

$$[A][X] = [b] \qquad (1)$$

where

$[A]$ = non-singular coefficient matrix of order n by n

$[X]$ = vector of unknown variables

$[b]$ = vector of known variables

n = number of variables or equations

Equation (1) can be solved by direct matrix inversion to give:

$$[X] = [A]^{-1}[b] \qquad (2)$$

However, this direct method involves n^3 arithmetical operations and although the matrix $[A]$ may have been very sparse, the inverse $[A]^{-1}$ is completely filled, that is the number of elements to be stored is n^2. Therefore, for large sparse networks this method is very inefficient and is justified only if the inverse matrix is required for other calculations.

One method which increases computational efficiency is Gauss-elimination. This is a systematised method based on the technique generally learned and used at school for solving simultaneous equations by hand. The method eliminates one variable at a time (by reducing its coefficient to zero) until one equation in one unknown remains, solving for this unknown variable and then back-substituting to solve for all the remaining variables. The method is best illustrated by the following example:

$$\begin{aligned} a_{11}x_1 + a_{12}x_2 + \ldots\ldots\; a_{1n}x_n &= b_1 \\ a_{21}x_1 + a_{22}x_2 + \ldots\ldots\; a_{2n}x_n &= b_2 \\ & \\ & \\ a_{n1}x_1 + a_{n2}x_2 + \ldots\ldots\; a_{nn}x_n &= b_n \end{aligned} \qquad (3)$$

The first equation can be conveniently divided by a_{11} and then used to eliminate x_1 from the remaining equations, resulting in (n-1) equations in (n-1) variables. The element a_{11} is called the pivot, and the equation containing the pivot, the pivotal equation. After eliminating the first variable (x_1), the remaining (n-1) equations can be written as:

$$\begin{array}{ccc} a'_{22}x_2 + \cdots\cdots & a'_{2n}x_n & = b'_2 \\ \vdots & \vdots & \\ a'_{n2}x_2 + \cdots\cdots & a'_{nn}x_n & = b'_n \end{array} \tag{4}$$

where

$$a'_{rc} = a_{rc} - a_{rp}a_{pp}^{-1}a_{pc}$$

$$b'_r = b_r - a_{rp}a_{pp}^{-1}b_p$$

r = row position
c = column position
p = pivot (row or column number)

The elimination process can be carried out until one equation with one variable is left. The various pivotal equations can be assembled together in the following triangular form:

$$\begin{array}{rl} c_{11}x_1 + c_{12}x_2 + \cdots\cdots c_{1n}x_n & = d_1 \\ c_{22}x_2 + \cdots\cdots c_{2n}x_n & = d_2 \\ & \\ c_{nn}x_n & = d_n \end{array} \tag{5}$$

where

$$c_{11} = a_{11} \qquad c_{12} = a_{12} \qquad \text{etc.}$$

$$c_{22} = a'_{22} \qquad d_2 = b'_2 \qquad \text{etc.}$$

The last equation containing one variable can be solved directly and the remaining variables follow in turn by substituting into the triangular form in reverse order.

In most cases it is sufficient to choose the diagonal coefficients as pivots unless they are small compared with off-diagonal coefficients or are zero. In this case, numerical accuracy can be improved by suitable modifications (interchange of rows and columns).

The order of elimination is arbitrary but it is important to note that some of the zero elements in the original matrix may become non-zero during the elimination. The number of new non-zero elements depends on the order of elimination.

Example

Consider the following equations:

$$x_1 - x_2 - x_3 = 1$$

$$-x_1 + 2x_2 = 2$$

$$-x_1 + 3x_3 - x_4 = 3$$

$$-x_3 + 2x_4 = 4$$

Choosing the first equation as pivot results in:

$$x_2 - x_3 = 3$$

$$-x_2 + 2x_3 - x_4 = 4$$

$$-x_3 + 2x_4 = 4$$

Choosing the top equation as pivot results in:

$$x_3 - x_4 = 7$$

$$-x_3 + 2x_4 = 4$$

Choosing the top equation again as pivot results in:

$$x_4 = 11$$

The pivotal equations are:

$$x_1 - x_2 - x_3 = 1$$

$$x_2 - x_3 = 3$$

$$x_3 - x_4 = 7$$

$$x_4 = 11$$

and the solution is:

$$x_4 = 11, \quad x_3 = 18, \quad x_2 = 21, \quad x_1 = 40$$

Matrix factorisation

Matrix factorisation methods are modifications of the basic Gauss elimination technique. In these methods the inverse matrix $[A]^{-1}$ is not calculated explicitly but exploits the property that matrices can be represented by the product of a number of factor matrices. Also, by exploiting sparsity in these methods, the number of operations and the computer storage requirement can be reduced drastically. There are several factorisation methods available, the best known being the triangular decomposition and the product form of the inverse.

By simple modification of the Gaussian elimination, it is possible to obtain the inverse of a given matrix as the product of a number of matrices which are very sparse.

$$[A]^{-1} = [T_n] \ldots\ldots\ldots [T_3][T_2][T_1] \tag{6}$$

This factorisation process can be obtained from the following considerations:

Let:

$$[T_1][A] = [A_1] \tag{7}$$

where:

$1/a_{11}$	•	•	•
$-a_{21}/a_{11}$	1	•	•
$-a_{31}/a_{11}$	•	1	•
$-a_{41}/a_{11}$	•	•	1

$[A_1]$ =

1	a'_{12}	a'_{13}	a'_{14}
•	a'_{22}	a'_{23}	a'_{24}
•	a'_{32}	a'_{33}	a'_{34}
•	a'_{42}	a'_{43}	a'_{44}

(•) represents element with zero value

$$a'_{pc} = a'_{pp} a_{pc}$$

$$a'_{rc} = a_{rc} - a_{rp} a_{pp}^{-1} a_{pc} \qquad (r \neq p,\ c \neq p)$$

Starting with the equation:

$$[A][X] = [b] \tag{8}$$

multiplying by $[T_1]$ and substituting from equation (7) for $[T_1][A]$:

$$[T_1][A][X] = [T_1][b]$$

$$[A_1][X] = [T_1][b] \tag{9}$$

Let

$$[T_2][A_1] = [A_2]$$

where

$$[T_2] = \begin{bmatrix} 1 & -a'_{12}/a'_{22} & & \\ & 1/a'_{22} & & \\ & -a'_{32}/a'_{22} & 1 & \\ & -a'_{42}/a'_{22} & & 1 \end{bmatrix} \qquad [A_2] = \begin{bmatrix} 1 & & a''_{13} & a''_{14} \\ & 1 & a''_{23} & a''_{24} \\ & & a''_{33} & a''_{34} \\ & & a''_{43} & a''_{44} \end{bmatrix}$$

Continuing the process and multiplying successively by [T] results in:

$$[A][X] = [b]$$

$$[T_1]\,[A][X] = [T_1][b]$$

$$[A_1][X] = [T_1][b] \qquad \text{since} \quad [T_1][A] = [A_1]$$

$$[T_2][A_1][X] = [T_2][T_1][b]$$

$$[A_2][X] = [T_2][T_1][b] \quad \text{since } [T_2][A_1] = [A_2]$$

$$[A_n][X] = [T_n] \;.......\; [T_3][T_2][T_1][b]$$

but $[A_n]$ reduces to a unit matrix, therefore:

$$[T_n] \;.......\; [T_3][T_2][T_1] = [A]^{-1}$$

Numerical example

The solution of the following simple example is given to illustrate the procedure.

$$[A] = \begin{bmatrix} 1 & -1 & -1 & \\ -1 & 2 & & \\ -1 & & 3 & -1 \\ & & -1 & 2 \end{bmatrix} \qquad [b] = \begin{bmatrix} 1 \\ 2 \\ 3 \\ 4 \end{bmatrix} \qquad [X] = \begin{bmatrix} x_1 \\ x_2 \\ x_3 \\ x_4 \end{bmatrix}$$

1st reduction step $[A_1] = [T_1][A]$

$$\begin{bmatrix} 1 & -1 & -1 & \\ & 1 & -1 & \\ & -1 & 2 & -1 \\ & & -1 & 2 \end{bmatrix} = \begin{bmatrix} 1 & & & \\ 1 & 1 & & \\ 1 & & 1 & \\ & & & 1 \end{bmatrix} \begin{bmatrix} 1 & -1 & -1 & \\ -1 & 2 & & \\ -1 & & 3 & -1 \\ & & -1 & 2 \end{bmatrix}$$

2nd reduction step

$$\begin{bmatrix} 1 & & -2 & \\ & 1 & -1 & \\ & & 1 & -1 \\ & & -1 & 2 \end{bmatrix} = \begin{bmatrix} 1 & 1 & & \\ & 1 & & \\ & 1 & 1 & \\ & & & 1 \end{bmatrix} \begin{bmatrix} 1 & -1 & -1 & \\ & 1 & -1 & \\ & -1 & 2 & -1 \\ & & -1 & 2 \end{bmatrix}$$

3rd reduction step

$$\begin{bmatrix} 1 & & & -2 \\ & 1 & & -1 \\ & & 1 & -1 \\ & & & 1 \end{bmatrix} = \begin{bmatrix} 1 & & 2 & \\ & 1 & 1 & \\ & & 1 & \\ & & 1 & 1 \end{bmatrix} \begin{bmatrix} 1 & & -2 & \\ & 1 & -1 & \\ & & 1 & -1 \\ & & -1 & 2 \end{bmatrix}$$

4th reduction step

$$\begin{bmatrix} 1 & & & \\ & 1 & & \\ & & 1 & \\ & & & 1 \end{bmatrix} = \begin{bmatrix} 1 & & & 2 \\ & 1 & & 1 \\ & & 1 & 1 \\ & & & 1 \end{bmatrix} \begin{bmatrix} 1 & & & -2 \\ & 1 & & -1 \\ & & 1 & -1 \\ & & & 1 \end{bmatrix}$$

and the product $[X] = [A]^{-1}[b]$ is:

$$\begin{bmatrix} 1 & & & 2 \\ & 1 & & 1 \\ & & 1 & 1 \\ & & & 1 \end{bmatrix} \begin{bmatrix} 1 & & 2 & \\ & 1 & 1 & \\ & & 1 & \\ & & 1 & 1 \end{bmatrix} \begin{bmatrix} 1 & 1 & & \\ & 1 & & \\ & 1 & 1 & \\ & & & 1 \end{bmatrix} \begin{bmatrix} 1 & & & \\ 1 & 1 & & \\ 1 & & 1 & \\ & & & 1 \end{bmatrix} \begin{bmatrix} 1 \\ 2 \\ 3 \\ 4 \end{bmatrix} = \begin{bmatrix} 40 \\ 21 \\ 18 \\ 11 \end{bmatrix}$$

Sparsity-directed elimination

Owing to the subtraction and addition of equations during the Gaussian elimination some of the zero elements may become non-zero. This causes an increase in required storage and computation time. To illustrate this, consider the following radial system and the corresponding matrix:

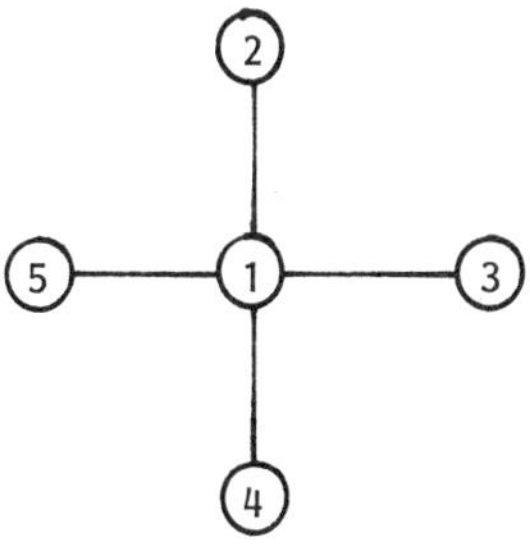

$$[A] = \begin{bmatrix} x & x & x & x & x \\ x & x & & & \\ x & & x & & \\ x & & & x & \\ x & & & & x \end{bmatrix}$$

where x represents non-zero element.

Eliminating in natural order starting with first row and column results in:

$$[T_1] = \begin{bmatrix} x & & & & \\ x & 1 & & & \\ x & & 1 & & \\ x & & & 1 & \\ x & & & & 1 \end{bmatrix} \qquad [A_1] = \begin{bmatrix} 1 & x & x & x & x \\ & x & x & x & x \\ & x & x & x & x \\ & x & x & x & x \\ & x & x & x & x \end{bmatrix}$$

The matrix $[A_1]$ is full except for the pivotal column, the sparsity of the original matrix is lost and the pivotal columns of all subsequent [T] matrices will also be full.

If the nodes are renumbered or the elimination is carried out in reverse order, for example, the sparsity can be preserved as shown below:

$[T_5]$ =

1				x
	1			
		1		
			1	
				x

$[A_5]$ =

x	x	x	x	
x	x			
x		x		
x			x	
				1

The transformation matrices will for this particular example retain the sparsity of the original matrix.

The simple example shown above clearly indicates the advantages that can be gained by a suitable order of elimination. To achieve optimum ordering for maximum sparsity it is necessary to enumerate all the possible alternatives. The extra computation time required will nullify all the gains of taking advantage of sparsity. In practice, therefore, the strategy is to use some simple rule which is efficient and accept a near optimum ordering sequence. There are many possible alternatives which can be divided into two categories, one which determines the order before elimination and the other known as the dynamic ordering which determines the next pivot after previous elimination.

The simplest and most common scheme is one where at each stage of elimination, the diagonal element in the row and column of the reduced matrix with the fewest number of elements is selected as the next pivot.

The solution by matrix factorisation can be applied to a large variety of problems; some may involve many different solutions for different right-hand side without changes in the system. In this case the ordering and factorisation has to be performed once only and the solution step repeated a number of times. Other problems like the solution of a non-linear problem using Newton-Raphson method have different requirements. During each iterative step it is necessary to solve a linear set of equations with a Jacobian coefficient matrix. Such matrices can be highly sparse and the sparsity remains the same from iteration to iteration although the numerical values of the non-zero elements change.

Optimal ordering can be determined during the course of the computation, but for problems like the one outlined above it is more efficient to determine it by simulating the reduction process beforehand. The optimal order of elimination can then be recorded and used for subsequent iterative steps. For practical purposes the computer program should be subdivided into four independent subroutines as shown in the flow chart below:

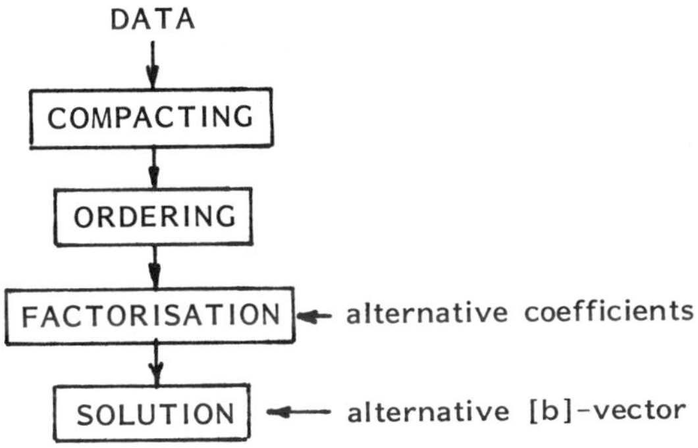

In order to exploit the benefits of sparsity the digital computer program must be written efficiently. For this purpose it is necessary to introduce tables of indexing information to identify the elements and to facilitate their addressing. There are many schemes that can be used and to illustrate the process involved consider storing the matrix [A]

[A] =

3.1	-1.3	-1.7	
-1.3	2.6		
-1.7		2.8	-1.2
		-1.2	3.7

Location	1	2	3	4	5	6	7	8	9
Value	-1.3	-1.7	-1.3	-1.7	-1.2	-1.2			
Row	2	3	1	1	4	3			
Next address	2	0	0	5	0	0			

Location	1	2	3	4
Diagonal value	3.1	2.6	2.8	3.7
Column address	1	3	4	6
No. of non-zeros	2	1	2	1
Order of elimination				

Note: 0 in the next address row indicate that there are no more elements in the column
* indicates next available location

The solution $[X] = [A]^{-1}[b]$ is obtained by multiplying the last factor matrix by the vector [b] to obtain another vector. The next factor matrix is then multiplied by this vector until the final solution is obtained. Because the factor matrices are all unit matrices except for one column, the actual multiplication is performed only on the non-zero elements. For example:

$$\begin{bmatrix} 1 & f_{12} & & \\ & f_{22} & & \\ & & 1 & \\ & & & 1 \end{bmatrix} \begin{bmatrix} b_1 \\ b_2 \\ b_3 \\ b_4 \end{bmatrix} = \begin{bmatrix} b'_1 \\ b'_2 \\ b'_3 \\ b'_4 \end{bmatrix}$$

$$\begin{aligned} b'_1 &= b_1 + f_{12}b_2 \\ b'_2 &= f_{22}b_2 \\ b'_3 &= b_3 \\ b'_4 &= b_4 \end{aligned}$$

Comparison of calculation time and storage for different networks on the CDC 7600 computer using sparsity programming can be seen from the following table:

Number of		No.of non-zeros		Comp-acting	Order-ing	Factor-isation	Solution	Total
Nodes	Bran-ches	Initial	Final					
30	41	112	85	.001	.003	.003	.001	.008
57	80	217	200	.003	.007	.007	.001	.018
118	184	486	390	.007	.020	.013	.003	.043
621	794	2209	2136	.030	.422	.082	.017	.551
1242	1593	4428	4430	.058	1.578	.182	.038	1.856

Table 1 Computation time (sec) and storage using CDC 7600 computer

Numerical example

To illustrate the process of factorisation and sparsity programming consider the numerical example of factorising the following matrix:

[A] =

2	−1	−1	
−1	1		
−1		2	−1
		−1	2

The efficiency of the solution depends very much on the skill and method of programming. The solutions for the particular example are shown in tabular form where the columns have the following values:

a = row position
b = location of next non-zero element
c = non-zero element value
d = starting position of columns
e = number of non-zero elements in the column
f = pivot sequence
0 = indicates last term in column

For the particular example the tables will change during the process of simulation and ordering as follows:

Initial

1	2	3	4
x	x	x	
x	x		
x		x	x
		x	x

2	−1	−1	
−1	1		
−1		2	−1
		−1	2

Step 1
Pivot 2

1	2	3	4
x	x	x	
x	x		
x		x	x
		x	x

1	1	−1	
	1		
−1		2	−1
		−1	2

Step 2
Pivot 1

1	2	3	4
x		x	
	1		
x		x	x
		x	x

1			
1		1	−1
		−1	2

Step 3
Pivot 3

1	2	3	4
1			
	1		
		x	x
		x	x

		1	
		1	1

Step 4
Pivot 4

1	2	3	4
1			
	1		
		1	
			x

			1

Column	Address d	Non-zeros e	Sequence f	Address d	Non-zeros e	Sequence f	Address d	Non-zeros e	Sequence f	Address d	Non-zeros e	Sequence f	Address d	Non-zeros e	Sequence f
1	1	3		1	2	←	1	2	2	1	2	2	1	2	2
2	4	2	←	4	2	1	4	2	1	4	2	1	4	2	1
3	6	3		6	3		7	2	←	7	2	3	7	2	3
4	9	2		9	2		9	2		10	1	←	10	1	4

	Row a	Next b	Value c	Row a	Next b	Value c	Row a	Next b	Value c	Row a	Next b	Value c	Row a	Next b	Value c
1	1	2	2	1	3	1	1	3	1	1	3	1	1	3	1
2	2	3	−1	0	11	0 *	0	11	0	0	11	0	0	11	0
3	3	0	−1	3	0	−1	3	0	−1	3	0	1	3	0	1
4	1	5	−1	1	5	−1	1	5	1	1	5	1	1	5	1
5	2	0	1	2	0	1	2	0	1	2	0	1	2	0	1
6	1	7	−1	1	7	1	0	2	0 *	0	2	0	0	2	0
7	3	8	2	3	8	2	3	8	−1	3	8	1	3	8	1
8	4	0	−1	4	0	−1	4	0	−1	4	0	1	4	0	1
9	3	10	−1	3	10	−1	3	10	−1	0	6	0 *	0	6	0 *
10	4	0	2	4	0	2	4	0	2	4	0	1	4	0	1
11	0	12	0 *	0	12	0	0	12	0	0	12	0	0	12	0
12	0	0	0 **	0	0	0 **	0	0	0 **	0	0	0 **	0	0	0 **

* = First empty position

** = Last empty position

3 ITERATIVE METHODS FOR SPARSE MATRICES

David J. Evans
Department of Computer Studies
Loughborough University of Technology
Loughborough, Leicestershire,
U.K.

ABSTRACT

After brief introductory sections covering basic graph theory, computer storage techniques for sparse matrix manipulation, direct methods of solution and the sparse eigenvalue problem, later sections will be concerned with iterative algorithms for the solution of sparse matrices of large order. Such matrices are derived from discretised problems involving electrical networks, structural analyses, partial differential equations, power distribution systems, nuclear physics and operational research.

The basic first order stationary matrix iterative methods i.e. Jacobi, Gauss Seidel will be initially introduced and their convergence characteristics analysed. Then, the basic ideas of the overrelaxation and preconditioning strategies are compared to obtain an improved method.

Further classes of iterative methods are also discussed i.e. variational methods such as the conjugate gradient method and its applicability to the new concepts of preconditioning emphasised.

Finally, the recently introduced approximate matrix factor-isation methods i.e. incomplete Choleski, etc. are presented and when coupled with preconditioning strategies it is shown that some of the most powerfully convergent sparse matrix iterative methods are derived.

1. INTRODUCTION

Sparse matrix research is an active and expanding area, Duff (1977). Although we concentrate on *iterative methods* mainly it is in the area of *direct methods* that most progress has been made in recent years. We also include brief sections on the sparse *eigenvalue problem. Iterative methods* were developed principally for the solution of finite difference equations derived from partial differential equations especially for higher dimensional problems ($n \geq 3$) but recently have been extended to cover other fields, i.e. finite element methods, etc. which for the method of conjugate gradients is of very wide applicability. Finally, we make some remarks on the relative merits of direct and iterative methods.

1.1 Sparse Matrices

A matrix is said to be sparse if it has sufficiently many zero elements for it to be worthwhile to use special techniques to avoid storing or operating with the zeros. Unfortunately, these techniques almost always imply an overhead on other operations so they are worthwhile only when the number of zeros is large. Sometimes it is not obvious that sparse techniques are appropriate but generally a matrix with more than 80% zeros is considered to be 'sparse' and would benefit from special treatment.

A sparse matrix is defined as a matrix with only a small percentage of non-zero elements. In many practical applications, a typical sparse matrix usually has five to ten non-zeros per row, (as ordinarily occurs in finite-difference or finite-element approximations to an elliptic partial differential equation), and the matrix to be considered is of the order of a few hundred, but orders in the range of thousands or tens of thousands can arise quite frequently in three-dimensional problems.

Matrices of this kind arise in such diverse fields as power systems analysis, surveying, circuit theory and structural analysis and the sparsity usually arises because the underlying problem consists of a network or graph. The equation associated with an individual node will involve the variable or variables belonging to that node and the variable or variables belonging to nodes to which it is connected in the network. The matrix is sparse because each node is likely to be connected only to a few neighbouring nodes.

1.2 Graph Theory Associated with Matrices

In this section we introduce a few basic graph theory notions, and establish their correspondence to matrix concepts. Although rather few results from graph theory have found direct application to the analysis of sparse matrix computations, the notation and concepts are convenient and helpful in describing algorithms and identifying or characterizing matrix structure.

For our purposes, a graph $G=(X,E)$ consists of a finite set of nodes or vertices together with a set E or edges, which are unordered pairs of vertices. An ordering {labelling} α of $G=(X,E)$ is simply a mapping of $\{1,2,\dots,N\}$ onto X, where N denotes the number of nodes of G. Unless we specifically state otherwise, a graph will be unordered; the graph G labelled by α will be denoted by $G^{\alpha}=(X,E)$.

Since our objective in introducing graphs is to facilitate the study of sparse matrices, we now establish the relationship between graphs and matrices. Let A be an N by N symmetric matrix. The ordered graph of A, denoted by $G^A=(X^A,E^A)$ is one for which the N vertices of G^A are numbered from 1 to N, and $\{x_i,x_j\} \in E^A$ if and only if $a_{ij}=a_{ji}\neq 0$, $i\neq j$. Here x_i denotes the node of X^A with label i. Figure 1.1 illustrates the structure of a matrix and its labelled graph. We denote the ith diagonal element of a matrix by ⓘ to emphasize its correspondence with node i of the corresponding graph. Off-diagonal nonzeros are depicted by "*".

FIGURE 1.1: A matrix and its labelled graph, with * denoting a nonzero entry of A.

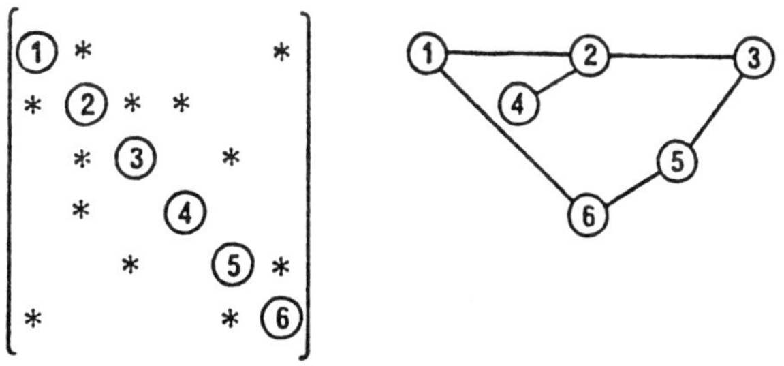

Matrix A Graph G^A

The directed graph associated with an unsymmetric matrix is usually called a *digraph* and the word *graph* is used for the undirected graph associated with a symmetric matrix.

FIGURE 1.2: An unsymmetric matrix and its associated graph

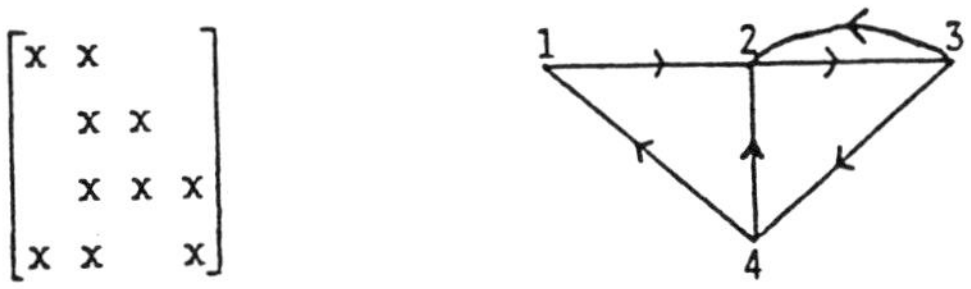

Another example is where the matrix can be generated from a computational molecule arising from the 5-point finite difference discretization of Laplace's equation as shown in Figure 1.3a,b,c.

FIGURE 1.3: The matrix and associated graph arising from the finite difference approximation of Laplace's equation on a square

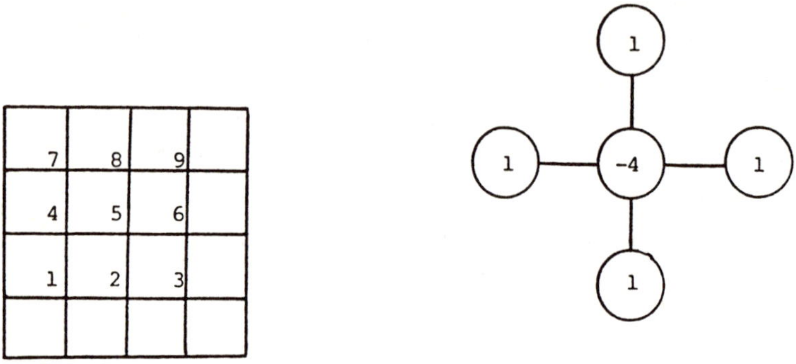

(a) (b)

$$\left[\begin{array}{ccc|ccc|ccc} -4 & 1 & & 1 & & & & & \\ 1 & -4 & 1 & & 1 & & & & \\ & 1 & -4 & & & 1 & & & \\ \hline 1 & & & 4 & -1 & & 1 & & \\ & 1 & & -1 & 4 & -1 & & 1 & \\ & & 1 & & -1 & 4 & & & 1 \\ \hline & & & 1 & & & 4 & -1 & \\ & & & & 1 & & -1 & 4 & -1 \\ & & & & & 1 & & -1 & 4 \end{array}\right]$$

(c)

Direct methods lead to additional nonzero elements (fill-ins) being created during the solution, and their success depends critically on how many such fill-ins take place whilst iterative methods do not suffer from this defect. Generally speaking, there are fewer fill-ins in cases where the associated graph can be separated into disjoint parts by the removal of small numbers of nodes. Thus, matrices arising from power transmission networks are likely to suffer little fill-in whereas those from partial differential equations, particularly in three dimensions, suffer much fill-in.

1.3 Storage Schemes for Sparse Matrices

The generated matrix shown in Figure 1.3 has a repetitive structure which can be exploited in order that only the network (Fig. 3a), and the associated logic of the computational molecule transversing the grid or the ordering of the points need to be stored. In general, however we need to store each non-zero explicitly along with some indexing information to indicate where each element belongs in the matrix. The most convenient way to do this when specifying a matrix is to store just the non-zeros as triples (a_{ij}, i,j) held contiguously in any order in a real array and two integer arrays. Unfortunately, when manipulating such data structures (for example, performing Gaussian elimination) we are likely to require access to the rows and/or columns, each of which can only be found by scanning the whole stored matrix.

There are many schemes for the storage of sparse matrices, of which not one is perfectly efficient and ideal. For example, if a method is said to be good because it uses fewer storage elements, then it will be more difficult to manipulate the matrices which are stored in this way. However, some storage methods are more efficient and effective in certain programming languages.

A few methods for the storage of sparse matrices are presented here.

1.3.1 Use of Coordinates

One simple and popular method of storage of sparse matrices is to use a one-dimensional vector for the storage of the non-zero elements and their coordinates in the matrix array. The method allows enough information to be stored in the vector so that the position of every existing nonzero element can be found quite easily. Consider the Example 1.1 and let us assume that vector $\underline{V}$ (say) has been used for the nonzero elements of the matrix to be stored.

$$\begin{bmatrix} 0 & a_{12} & 0 & a_{14} & 0 \\ 0 & a_{22} & 0 & 0 & a_{25} \\ 0 & 0 & a_{33} & 0 & 0 \\ 0 & 0 & 0 & 0 & 0 \\ a_{51} & 0 & 0 & a_{54} & a_{55} \end{bmatrix}$$

Example 1.1

In the matrix the first element which is nonzero lies on the first row and on the second column, so the first location of vector $\underline{V}$

contains 1, indicating the row which the element lines on. The second location of vector $\underline{V}$ contains 2, indicating the column number and the third location contains the element itself. Generally if we denote 3 locations of vector $\underline{V}$ for each non-zero element, then the first of these locations indicates the row position, the second one indicates the column position and the last one contains the element itself (see Figure 1.4

FIGURE 1.4:

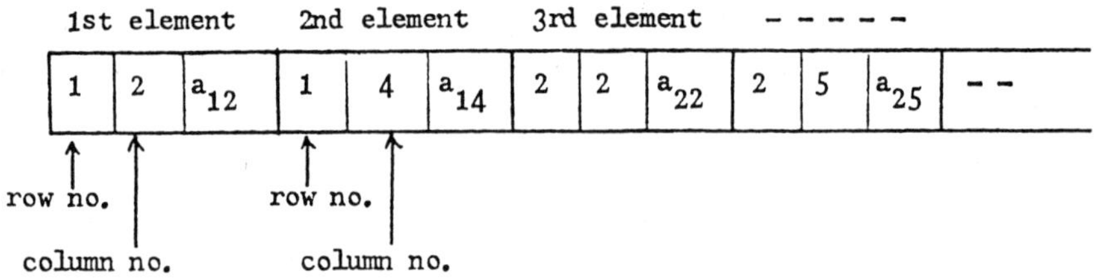

If a matrix of the size n*m contains p nonzero elements, then the amount of storage we should need is 3*p. However the insertion of new values in the vector $\underline{V}$ above could involve a lot of shuffling operations which could be very time consuming.

An extension of the above method is to use an extra storage element to actually link each nonzero element to the next one and to arrange for the last element to point to either the first element or a terminating symbol, indicating that there are no more nonzero elements. The vector $\underline{V}$ for Example 1.1 would now look like:

FIGURE 1.5:

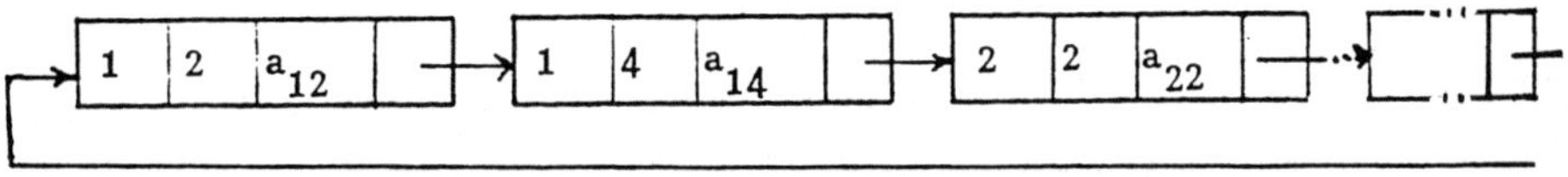

As can be seen, the storage elements for the nonzero elements do not now need to be contiguous and the insertion of new values can be done easily.

Here, although we have more control on the vector $\underline{V}$ from a manipulation point of view, the amount of storage has increased to 4*X for X nonzero elements in the matrix.

1.3.2 Bit Pattern

The next method of storage which we consider is to use a bit pattern in such a way that if the element (i,j) in the matrix is nonzero, then the (i,j)th element in the bit pattern is one and zero otherwise. Considering Example 1.1, our bit pattern would be:

$$\begin{bmatrix} 0 & 1 & 0 & 1 & 0 \\ 0 & 1 & 0 & 0 & 1 \\ 0 & 0 & 1 & 0 & 0 \\ 0 & 0 & 0 & 0 & 0 \\ 1 & 0 & 0 & 1 & 1 \end{bmatrix}$$

The bit pattern is usually followed by a one-dimensional vector $\underline{V}$ (say) which holds the nonzero elements of the matrix.

Vector $\underline{V}$:

a_{12}	a_{14}	a_{22}	a_{25}	a_{33}	a_{51}	a_{54}	a_{55}

The amount of storage in this method depends on the machine and its software. Some systems allow each boolean value to be stored in one bit only whilst others require one byte or even a word.

If we are allowed to store a boolean value in a bit, then the amount of storage for an n*m matrix which contains $\underline{p}$ nonzero elements will be

$$p + \frac{n*m}{y}$$

words of computer memory where y is the number of bits per word. It is obvious that the amount of storage is considerably less than the methods in Section 1.1 but the manipulation on matrices stored in this way is more difficult.

1.3.3 Uncompressed Storage Scheme

Another storage scheme for sparse matrices is called aptly an uncompressed storage of sparse matrices.

The scheme uses three one-dimensional vectors. One vector called $\underline{a}$ (say) contains all the nonzero elements in the matrix in lexicographical order. The second vector called $\underline{ja}$ (say) is used for storing the column positions of the nonzero elements which have already been stored in vector $\underline{a}$ and in the same order, i.e. lexicographical order. The third vector $\underline{ia}$ (say) has the property that if we wanted to know how many nonzero elements existed in the ith row, then ia(i+1)-ia(i) would be the answer. Let us see what it all means by applying these rules on an example. Consider Example 1.2:

$$\begin{bmatrix} 0 & 0 & a_{13} & 0 & 0 \\ a_{21} & 0 & 0 & 0 & a_{25} \\ 0 & 0 & 0 & 0 & 0 \\ a_{41} & a_{42} & 0 & 0 & 0 \\ 0 & 0 & 0 & a_{54} & a_{55} \end{bmatrix}$$

Example 1.2

The vectors a and ja in description above now are:

a:	a_{13}	a_{21}	a_{25}	a_{41}	a_{42}	a_{54}	a_{55}
	1	2	3	4	5	6	7

ja:	3	1	5	1	2	4	5

Vector ia needs n+1 locations for a matrix with n rows. If we assume that ia(1)=1, then vector ia for Example 1.2, will be:

	1	2	3	4	5	6
ia:	1	2	4	4	6	8

Now, ia(2)-ia(1) = 2-1 = 1 indicates that there is one nonzero element in row 1, whilst,

ia(3)-ia(2) = 4-2 = 2 says that in row 2 there are two nonzero elements , and,

ia(4)-ia(3) = 4-4 = 0 indicates that there is not any nonzero element in row three, and so on.

So, for a matrix of the size n*m with X number of nonzero elements, we would require n+1 memory words for vector ia and 2X memory words for vectors a and ja, which all together is 2X+n+1 memory words.

Note that it would not make any difference if we put other numbers rather than 1 in ia(1), but one should always remember this number and build the vector ia such that the property of ia is not changed. For example, if we had chosen L instead of 1, then the vector ia would have looked like:

	1	2	3	4	5	6
ia:	L	L+1	L+3	L+3	L+5	L+7

All the storage schemes which have been developed for storing sparse matrices are effective only when the percentage of nonzero elements existing in the matrices is low and certainly not more than 40% because as can be seen in the uncompressed storage method, not only do we need n storage elements for n non-zero elements, but we need n storage elements for the vector indicating the columns on which the nonzero elements lie (ja). For example, if we had a matrix of the size 200*200, we would need 40000 memory locations (if real, then 80000) to store this matrix using the conventional matrix storage. But if 50% of the elements of this matrix were nonzero, then using the uncompressed storage scheme, we would need 20000 storage elements for the nonzero elements to be stored in vector a and another 20000 storage elements for vector ja (which is already 40000 and is equal to the amount of storage using the conventional matrix storage) and yet we would still need 201 storage elements for vector ia. Therefore, we emphasize that if any advantage is going to be gained, the number of nonzero elements should be low.

2. DIRECT METHODS FOR SPARSE LINEAR SYSTEMS

By direct methods, we mean methods which give the exact solution in a finite number of elementary arithmetic operations, provided there were no rounding errors. The principal direct methods for general computational problems fall into three categories:

(1) Gaussian elimination with interchanges

(2) Triangular factorization without interchanges, including Choleski factorization for symmetric and positive-definite matrices.

(3) Householder reduction to upper triangular form.

Detailed analyses and discussion of the numerical stability for these methods have been well established, and the reader may consult the literature.

We devote this section to a full description of sparse elimination techniques and in particular symmetric Gaussian elimination for linear systems and eigenvalues.

2.1 Gaussian Elimination

We first describe in some detail the theory of Gaussian-elimination even though it is a widely known process. However, a detailed description of the method and some satisfactory error bounds can be found in Wilkinson (1965).

Gaussian elimination is generally used to solve a set of linear equations of the form,

$$A\underline{x} = \underline{b} \ , \tag{2.1.1}$$

by reducing A to an upper triangular matrix. The elimination in the matrix takes place in (n-1) stages, evaluating matrices $A^{(k)}$, k=1,2,..., n-1 where in general after the (k-1)th stage of elimination, the matrix $A^{(k-1)}$ has the form,

$$A^{(k-1)} = \begin{bmatrix} a_{1,1} & a_{1,2} & a_{1,3} & \cdots & & & a_{1,n} \\ 0 & a_{2,2}^{(1)} & a_{2,3}^{(1)} & \cdots & & & a_{2,n}^{(1)} \\ & 0 & a_{3,3}^{(2)} & \cdots & & & a_{3,n}^{(2)} \\ & & \ddots & \ddots & & & \vdots \\ & & & a_{k,k}^{(k-1)} & a_{k,k+1}^{(k-1)} & \cdots & a_{k,n}^{(k-1)} \\ & 0 & & a_{k+1,k}^{(k-1)} & a_{k+1,k+1}^{(k-1)} & \cdots & a_{k+1,n}^{(k-1)} \\ & & & \vdots & & & \vdots \\ & & & a_{n,k}^{(k-1)} & a_{n,k+1}^{(k-1)} & \cdots & a_{n,n}^{(k-1)} \end{bmatrix} \tag{2.1.2}$$

where the elements $a_{i,k}^{(k-1)}$, i=k+1,k+2,...,n are now eliminated by calculating

$$a_{i,j}^{(k)} = a_{i,j}^{(k-1)} - a_{i,k}^{(k-1)} * a_{k,j}^{(k-1)} / a_{k,k}^{(k-1)} \ , \quad i,j=k+1,k+2,\ldots,n. \tag{2.1.3}$$

It can be seen that after the (k-1) stages of elimination any one of the elements $a_{i,j}^{(k-1)}$, i,j=k,k+1,...,n could be chosen as the new pivot. Pivotal strategies are employed in order to maintain accuracy and retain sparsity when A is originally sparse. These strategies are fully discussed in later sections.

It usually happens that during the elimination process new nonzero elements (fill-ins) are created and therefore, the major purpose of this chapter is to study the effect of the elimination process on the sparsity of the matrices in question and to minimise the amount of fill-ins by using some types of pivotal strategies. The following example illustrates the important role of a suitable choice of pivoting in order to minimise the number of nonzero elements created during the process. Suppose that the matrix A is of the form as follows:

$$A \equiv \begin{bmatrix} X & X & X & X & X & \cdots & X \\ X & X & & & & & \\ X & & X & & & 0 & \\ X & & & X & & & \\ X & & & & X & & \\ \vdots & & 0 & & & \ddots & \\ X & & & & & & X \end{bmatrix} \tag{2.1.4}$$

where (X) denotes the position of a nonzero element.

If we apply the Gaussian elimination method (2.1.3) on A without any re-ordering (rows or columns) we will have a full matrix by the end of the process. However, if we permute the matrix to have the form as follows:

$$B=PAP^T \equiv \begin{bmatrix} X & & & & X \\ & X & & 0 & \vdots \\ & & \ddots & & \vdots \\ & 0 & & X & X \\ X & X & \cdots & \cdots & X \end{bmatrix} \tag{2.1.5}$$

where P is a permutation matrix defined as,

$$P \equiv \begin{bmatrix} & & & & 1 \\ & 0 & & 1 & \\ & & \iddots & & \\ & 1 & & 0 & \\ 1 & & & & \end{bmatrix} \tag{2.1.6}$$

then there will be no fill-ins during the elimination process.

The above situation where we have an ordering beforehand, which when operated on the matrix minimises the number of new nonzero elements (fill-ins) occurs rarely in practice. However, an ordering which locally minimises the amount of fill-ins, as described in later sections, can be used.

2.2 Symmetric Gaussian Elimination and Local Minimisation Algorithms

In this section we describe the Crout form of factorisation for a symmetric positive definite matrix, $A=U^TDU$. (Isaccson & Keller (1966) Rose, (1970), Sherman, (1975)).

The U^TDU factorisation of the matrix A is obtained in a recursive procedure where at the first step of the process, we write A as:

$$A \equiv A^{(1)} \equiv \left[\begin{array}{c|c} a_1 & r_1^t \\ \hline r_1 & C_1 \end{array}\right]$$

where a_1 is a single element, r_1 is a $(n-1)*1$ vector and C_1 is a $(n-1)*(n-1)$ matrix. The first element of the diagonal matrix D, i.e. $d_{1,1}$ and the first row of the upper triangular matrix U is now computed as:

$$u_{1,j} = [r_1^t/a_1]_j \;, \quad j=2,3,\ldots,n$$

and

$$d_{1,1} = a_1 \;.$$

We then compute $A^{(2)}$ as:

$$A^{(2)} = \left[\begin{array}{c|c} a_1 & r_1^t \\ \hline r_1 & C_1 - \dfrac{r_1 * r_1^t}{a_1} \end{array}\right]$$

where the matrix $\dfrac{r_1 * r_1^t}{a_1}$ is subtracted element by element from C_1.

In general, at the kth step $(1 \leq k \leq n-1)$ we have $A^{(k)}$ as:

$$A^{(k)} = \left[\begin{array}{c|cc} A_k & \multicolumn{2}{c}{B_k^t} \\ \hline & a_k & r_k^t \\ \cline{2-3} B_k & r_k & C_k \end{array}\right]$$

where A_k is a $(k-1)*(k-1)$ matrix, B_k is a $(n-k+1)*(k-1)$ array, a_k is a single element, r_k is $(n-k)*1$ vector and C_k is a $(n-k)*(n-k)$ matrix. We now have $d_{k,k}=a_k$ and the kth row of the matrix U is replaced by r_k^t/a_k. Finally, we compute $A^{(k+1)}$ as:

$$A^{(k+1)} = \left[\begin{array}{c|cc} A_k & \multicolumn{2}{c}{B_k^t} \\ \hline & a_k & r_k^t \\ \cline{2-3} B_k & r_k & C_k - \dfrac{r_k * r_k^t}{a_k} \end{array}\right] .$$

For our further discussion, it is of great convenience to visualise the elimination process by representing the structure of the matrix in a graph since the process of vertex elimination on graphs models the process of performing Gaussian elimination on the matrix, c.f. Harary, (1969), Rose, (1972).

The structure of an $(n*n)$ symmetric matrix A is given by the graph

$$G(A) = (X(A), E(A))$$

where

$$X(A) = \{x_1, x_2, \ldots, x_n\}$$

is the set of vertices and

$$E(A) = \left\{\{x_i, x_j\} \quad i \neq j, \quad a_{i,j} \neq 0\right\}$$

is the set of edges. For example, if A is

$$A \begin{bmatrix} a_{1,1} & 0 & a_{1,3} & a_{1,4} & 0 & a_{1,6} \\ 0 & a_{2,2} & a_{2,3} & a_{2,4} & 0 & 0 \\ a_{3,1} & a_{3,2} & a_{3,3} & 0 & a_{3,5} & a_{3,6} \\ a_{4,1} & a_{4,2} & 0 & a_{4,4} & a_{4,5} & 0 \\ 0 & 0 & a_{5,3} & a_{5,4} & a_{5,5} & a_{5,6} \\ a_{6,1} & 0 & a_{6,3} & 0 & a_{6,5} & a_{6,6} \end{bmatrix}$$

then the corresponding graph is shown below in Figure 2.1.

FIGURE 2.1:

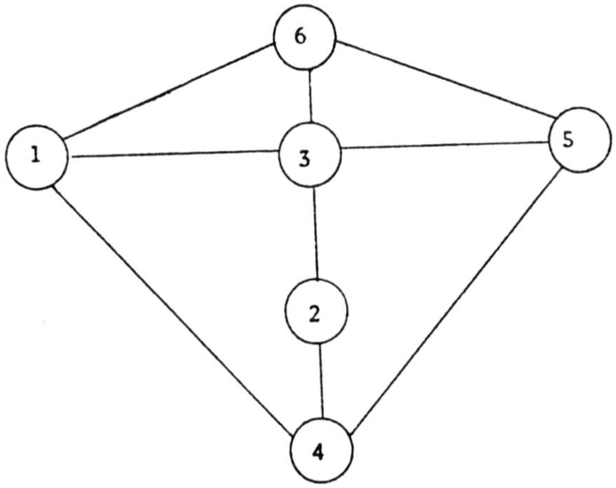

In the above graph, the vertex x_i corresponds to row i (or column i) and the set of edges i.e.

$$E(A) = \left\{ \{x_1,x_6\},\{x_1,x_3\},\{x_1,x_4\},\{x_6,x_3\},\{x_3,x_2\},\{x_3,x_4\} \right.$$
$$\left. \{x_6,x_5\},\{x_3,x_5\},\{x_5,x_4\} \right\}$$

corresponds to the nonzero elements of A.

For any vertex $x_i \in X(A)$, we define the set of vertices which are adjacent to x_i as,

$$\text{Adj}(x_i) = \{x_j \mid x_j \in X(A), \; x_j \text{ is adjacent to } x_i\}$$

and the degree of x_i as

$$\text{Deg}(x_i) = |\text{Adj}\ (x_i)|.$$

The deficiency of a vertex x_i, def(x_i), is the set of all pairs of distinct vertices $\{x_j,x_k\}$ where x_j and x_k are not adjacent to each other and are both adjacent to x_i. For example, in Figure 2.1 we have

$$\text{Adj}(x_1) = \{x_3,x_4,x_6\} \rightarrow \text{Deg}(x_1) = 3$$
$$\text{Adj}(x_3) = \{x_2,x_6\} \quad \rightarrow \text{Deg}(x_3) = 2$$

and

$$\mathrm{Def}(x_3) = \left\{\{x_1,x_5\},\{x_2,x_5\},\{x_1,x_2\}\right\}$$
$$\mathrm{Def}(x_1) = \left\{\{x_3,x_4\},\{x_4,x_6\}\right\}.$$

If in the Crout method as described above, the element $a_{i,i}$ is reordered so that it is chosen as the pivot in the first elimination step then the x_1-elimination graph of G as graph $G^{(2)}$ is formed by:

i) deleting vertex x_1 and all edges upon it,

ii) joining all pairs of vertices each of which belonged to $\mathrm{Adj}(x_1)$ in $G^{(1)}$.

For the Graph G shown in Figure 2.1 if the vertex x_5 (corresponding to the element $a_{5,5}$ of the matrix A) is reordered as x_1, then the graph $G^{(2)}$ which is formed by eliminating the vertex x_1 according to the above algorithm is given below.

FIGURE 2.2:

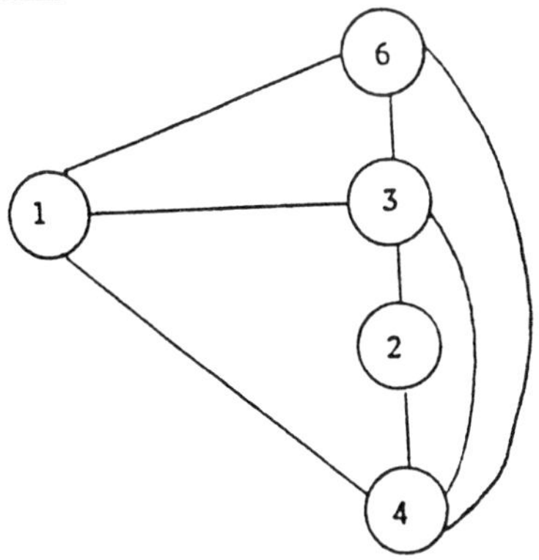

It can be seen that graph $G^{(2)}$ is not a subgraph of G, c.f. Harary, 1969. Indeed the graph elimination process can be defined as a sequence of graphs $G^{(1)}, G^{(2)}, \ldots, G^{(n)}$ where,

$$G^{(k)} = (X_k, E_k)$$

with $$X_k = \{x_i \mid x_i \in X \ \&\ i \geq j\}, \quad E_k = E_{k-1} \cup \mathrm{Def}(x_k) - \frac{E_{k-1}}{x_k}$$

where, $$\frac{E_{k-1}}{x_k} = \left\{\{x_k,x_j\} \in E_{k-1} \mid x_j \in \mathrm{adj}(x_k)\right\}.$$

Comparing the two graphs, $G^{(1)}$, $G^{(2)}$, we can see that there are two edges i.e. $\{x_3,x_4\},\{x_4,x_6\}$, which have been created after the first step of the elimination procedure and their set is the deficiency of vertex x_5 in $G^{(1)}$. In general, at the kth step, the number of such vertices (fill-ins) which are created as a result of eliminating x_k from the graph $G^{(k)}$ is the $\mathrm{Def}(x_k)$ in $G^{(k)}$. This fact is the basis of the first ordering we shall discuss in this section in order to minimise the storage, by creating as few non-zero elements as possible at each step of elimination.

This ordering is called the minimum deficiency algorithm, Rose, (1972). In this algorithm at the kth step of the elimination process, we search among the remaining unordered vertices and a vertex with the smallest deficiency is ordered as x_k and the elimination proceeds using x_k as the new pivot.

The second ordering which is described here, locally minimises the computational cost by reducing the number of arithmetic operations. The method is called the minimum degree algorithm, see Rose, (1970). It can be shown, Sherman, (1975), that the total number of multiplications or divisions i.e. θ_M and the total number of additions or subtractions, i.e. θ_A required for the U^TDU factorisation of A are:

$$\theta_M = \sum_{k=1}^{n-1} \frac{\text{Deg}(x_k)(\text{Deg}(x_k)+3)}{2}$$

$$\theta_A = \sum_{k=1}^{n-1} \frac{\text{Deg}(x_k)(\text{Deg}(x_k)+1)}{2}$$

and to minimise this computational cost, at the kth step of the elimination process, a vertex is reordered as x_k which has the minimum degree among the remaining vertices in $G^{(k)}$.

In both algorithms described above, tie-breaking strategies must be applied when searching for a vertex of minimum deficiency or minimum degree, if there are several vertices which all have the same smallest deficiency or minimum degree.

2.3 Some Pivotal Strategies for Arbitrary Matrices

In the previous section we discussed two pivotal strategies for sparse symmetric matrices which locally minimise either the storage requirement or computational cost of the factorisation process. In this section it is assumed that $A \equiv (a_{i,j})$, $i,j=1,2,\ldots,n$ is a general matrix and that (k-1) steps of the elimination process have been completed and the kth pivotal element is to be found.

We define the number of nonzero elements in row i $(i=k,k+1,\ldots n)$ of A with their column number greater than or equal to k as r(i,k), i.e.

$$r(i,k) = |\{a_{i,j} \mid j \geq k \ \&\ a_{i,j} \neq 0\}|, \quad i=k,k+1,\ldots,n \qquad (2.3.1)$$

and the number of nonzero elements in column j $(j=k,k+1,\ldots,n)$ with their row number greater than or equal to k as c(j,k), i.e.,

$$c(j,k) = |\{a_{i,j} \mid i\geq k \ \& \ a_{i,j}\neq 0\}| \ , \ j=k,k+1,\ldots,n. \quad (2.3.2)$$

We also define the submatrix A_k of A which contains all the elements $a_{i,j}$ where $i,j\geq k$ as the active part of A at stage k.

Perhaps the most widely used pivotal strategy is that proposed by Markowitz, (1957), whereby the nonzero element $a_{i,j}$ of the matrix A_k which minimises the quantity:

$$(r(i,k)-1) * (c(j,k)-1) \ , \quad (2.3.3)$$

is chosen as the pivot at the kth stage. This method locally minimises the number of new nonzero elements (fill-ins) created at step k of the elimination process since the quantity (2.3.3) is an upper bound on the fill-in. On the other hand, the choice of a nonzero element $a_{i,j}$ of the matrix A_k which minimises the quantity:

$$r(i,k)(c(j,k)-1)+1 \ , \quad (2.3.4)$$

locally reduces the number of arithmetic operations since quantity (2.3.4) represents the number of multiplications and divisions required at stage k of the elimination process.

It can be shown, Duff & Reid, 1974, that the use of expression (2.3.3) gives virtually identical results to the use of expression (2.3.4).

Zlatev, 1978, has recently developed the new pivotal strategies known as the generalised Markowitz algorithm which depend on the following two parameters:

1. U - the stability factor
2. p(k) - the number of rows that will be investigated for a pivotal element at stage k of the elimination process,

and shows that pivotal strategies with small p(k) may be more profitable than the original Markowitz algorithm.

Since pivotal strategies of this type have been used for numerical experiments on sparse matrices we shall described these methods in more detail. However, a full description of the strategies can be found in Zlatev, (1978).

We define B_k as a set of the nonzero elements of A_k at stage k as:

$$B_k = \{a_{i,j}^{(k)} \mid a_{i,j} \in A_k \ \& \ |a_{i,j}^{(k)}| > \max_{k\leq p\leq n} |a_{i,p}^{(k)}|/U, \ i \in I_k\} \quad (2.3.5)$$

where, $U > 1$,

$$I_k = \{i_1,i_2,\ldots,i_{p(k)}\},$$

for $k \leq i_m \leq n$, $m=1,2,\ldots,p(k)$

with,

$$r(i_1,k) \leq r(i_2,k) \leq \ldots \leq r(i_{p(k)},k) ,$$

and if $i \notin I_k$ and $k \leq i \leq n$ then $r(i_{p(k)},k) \leq r(i,k)$.

The parameter U is called a stability factor and if U>1 and A is a non-singular matrix then the set B_k whose elements are said to satisfy the stability condition will not be empty.

In the generalised Markowitz algorithm any member of the set B_k for which the quantity (2.3.3) is minimum can be chosen as the pivotal element at stage k. We define the set of such elements as C_k.

As we mentioned above, the choice of a pivotal element according to the generalised Markowitz algorithm depends on two parameters U and p(k) and when $U=\infty$ and $p(k)=n-k+1$ $(k=1,2,\ldots,n)$ the algorithm is equivalent to the original Markowitz strategy. The suggested values for the two parameters U and p(k) in two sparse elimination programs are as follows:

i) MA28 (c.f. Duff,(1977), and Duff & Reid,(1977) - $4 \leq U \leq 10$ and $p(k)=n-k+1$

ii) The FORTRAN version of SSLEST (c.f. Zlatev et al,(1978), - $4 \leq U \leq 16$ and $p(s)=\min(RPIV,n-k+1)$, where RPIV may be 1,2,... n.

It can be shown (c.f. Zlatev,(1978), that if the above pivotal strategy is employed to determine the pivot at each stage of the elimination process then the following bound can be derived:

$$b \leq (1+u)^{n-1} a , \qquad (2.3.6)$$

where a is the largest element in absolute value of the original matrix A, and b is the largest element in absolute value created at any stage of the elimination. On the other hand, it can be proved (c.f. Reid, (1971), that if an approximation LU-factorisation of A where L and U are lower and upper triangular matrices respectively is carried out so that:

$$LU = PAQ + E , \qquad (2.3.7)$$

then the elements of the perturbation matrix E, i.e. $e_{i,j}$ satisfy the inequality,

$$|e_{i,j}| \leq 3.01\ nb\varepsilon \qquad (2.3.8)$$

where ε is the machine accuracy. Therefore, to keep the elements $e_{i,j}$ small enough to obtain an accurate LU-factorisation of A, the value b must be kept small. Obviously, the right-hand side of (2.3.6) may be reduced by the choice of a smaller value of the stability parameter U. However, decreasing the value of U may cause the number of elements in B_k to decrease too, and this may force one to choose an element of B_k as the pivot for which the quantity (2.3.3) is not small enough to keep the amount of fill-ins down.

Another pivotal strategy which sometimes gives a possibility of reducing the right-hand side of (2.3.6) is to choose the largest element of the set C_k as a pivotal element. This is called the improved generalised Markowitz strategies. The reasons for using the term "improved" for this strategy and some examples of the application of the above two strategies can be found in Zlatev, (1978).

2.4 Nested Dissection Algorithms

In this section we consider the ordering strategies for a finite element system of equations associated with a regular n*n mesh or grid. We start by giving an example which illustrates how the way we number or order the vertices of an n*n mesh, D_n, can affect the amount of computation and storage requirements for the Gaussian elimination of the corresponding matrix.

Suppose that the internal points of a 3*3 mesh, D_3, are ordered as follows:

FIGURE 2.3:

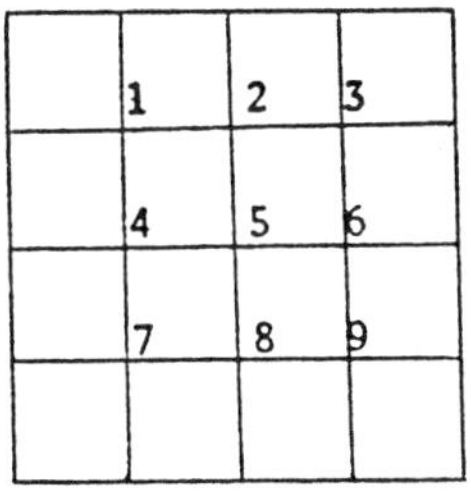

then, the nonzero elements of the matrix derived from the 5 point finite difference analogue of the Laplace equation are marked by an 'X' in Figure 2.4.

<u>FIGURE 2.4</u>:

	1	2	3	4	5	6	7	8	9
1	X	X		X					
2	X	X	X	.	X		0		
3		X	X	.	.	X			
4	X	.	.	X	X	.	X		
5		X	.	X	X	X	.	X	
6			X	.	X	X	.	.	X
7				X	.	.	X	X	.
8		0			X	.	X	X	X
9						X	.	X	X

Ignoring the accidental creation of zero elements during the elimination process, it can easily be seen that the number of the nonzero elements created during the elimination process whose positions are marked by a (.) in Figure 2.4 is sixteen.

However, if we re-number the above region as follows

<u>FIGURE 2.5</u>:

	1	2	4
	3	5	7
	6	8	9

then Figure 2.6 shows the structure of the corresponding matrix and the occurrence of fill-ins. It can be seen that the number of nonzero elements which are created as a result of the elimination is now thirteen. For larger mesh problems, this difference would be very much greater.

FIGURE 2.6:

	1	2	3	4	5	6	7	8	9
1	X	X		X					
2	X	X		X	X				
3	X	.	X	.	X	X			
4		X	.	X	.	.	X		
5		X	X	.	X	.	X		
6			X	.	.	X	.	X	
7				X	X	.	X	.	X
8					X	X	.	X	X
9							X	X	X

We now turn to near optimal orderings of the vertices of a mesh produced by George,(1973), e.g. for a nine-point model problem (see also Woo, et al,(1973), Birkhoff and George,(1973),) and briefly describe the algorithm for an (n*n) mesh where $n=2^t-1$ for some positive integer t, c.f. Sherman,(1975). For other values of n, see Duff et al, (1974), Rose & Whitten,(1974).

The nested dissection algorithm essentially involves successive dissections of successive subregions where the basic step of the process consists of first choosing a separating cross which divides the submesh into identical m*m quadrants, i.e.,

FIGURE 2.7: Separating cross in D_7

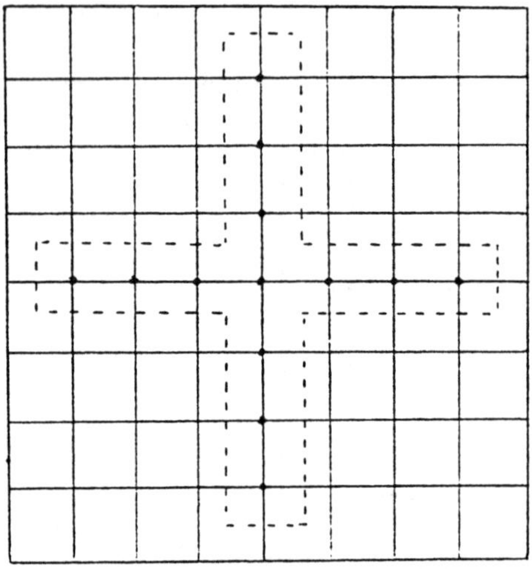

then ordering the vertices on the separating cross is completed as follows.

If the vertices $x_{k+1}, x_{k+2}, \ldots, x_n$ have already been determined then the 2m-1 vertices along the separating cross in any submesh D_m are ordered as $x_{k-2m+2}, x_{k-2m+3}, \ldots, x_k$. See Figure 2.8. Note that the ordering within each separating cross is relatively unimportant.

The algorithms start by applying the basic step as described above to the mesh D_n and divides it into four identical (m*m) submeshes where $m=2^{t-1}-1$, see Figure 2.9. The process is now repeated by applying the basic step to each of the submeshes D_m to obtain 16 square submeshes $D_{m'}$, where $m'=2^{t-2}-1$ and continues in this way until all the vertices on the mesh have been ordered. Figure 2.8 illustrates the nested dissection ordering on a (7*7) grid.

FIGURE 2.8: Numbering of vertices in separating cross in D_7

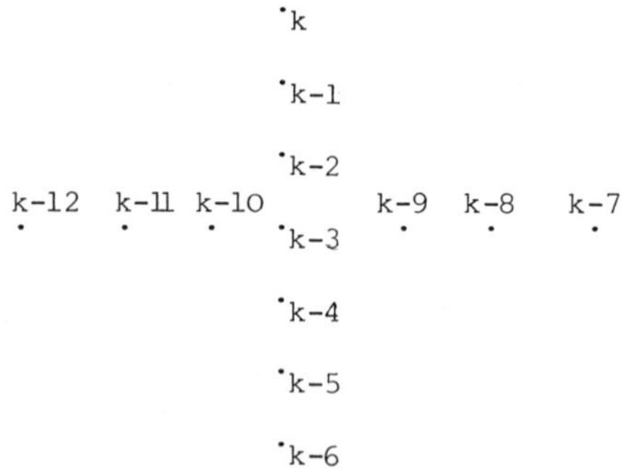

FIGURE 2.9:

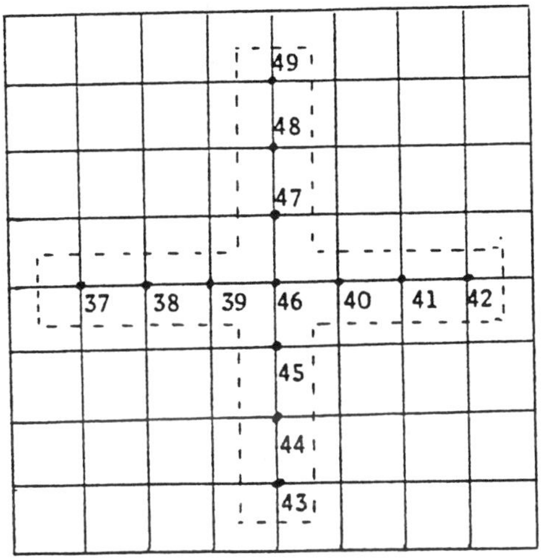

FIGURE 2.10:

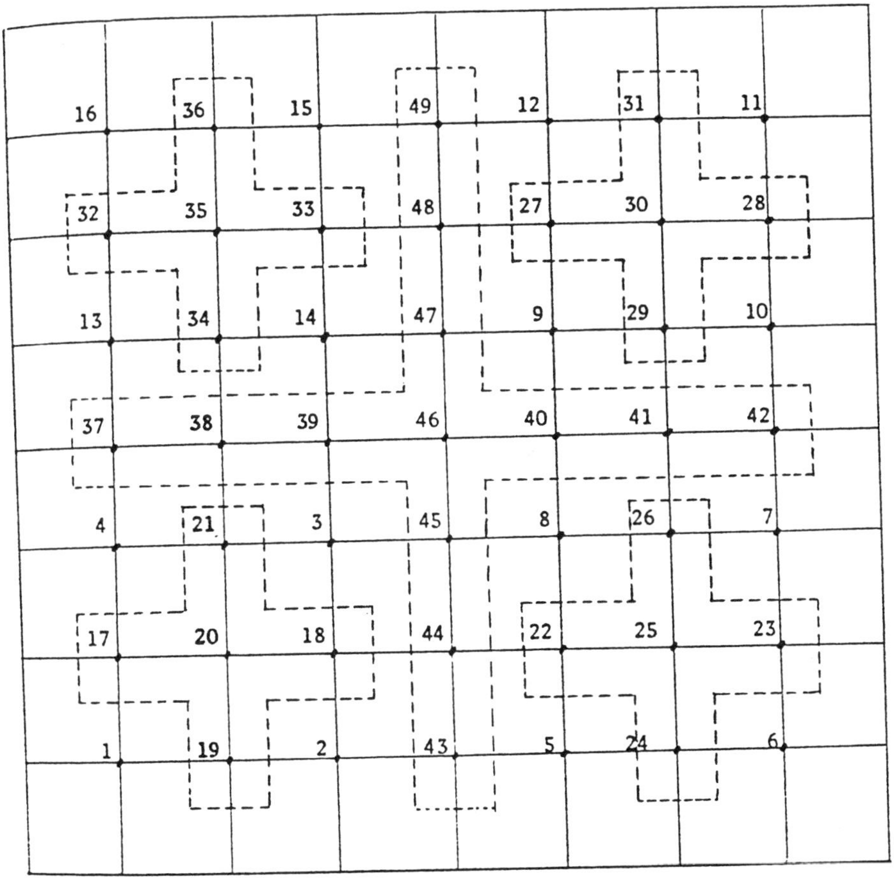

Note that in the above algorithm, the basic ordering step is applied to all the submeshes of equal size before operating on any smaller submeshes. However Rose and Whitten, (1974), have produced a version of the nested dissection algorithm whereby at the first step we divide the mesh D_n into four identical m*m submeshes D_m where $m=2^{t-1}-1$ by applying the basic step to D_n. Next, the basic step is applied to one of the generated D_m's to obtain four smaller square submatrices $D_{m'}$, where $m'=2^{t-2}-1$, then one of the $D_{m'}$ is further subdivided and the algorithm continues in this way recursively until all the vertices in D_n have been ordered. This strategy is called the <u>depth-first nested</u> dissection whereas the first method is called the <u>breadth-first nested</u> dissection. The two algorithms differ only in the order in which they possess the submeshes. However both have an equivalent set of separating crosses. Figure 2.11 illustrates the depth-first nested dissection ordering on a (7*7) grid.

<u>FIGURE 2.11</u>:

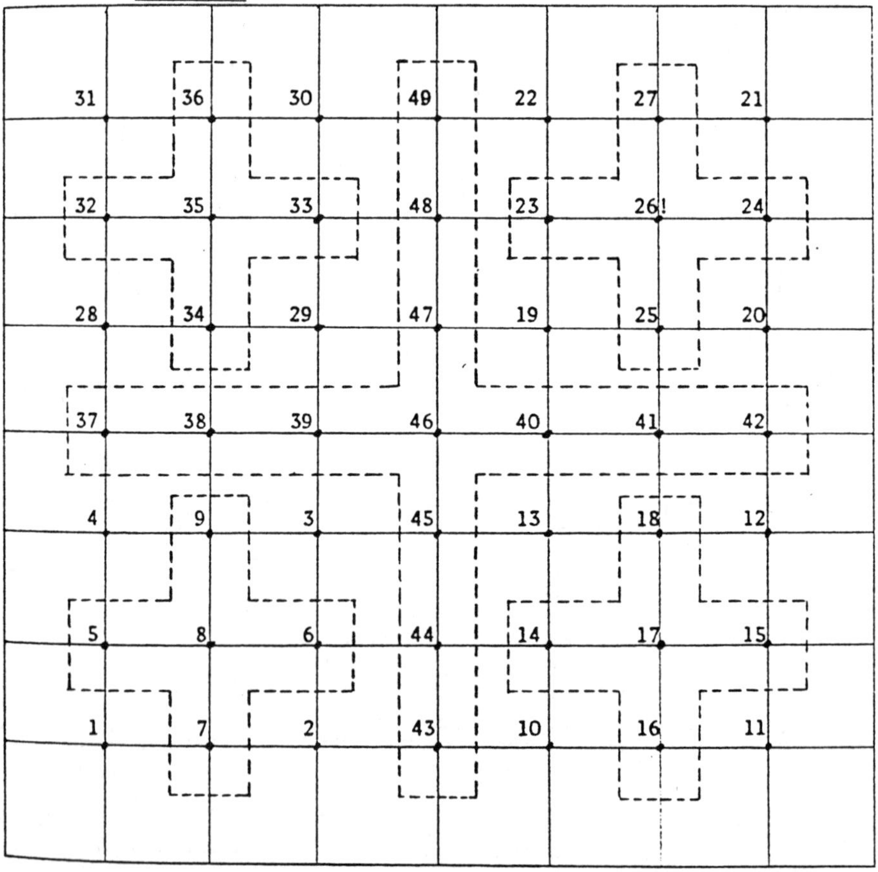

It can be shown, George, (1973), that the number of nonzero elements of the final upper triangular matrix of the elimination process is $O(n^2 \ell g_2 n)$ for an n*n mesh, and the amount of arithmetic operations required is $O(n^3)$. In the natural ordering of the vertices these numbers are, $O(n^3)$ and $O(n^4)$ respectively.

2.5 Determination of the Eigenvalues of a Matrix by the Elimination Method

Now that sparse elimination techniques and the significant role of pivotal strategies have been fully explained, we outline the procedure for determining the eigenvalues of a matrix A by a suitable modification to the Gaussian elimination process together with the use of polynomial solvers.

In this section we describe how the eigenvalues of a matrix can be obtained by the elimination method and therefore, the modifications required to convert a linear system solver program package to an eigenvalue system solver.

We first prove that the determinant of an upper triangular matrix A, is equal to the product of the elements along the leading diagonal, i.e.,

$$\det(A) = \prod_{i=1}^{n} a_{i,i} , \tag{2.5.1}$$

The proof is obtained easily by induction, where for n=2, using the expansion of $\det(A_2)$ by its second row we have,

$$\det(A_2) = \begin{bmatrix} a_{1,1} & a_{1,2} \\ 0 & a_{2,2} \end{bmatrix} = a_{1,1}*a_{2,2} . \tag{2.5.2}$$

We now assume that for n=m we have,

$$\det(A_m) = \prod_{i=1}^{m} a_{i,i} , \tag{2.5.3}$$

and prove that for n=m+1 the following relationship holds:

$$\det(A_{m+1}) = \prod_{i=1}^{m+1} a_{i,i} . \tag{2.5.4}$$

The matrix A_{m+1} can be written as:

$$A_{m+1} = \left[\begin{array}{ccc|c} & & & a_{1,m+1} \\ & A_m & & a_{2,m+1} \\ & & & \vdots \\ & & & a_{m,m+1} \\ \hline 0 & 0 \ldots & 0 & a_{m+1,m+1} \end{array}\right] \tag{2.5.5}$$

Then using the expansion of $\det(A_{m+1})$ by its $(m+1)^{st}$ row we have:

$$\begin{aligned} \det(A_{m+1}) &= a_{m+1,m+1} * (-1)^{(m+1)+(m+1)} * \det(A_m) \\ &= a_{m+1,m+1} * (-1)^{2(m+1)} * \det(A_m) = a_{m+1,m+1} \det(A_m) \end{aligned} \tag{2.5.6}$$

and substituting for $\det(A_m)$ in the above expression its equivalent value in (2.5.3) we have:

$$\det(A_{m+1}) = a_{m+1,m+1} * \prod_{i=1}^{m} a_{i,i} = \prod_{i=1}^{m+1} a_{i,i} , \tag{2.5.7}$$

and the result (2.5.4) is obtained.

From the above discussion and the properties of determinants we can conclude that once a matrix A has been factorised and the pivotal elements together with the total number of interchanges made during the factorisation process are known then,

$$\det(A) = (-1)^p \prod_{i=1}^{n} \text{pivot}(i) , \tag{2.5.8}$$

where p is the total number of interchanges and pivot(i), i=1,2,...,n denote the pivotal elements of A.

We are now able to compute the determination of an eigenvalue λ of a matrix A by the elimination method with the help of polynomial solvers (i.e. Newton-Raphson iteration, Muller's method, etc.). The problem is to find a value of λ for which we have

$$F(\lambda) \equiv \det(A-\lambda I) = 0 . \tag{2.5.9}$$

To determine such a value we proceed as follows:

1. An initial guess to λ (i.e. λ_0) is made and the determinant of $(A-\lambda_0 I)$ is evaluated by the elimination procedure and multiplying all the pivots together as described above.
2. If for this value of λ the $\det(A-\lambda_0 I)$ is equal to zero then

λ_0 is the eigenvalue required and the process is terminated.

3. Otherwise, an improved estimate to λ (i.e. λ_1) is obtained from λ_0 and the determinantal value $F(\lambda_0)$ and its derivative $F'(\lambda_0)$, if required, together with previous evaluations, i.e. $\lambda_{-1}, F(\lambda_{-1})$, $F'(\lambda_{-1}), \lambda_{-2}(\lambda_{-2})$, $F'(\lambda_{-2})$, etc. by using one of the many well known polynomial solvers.

4. The process is repeated by replacing λ_1 for λ_0 in the above order until the difference between two successive values of (i.e. λ_i, λ_{i+1}) is smaller than some predefined tolerance ε.

Note that the choice of the polynomial solver for the above algorithm is important because of the arbitrary interchanges made during the elimination process. In the above two methods (i.e. the Secant and Muller's methods) the correct sign for the evaluated determinant after the completion of the elimination process can be obtained easily from the formula (2.5.8), however, during the row/column interchanging process it can be shown that the sign of successive minors of the matrix A may be disturbed and consequently their use in a Sturm sequence procedure using the well known bisection algorithm (which is less costly than the above two methods) can lead to erroneous results. Further research into this topic is necessary before any firm conclusions can be given.

3. ITERATIVE METHODS FOR PARTIAL DIFFERENTIAL EQUATIONS

3.1 Introduction

Mathematical models of physical phenomena often require the solution of partial differential equations. These equations frequently involve complex geometries and awkward boundary conditions. Because of these restrictions, analytic solutions of the partial differential equations are frequently impossible to achieve.

In this section we shall be concerned with a numerical solution of a linear elliptic partial differential equation. The typical form of the equation is given by,

$$Lu = -\sum_{i=1}^{n} \frac{\partial}{\partial x_i}\left(a_i(x)\frac{\partial u}{\partial x_i}\right) + q(x)u = f(x) \quad , \quad \forall x \in D \tag{3.1.1}$$

where $x=(x_1,x_2,\ldots,x_n)$, $a_i(x)>0$, $\forall i$, $q(x)\geq 0$ and D is a closed bounded region with boundary ∂D. Equation (3.1.1) is solved under given boundary conditions, which often take one of the three following forms:-

(1) Dirichlet conditions, where the value of the function u(x) is specified on the boundary, that is,

$$u(x) = g(x) \ , \quad \forall x \in \partial D$$

(2) Neumann conditions, where the normal derivative is specified, such that,

$$\frac{\partial u}{\partial n} = g(x) \ , \quad \forall x \in \partial D$$

where $\partial u/n$ denotes the outward normal derivative of u evaluated on the boundary.

(3) Robbin's conditions, which are mixed boundary conditions of the form,

$$\alpha(x)u + \beta(x)\frac{\partial u}{\partial n} = g(x) \ ,$$

where $\alpha(x)>0, \ \beta(x)>0 \ , \quad \forall x \in \partial D.$

By using either finite-difference approximations or finite-element methods, we then reduce the solution of equation (3.1.1) to the solution of systems of simultaneous linear equations,

$$Au = b \ , \tag{3.1.2}$$

where A is a square (large and sparse) matrix, b is a known column vector, and u is the unknown column vector.

In this paper, however, we shall mainly study a numerical solution of large systems of linear equations. The question of how these equations arise from the discretization of a partial differential equation will not be discussed, and the reader may consult the literature (Varga, (1962), Strang and Fix, (1973)) as necessary.

Iterative methods, consist of repeated application of an often simple algorithm until the required accuracy is achieved. The methods have been preferred to direct methods for solving large sparse systems of linear equations, because they are easier to program and require less computer storage as well as less rounding error growth in the solution.

It must be emphasized although there is much discussion on iterative versus direct methods as soon as we come to consider very large systems (as arising for problems in three or higher dimensions), there is little doubt that iterative methods are likely to be more favourable than direct methods.

3.2 General Principles for Iterative Processes

In any iterative method we begin with an arbitrary initial

approximation according to some scheme. The rate of convergence of the iteration is, of course, important, since no method can be regarded as practical and efficient unless the convergence rate is rapid. In this section, we shall present the general principles of analysis of convergence for iterative processes, and the rate of convergence for various methods will be stated without giving proof. Excellent accounts of matrix iterative techniques are given in Varga, (1962) and Young, (1971).

The most general form of an iterative process for solving linear systems of equations

$$Ax = b\ , \tag{3.2.1}$$

may be described by a sequence of vectors $x^{(1)}, x^{(2)}, \ldots, x^{(n)}$, which is constructed according to the formula

$$x^{(k+1)} = C_k x^{(k)} + d_k\ , \tag{3.2.2}$$

where C_k is defined as the iteration matrix, which depends upon the iteration number k, the matrix A and the vector b. If C_k is independent of k, the iteration is said to be <u>stationary</u>. A different choice of the matrices C_k leads to a different iterative process. The iterative process arising from the recurrence relation (3.2.2) possesses the property that successive vectors $x^{(k)}$ satisfy the fixed point solution x^*. This means that if the process converges, then all subsequent approximations will also equal x^*, such that

$$x^* = C_k x^* + d_k \tag{3.2.3}$$

At this point, two questions while exceedingly important, must be answered before we go on to discuss various methods based on equation (3.2.2).

(1) How can we tell whether the iterative scheme (3.2.2) converges?

(2) If the scheme does converge, how fast will the convergence rate be?

It is easy to derive the necessary and sufficient condition for the iterative process (3.2.2) to converge to the solution for any initial vector $x^{(0)}$. In fact, equations (3.2.2) and (3.2.1) give

$$\begin{aligned} x^* - x^{(k+1)} &= C_k(x^*-x^{(k)}) \\ &= C_k\ C_{k-1}, \ldots, C_1\ C_0(x^*-x^{(0)}) \\ &= \prod_{i=0}^{k} C_i(x^*-x^{(0)}) \end{aligned} \tag{3.2.4}$$

In order for $(x^*-x^{(0)})\to 0$ for any initial vector $x^{(0)}$, it is necessary and sufficient that $\prod_{i=0}^{k} C_i \to 0$ as $k\to\infty$. When the iteration is stationary, $C_i \equiv C$, so thus for convergence it requires $C^k \to 0$ and $k\to\infty$.

The quantity $(x^*-x^{(k)})$ is generally known as the error vector $e^{(k)}$ at the kth iterate. In terms of matrix and vector norms, we can show that for a stationary process

$$||e^{(k)}|| \leq ||C^k|| \; ||e^{(0)}||, \; k \geq 0 \quad . \tag{3.2.5}$$

Thus, if the eigenvalues of C are given by λ_i, then a necessary and sufficient condition for convergence is that the spectral radius of the iteration matrix $\rho(C)$ must be strictly less than unity, where $\rho(C)=\max_i|\lambda_i|$.

To a considerable extent we can answer the second question by a knowledge of the asymptotic behaviour of the error vector $e^{(k)}$ as $k\to\infty$. For a fixed number of iterations k, the quantity

$$\sigma = \left[\frac{||e^{(k)}||}{||e^{(0)}||}\right]^{1/k} \tag{3.2.6}$$

is the average reduction in the error norm per iteration. For a stationary process and if $||C^{(k)}||<1$, (3.2.6) becomes,

$$\sigma \leq (||C^k||)^{1/k} \tag{3.2.7}$$

As mentioned earlier, in constructing an iterative procedure, the speed of convergence is a critical quantity. In order to compare various iterative processes, we need a measure for the rate of convergence. Following the definition by Varga (1962), if C_1 and C_2 are two iteration matrices of the same order, and if for some positive integer k, $||C^k||<1$, then

$$R(C_1^k) = -\ln[(||C_1^k||)^{1/k}] = -\frac{1}{k}\ln||C_1^k|| \tag{3.2.8}$$

is the average rate of convergence for k iterations of the matrix C_1. If $R(C_1^k)<R(C_2^k)$, then C_2 is iteratively faster for k iterations than C_1. In the limit $k\to\infty$, the average rate of convergence $R(C_1^k)$ satisfies

$$\lim_{k\to\infty} R(C_1^k) = -\ln\rho(C_1) = R_\infty(C_1) \tag{3.2.9}$$

where $R_\infty(C)$ is called the asymptotic rate of convergence for C. It should be noted that although the asymptotic rate of convergence provides a simple and practical measure for the convergence rate of an iterative method, this measure must be used with discretion.

3.3 Basic Iterative Methods

We now consider the problem of constructing an iterative procedure for solving equation (3.2.1), where A is a given nonsingular $n \times n$ matrix, b is a known column vector, and x is the unknown solution vector. Let us consider A to be partitioned into the sum of two $n \times n$ matrices M and N, such that,

$$A = M - N , \tag{3.3.1}$$

where M is nonsingular. In general we can express a stationary iterative scheme in the form,

$$Mx^{(k+1)} = Nx^{(k)} + b , \tag{3.3.2}$$

where $x^{(k)}$ is the kth approximation to the solution vector. It should be noted that in splitting the matrix A, the most important consideration is to choose the matrix M so that it can be easily inverted. Depending on the splitting of the matrix, various iterative methods are constructed. A few of the best known iterative procedures are built by partitioning the matrix A into the form,

$$A = D - L - U , \tag{3.3.3}$$

where D is the diagonal $n \times n$ matrix with the diagonal entries a_{ii} of A, and $a_{ii} > 0$, $\forall i$. L and U are respectively strictly lower and upper triangular $n \times n$ matrices, whose entries are the negative of the entries of A below and above the main diagonal.

(1) Jacobi Method

The method is also known as the method of simultaneous displacements or iteration by total steps. In terms of actual computational procedure given by (3.3.2) the method is characterized by the choice

$$M = D , \quad N = (L + U) . \tag{3.3.4}$$

The Jacobi iteration is little used today, because it has comparatively a slow convergence rate. However its explicit virtue can be exploited by parallel computers of today.

(2) Gauss-Seidel (G-S) Method

The method was first used by Gauss and was independently discovered by Seidel. The method is also known as the method of successive displacements or iteration by single steps. In matrix form equation (3.3.2), is characterised by the choice

$$M = (D - L) , \quad N = U . \tag{3.3.5}$$

It can be proved that a sufficient condition for the methods of Jacobi and

G-S to converge is that the matrix A is diagonally dominant, that is

$$\max_{i} \sum_{j=1}^{n}{}' \left|\frac{a_{ij}}{a_{ii}}\right| < 1 \qquad (3.3.6)$$

where the prime on the summation sign means that the term for j=i is omitted. The G-S method converges twice as fast as the corresponding Jacobi method, provided the Jacobi matrix B is non-negative with $0<\rho(B)<1$. It is not hard to understand this, since the G-S method is based upon the use of the latest estimated values as soon as they are calculated.

(3) Successive Over-relaxation (SOR) Method

This method is closely related to the G-S method, but with the application of an acceleration technique so as to increase the rate of convergence of the iterative process. Let $\bar{x}^{(k)}$ be the vector calculated from the G-S method at the kth iteration, then the actual component $x^{(k)}$ of SOR is defined according to the scheme

$$x^{(k)} = x^{(k-1)} + \omega(\bar{x}^{(k)} - x^{(k-1)}) = (1-\omega)x^{(k-1)} + \omega\bar{x}^{(k)} \qquad (3.3.7)$$

The quantity ω is called the relaxation factor, and when $\omega=1$ the method becomes the G-S method. Expressed in matrix notation equation (3.2.16), the SOR method takes the form

$$M = \omega^{-1}(D-\omega L), \quad N=\omega^{-1}[(1-\omega)D+\omega U] \qquad (3.3.8)$$

Convergence can be proved if A is positive definite and $0<\omega<2$. In practice, the most difficult part of the application of the SOR method lies in the determination of the optimal relaxation factor ω, for which the rate of convergence can be greatly affected by a small deviation of the parameter ω.

The SOR method is particularly useful if the matrix A possesses "Property A", that is, if there exists a permutation matrix P, such that

$$PAP^T = \begin{bmatrix} D_1 & F \\ G & D_2 \end{bmatrix} \qquad (3.3.9)$$

where D_1 and D_2 are square diagonal matrices, and G and F are rectangular matrices. It should be noted that "Property A" is in fact the special case of the matrix which is consistently ordered and of 2-cyclic (Varga, 1962). An important consequence of this is that the optimal relaxation factor ω_{opt} can be theoretically determined according to

$$\omega_{opt} = 2/\{1-\sqrt{1-[\rho(C)]^2}\} \qquad (3.3.10)$$

where $\rho(C)$ is the spectral radius of the Jacobi matrix $C=D^{-1}(L+U)$.

(4) Generalisations of SOR

(i) SSOR

The SOR iteration described above lacks symmetry, which can be obtained if we perform one step of an SOR iteration on the equation $Ax=b$ to obtain $x_{\frac{1}{2}}$, and then follow with an SOR iteration on the equations in the reverse order. That is, we apply one step of the iteration to $\Pi Ax=\Pi b$ where,

$$\Pi = \begin{bmatrix} & & & 1 \\ & O & 1 & \\ & \cdot^{\cdot^{\cdot}} & & \\ 1 & & O & \end{bmatrix} \qquad (3.3.11)$$

using $x_{\frac{1}{2}}$ as a starting vector. The process is then repeated.

For the work performed, the symmetric SOR (SSOR) iteration gives about the same rate of convergence as SOR but has some attractions due to its symmetry, i.e. the eigenvalues of the iteration matrix are real and f.

(ii) Line and Block SOR

The SOR iteration (sometimes called point SOR to distinguish it from the methods under discussion here) has the form,

$$x^{(i+1)} = \mathcal{L}_{\omega} x^{(i)} + (D - \omega L)^{-1}\omega b \qquad (3.3.12)$$

where,

$$\mathcal{L}_{\omega} = (D - \omega L)^{-1}\{(1-\omega)D+\omega U\}, \qquad (3.3.13)$$

and $A = D - L - U$.

We may generalise the iteration by replacing each component of $x^{(i)}$ by a vector $x_s^{(i)}$ so that $x^{(i)}$ becomes

$$x_i^{(s)} = [x_1^{(i)}, x_2^{(i)}, \ldots, x_m^{(i)}]^T .$$

We replace the diagonal matrix D by a block diagonal matrix,

$$\begin{bmatrix} D_1 & & & \\ & D_2 & & O \\ & & \ddots & \\ O & & & D_m \end{bmatrix}$$

and replace L and U by (block) upper and lower triangular matrices such that $A = D - L - U$. (Each matrix D_j must be non-singular).

At the ith stage we calculate in succession the (sub)-vectors $x_1^{(i)}$,

$x_2^{(i)},\dots,x_m^{(i)}$. The rth vector $x_r^{(i)}$ must be obtained by the solution of a system of equations,

$$D_r x_r^{(i+1)} = \{\omega \sum_{j=1}^{r-1} L_{rj} x_j^{(i+1)} + \sum_{j=r+1}^{m} U_{rj}\, x_j^{(i)} + b_r\} + (1-\omega) D_r x_r^{(i)} ,$$

$$r=1,2,\dots,m , \qquad (3.3.14)$$

where $b=[b_1,b_2,\dots,b_m]^T$.

In the context of the finite difference solution of partial differential equations, the component of the vector $x_r^{(i)}$ will usually correspond to a set of neighbouring points in some geometrical sense. In particular, we may find the solution as a group for a whole line of mesh points (Line SOR).

Example

$$\begin{matrix} {}^1x & {}^2x & {}^3x \\ {}^4x & {}^5x & {}^6x \\ {}^7x & {}^8x & {}^9x \end{matrix}$$

If we consider the same mesh we considered in examples of Property (A), we may group the points in the form,

$$x_1 = [x_1,x_2,x_3]^T , \quad x_2 = [x_4,x_5,x_6]^T , \quad x_3 = [x_7,x_8,x_9]^T .$$

The corresponding example in three dimensions would lead to a block SOR method.

Generalisations of property (A) and consistent ordering to line and block SOR have been described in Forsythe and Wasow (1960) and Ames (1969).

The block iterative method have the advantage that they make the iterative process converge more rapidly than the corresponding point methods. It can be shown that the Jacobi method with single line converges at the same rate as the point G-S method and twice as fast as the point Jacobi matrix, if the matrix has "Property A". Furthermore, an improvement by the factor $(2k)^{\frac{1}{2}}$ over the corresponding optimum point SOR method can be achieved for the optimum k line SOR method for solving an elliptic partial differential equation.

(5) Explicit Block Iterative Methods

In this section we study alternative partitioning strategies for a system of linear equations arising from the five-point finite difference equations and compare them with the well known 1-line and 2-line block iterative methods.

For purposes of illustration, let us consider the following model problem. Given a function g(x,y) defined and continuous on the boundary $\partial\Omega$ of the unit square Ω, $0\leq x,y\leq 1$, and given the function f(x,y) defined and continuous in Ω, we seek a function u(x,y) which is continuous and twice differentiable in the interior R of Ω and which satisfies in R the Poisson differential equation

$$\frac{\partial^2 u}{\partial x^2} + \frac{\partial^2 u}{\partial y^2} = f(x,y) , \qquad (3.3.15)$$

the function u(x,y) is required to satisfy on $\partial\Omega$ the Dirichlet condition

$$u(x,y) = g(x,y) . \qquad (3.3.16)$$

To apply the method of finite differences, we replace the differential equation (3.3.15) by their usual five-point difference approximations. With the values of u(x,y) at the interior points of R considered as unknowns and with the values on $\partial\Omega$ determined by the boundary condition (3.3.16) we get a system of linear algebraic equations and by collecting all known terms to the right hand side, we obtain a linear system of the form,

$$A\underline{u} = \underline{b} . \qquad (3.3.17)$$

Consider the region in Figure 3.1.

FIGURE 3.1:

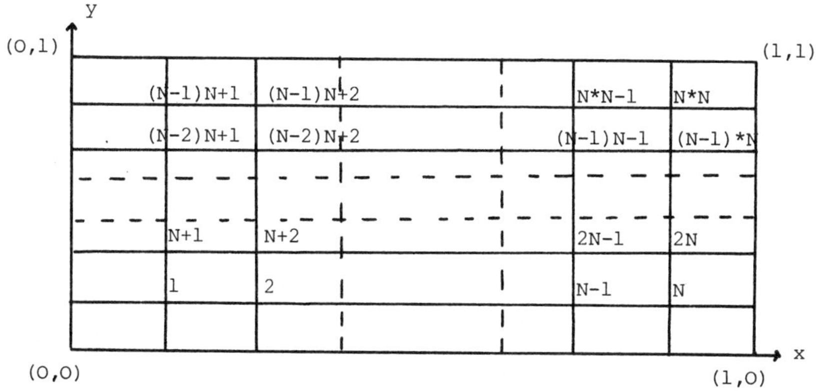

Now construct two novel groupings of mesh points in the following manner. Consider the net region depicted in Figure 3.1.

1) Place the 4 mesh points as shown in Figure 3.2 in each group where $k=(\ell N+1)(2)(\ell+1)N-1$ and $\ell=0(2)N-2$. In this case, where N is an even number we have the subset G_M, $M=1,2,\ldots,N^2/4$, such that each G_M

consists of four elements, $\{k,k+1,N+k,N+k+1\}$. However, the matrix $A_{M,M}$ is of order 4 and of the form,

FIGURE 3.2:

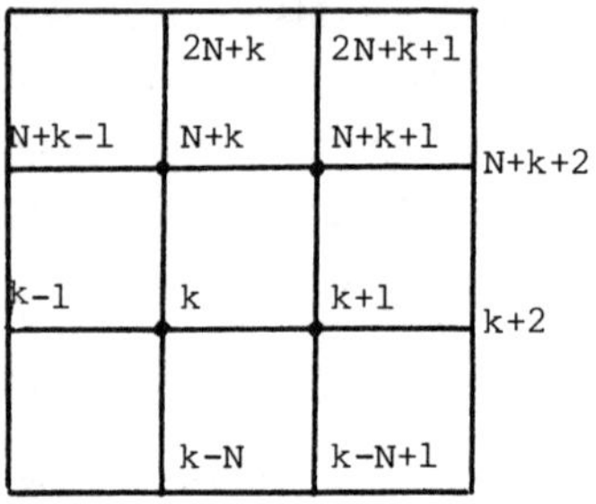

$$A_{M,M} = \begin{bmatrix} 4 & -1 & -1 & 0 \\ -1 & 4 & 0 & -1 \\ -1 & 0 & 4 & -1 \\ 0 & -1 & -1 & 4 \end{bmatrix}, \quad M=1,2,\ldots,N^2/4 . \tag{3.3.18}$$

Thus, for the model problem (3.3.15) we have the group Jacobi method given by,

$$\left.\begin{aligned} 4u_k^{(n+1)}-(u_{k+1}^{(n+1)}+u_{N+k}^{(n+1)}) &= u_{k-N}^{(n)}+u_{k-1}^{(n)}+f_k \\ 4u_{k+1}^{(n+1)}-(u_k^{(n+1)}+u_{N+k+1}^{(n+1)}) &= u_{k+2}^{(n)}+u_{k-N+1}^{(n)}+f_{k+1} \\ 4u_{N+k}^{(n+1)}-(u_k^{(n+1)}+u_{N+k+1}^{(n+1)}) &= u_{N+k-1}^{(n)}+u_{2N+k}^{(n)}+f_{N+k} \\ 4u_{N+k+1}^{(n+1)}-(u_{k+1}^{(n+1)}+u_{N+k}^{(n+1)}) &= u_{2N+k+1}^{(n)}+u_{N+k+2}^{(n)}+f_{N+k+1} \end{aligned}\right\} \tag{3.3.19}$$

such that, each iteration involves the solution of $N^2/4$ systems of order 4 with the matrix given in (3.3.18). Evidently, it can be seen that the matrix $A_{k,k}$ can be easily inverted. Therefore, in practice, group systems of the type (3.3.19) are solved beforehand and the iteration completed explicitly. As an example, the solution of the model problem (3.3.15) in the case of the Laplace equation (i.e. $f(x,y)=0$), gives the system of equations (3.3.19) which may be rewritten explicitly, as,

$$\begin{aligned} u_k^{(n+1)} &= \frac{1}{24}[7(u_{k-N}^{(n)}+u_{k-1}^{(n)}+2(u_{k-N+1}^{(n)}+u_{k+2}^{(n)}+u_{2N+k}^{(n)}+u_{N+k-1}^{(n)})+u_{N+k+2}^{(n)}+ \\ &\qquad u_{2N+k+1}^{(n)}] , \\ u_{k+1}^{(n+1)} &= \frac{1}{24}[7(u_{k-N+1}^{(n)}+u_{k+2}^{(n)})+2(u_{k-N}^{(n)}+u_{k-1}^{(n)}+u_{N+k+2}^{(n)}+u_{2N+k+1}^{(n)})+u_{2N+k}^{(n)} \\ &\qquad +u_{N+k-1}^{(n)}] , \end{aligned}$$

$$\left.\begin{aligned} u_{N+k}^{(n+1)} &= \frac{1}{24}[7(u_{2N+k}^{(n)}+u_{N+k-1}^{(n)})+2(u_{k-N}^{(n)}+u_{k-1}+u_{N+k+2}^{(n)}+u_{2N+k+1}^{(n)}) \\ &\qquad +u_{k-N+1}^{(n)}+u_{k+2}^{(n)}] , \\ u_{N+k+1}^{(n+1)} &= \frac{1}{24}[7(u_{N+k+2}^{(n)}+u_{2N+k+1}^{(n)})+2(u_{k-N+1}^{(n)}+u_{k+2}^{(n)}+u_{2N+k}^{(n)}+u_{N+k-1}^{(n)})+ \\ &\qquad u_{k-N}^{(n)}+u_{k-1}^{(n)}] , \end{aligned}\right\} \quad (3.3.20)$$

where $k=(\ell N+1)(2)(\ell+1)N-1$ and $\ell=0(2)N-2$. Note that, if any of $u_M^{(n)}$ in the right hand sides of (3.3.20) is a point of the boundary $\partial\Omega$ then we compute the appropriate value of u at this point by (3.3.16). Obviously similar sets of equations can be written down for the 4 pt. explicit Gauss-Seidel and S.O.R. methods.

2) The second grouping of the mesh points which we are concerned with is as follows. Consider each group as being formed from 9 points of the net region Figure 3.1 in accordance with Figure 3.3, where $k=(\ell N+1)(3)(\ell+1)N-2$ and $\ell=0(3)N-3$. For this scheme of grouping of the mesh points N must be divisible by 3. In this case, each subset G_M, $M=1,2,\ldots,N^2/9$ consists of 9 elements and the submatrices $A_{M,M}$, $M=1,2,\ldots,N^2/9$, of the original matrix A of order 9, are of the form,

FIGURE 3.3:

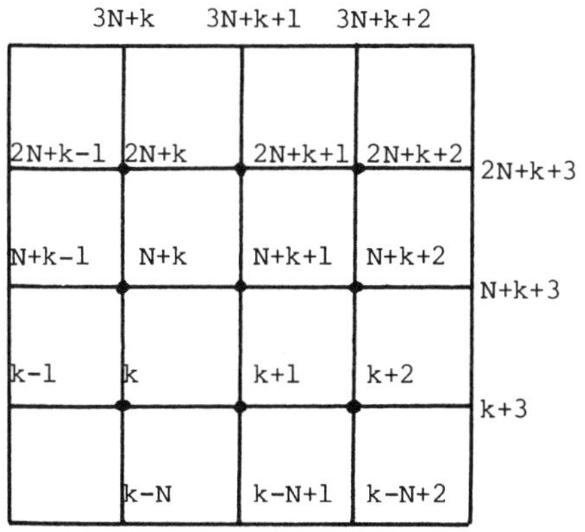

$$A_{M,M} = \begin{vmatrix} 4 & -1 & 0 & -1 & \cdot & \cdot & \cdot & \cdot & \cdot \\ -1 & 4 & -1 & 0 & -1 & \cdot & \cdot & \cdot & \cdot \\ 0 & -1 & 4 & 0 & 0 & -1 & \cdot & \cdot & \cdot \\ -1 & 0 & 0 & 4 & -1 & 0 & -1 & \cdot & \cdot \\ \cdot & -1 & 0 & -1 & 4 & -1 & 0 & -1 & \cdot \\ \cdot & \cdot & -1 & 0 & -1 & 4 & 0 & 0 & -1 \\ \cdot & \cdot & \cdot & -1 & 0 & 0 & 4 & -1 & \cdot \\ \cdot & \cdot & \cdot & \cdot & -1 & 0 & -1 & 4 & -1 \\ \cdot & \cdot & \cdot & \cdot & \cdot & -1 & 0 & -1 & 4 \end{vmatrix}, \quad M=1,2,\ldots,N^2/9. \tag{3.3.21}$$

The matrix (3.3.21) again, can be easily inverted, and for the solution of the model problem (3.3.15) in the case $f(x,y)=0$, the Laplace equation by the group Jacobi method where each group will be solved by the following formulae,

$$\left.\begin{aligned}
u_k^{(n+1)} &= \frac{1}{224}[67(u_{k-N}^{(n)}+u_{k-1}^{(n)})+22(u_{k-N+1}^{(n)}+u_{N+k-1}^{(n)})+7(u_{k-N+2}^{(n)}+u_{k+3}^{(n)}+u_{3N+k}^{(n)}+\\
&\qquad u_{2N+k-1}^{(n)})+6(u_{N+k+3}^{(n)}+u_{3N+k+1}^{(n)})+3(u_{2N+k+3}^{(n)}+u_{3N+k+2}^{(n)})],\\
u_{k+1}^{(n+1)} &= \frac{1}{112}[37u_{k-N+1}^{(n)}+11(u_{k-N}^{(n)}+u_{k-N+2}^{(n)}+u_{k+3}^{(n)})+7(u_{N+k+3}^{(n)}+u_{N+k-1}^{(n)})+5u_{3N+k+1}^{(n)}+\\
&\qquad +11u_{k+1}^{(n)}+3(u_{2N+k+3}^{(n)}+u_{3N+k+2}^{(n)}+u_{3N+k}^{(n)}+u_{2N+k-1}^{(n)})],\\
u_{k+2}^{(n+1)} &= \frac{1}{224}[67(u_{k-N+2}^{(n)}+u_{k+3}^{(n)})+22(u_{k-N+1}^{(n)}+u_{N+k+3}^{(n)})+7(u_{k-N}^{(n)}+u_{2N+k+3}^{(n)}+u_{3N+k+2}^{(n)}+\\
&\qquad +u_{k-1}^{(n)})+6(u_{3N+k+1}^{(n)}+u_{N+k-1}^{(n)})+3(u_{3N+k}^{(n)}+u_{2N+k-1}^{(n)})],\\
u_{N+k}^{(n+1)} &= \frac{1}{112}[37u_{N+k-1}^{(n)}+11(u_{k-N}^{(n)}+u_{3N+k}^{(n)}+u_{2N+k-1}^{(n)}+u_{k-1}^{(n)})+7(u_{k-N+1}^{(n)}+u_{3N+k+1}^{(n)}+\\
&\qquad +5u_{N+k+3}^{(n)}+3(u_{k-N+2}^{(n)}+u_{k+3}^{(n)}+u_{2N+k+3}^{(n)}+u_{3N+k+2}^{(n)})],\\
u_{N+k+1}^{(n+1)} &= \frac{1}{16}[2(u_{k-N+1}^{(n)}+u_{N+k+3}^{(n)}+u_{3N+k+1}^{(n)}+u_{N+k+1}^{(n)})+u_{k-1}^{(n)}+u_{k-N+2}^{(n)}+u_{2N+k+3}^{(n)}+\\
&\qquad u_{k+3}^{(n)}+u_{k-N}^{(n)}u_{3N+k+2}^{(n)}+u_{3N+k}^{(n)}+u_{2N+k-1}^{(n)})],\\
u_{N+k+2}^{(n+1)} &= \frac{1}{112}[37u_{N+k+3}^{(n)}+11(u_{k-N+2}^{(n)}+u_{k+3}^{(n)}+u_{2N+k+3}^{(n)}+u_{3N+k+2}^{(n)})+7(u_{k-N+1}^{(n)}+u_{3N+k+1}^{(n)})\\
&\qquad +5u_{N+k-1}^{(n)}+3(u_{k-N}^{(n)}+u_{3N+k}^{(n)}+u_{2N+k-1}^{(n)}+u_{k-1}^{(n)})],\\
u_{2N+k}^{(n+1)} &= \frac{1}{224}[67(u_{3N+k}^{(n)}+u_{2N+k-1}^{(n)})+22(u_{3N+k+1}^{(n)}+u_{N+k-1}^{(n)})+7(u_{k-N}^{(n)}+u_{k-1}^{(n)}+u_{2N+k+3}^{(n)}+\\
&\qquad +u_{3N+k+2}^{(n)})+6(u_{k-N+1}^{(n)}+u_{N+k+3}^{(n)})+3(u_{k-N+2}^{(n)}+u_{k+3}^{(n)})],
\end{aligned}\right|$$

$$\left.\begin{aligned}u_{2N+k+1}^{(n+1)} &= \frac{1}{112}[37u_{3N+k+1}^{(n)}+11(u_{2N+k+3}^{(n)}+u_{3N+k+2}^{(n)}+u_{3N+k}^{(n)}+u_{2N+k-1}^{(n)})+5u_{k-N+1}^{(n)}\\ &\quad +7(u_{N+k+3}^{(n)}+u_{N+k-1}^{(n)})+3(u_{k-N}^{(n)}+u_{k-1}^{(n)}+u_{k-N+2}^{(n)}+u_{k+3}^{(n)})] ,\\ u_{2N+k+2}^{(n+1)} &= \frac{1}{224}[67(u_{2N+k+3}^{(n)}+u_{3N+k+2}^{(n)})+22(u_{N+k+3}^{(n)}+u_{3N+k+1}^{(n)})+7(u_{k-N+2}^{(n)}+u_{k+3}^{(n)}\\ &\quad +u_{3N+k}^{(n)}+u_{2N+k-1}^{(n)})+6(u_{k-N+1}^{(n)}+u_{N+k-1}^{(n)})+3(u_{k-N}^{(n)}+u_{k-1}^{(n)})] ,\end{aligned}\right\} \tag{3.3.22}$$

where $k=(\ell N+1)(3)(\ell+1)N-2$ and $\ell=0(3)N-3$. Again, similar sets of equations can be written down for the 9 pt. explicit Group Gauss-Seidel and S.O.R. methods.

The novelty of these types of grouping of the mesh points lie in the following facts. As mentioned before, in the solution of the problem by group iterative methods, each group of mesh points can be considered as a single point, thus if we consider here each group of 4 (or 9) mesh points as shown in Figure 3.2 or Figure 3.3 as a single point P_M, then, we have different methods of labelling the point P_M according to the choice of solving the groups successively; for example, if we choose the groups of mesh points corresponding to the first two (or three for 9 points grouping) rows from left to right and then the groups corresponding to third and fourth (or fourth, fifth and sixth in the case of 9 points) row and so on. Therefore, the labelling points P_M is known as the "natural" ordering and it can be seen that the matrix $Z=(z_{k,\ell})$ of the Definition (3.2),(Young, 1971) has property (A) and is consistently ordered. Thus, the matrix A has property $(A^{(\pi)})$ and is π-consistently ordered. By this consideration we have a different choice of labelling the points P_M which could result in the matrix A having property $(A^{(\pi)})$ and be π-consistently ordered, such as "red-black" ordering or "diagonal" ordering. Therefore, in each case the theory of group successive over-relaxation is valid.

Another advantage with these types of grouping of mesh points is that, the resulting matrices of the groups are easily invertible and since all the matrices are the same for linear p.d.e.'s therefore, only one matrix needs to be inverted and used thereafter. The inverse of these matrices can be seen to be obtained only with few operations, especially when the group size is small.

We now describe some numerical experiments in order to compare the new block methods with the existing methods i.e. the solution of the model problem,

$$\frac{\partial^2 u}{\partial x^2} + \frac{\partial^2 u}{\partial y^2} = 0 \text{ in } \Omega, \tag{3.3.23}$$

and
$$u = 0 \text{ on } \partial\Omega, \tag{3.3.24}$$

where Ω is a unit square with boundary $\partial\Omega$.

We use a red-black ordering of points of the mesh (Figure 3.1) and also of blocks of 4 points and 9 points where each point on the mesh can represent a single point, a block of four points or a block of nine points depending on the size of the grouping scheme. Therefore, the resulting matrices are π-consistently ordered, thus the optimum ω, ω_b (say) are calculated by the formula,

$$\omega_b = 2/(1+\{1-\rho(G)\}^{\frac{1}{2}}) , \tag{3.3.25}$$

where $\rho(G)$ is the spectral radius of the Gauss-Seidel iteration matrix and is estimated by use of the Power method from,

$$\rho(G) = \lim_{n\to\infty} \frac{||d^{(n)}||}{||d^{(n-1)}||} , \tag{3.3.26}$$

where $d^{(n)}=u^{(n)}-u^{(n-1)}$ and the norm $||d^{(n)}||$ can be defined as,

$$||d^{(n)}|| = [\sum_{i=1}^{N} (u_i^{(n)} - u_i^{(n-1)})^2]^{\frac{1}{2}} . \tag{3.3.27}$$

An experimental optimum value of ω is also found by solving the problem using different values of ω. The condition for convergence of the S.O.R. methods is,

$$|u_{i,j}^{(n+1)}| < \varepsilon, \text{ for all } i,j , \tag{3.3.28}$$

where $\varepsilon=5*10^{-6}$, and similarly for the Power method where the difference between successive estimates of $\rho(G)$ is chosen to be less than the same value of ε. Different mesh sizes are used which produce (20×20), (32×32), (44×44), (56×56) and (62×62) networks. The results obtained from these experiments are recorded in Table 3.1 where n is the minimum number of iterations required for convergence.

method / h^{-1} mesh spacing	4 point explicit block SOR			9 point explicit block SOR			SOR	SLOR	S2LOR
	ρ(B)	ω	n	ρ(B)	ω	n	n	n	n
19	0.9731	1.635	30	0.9602	1.573	24	40	30	21
31	0.9898	1.755	46	0.9848	1.709	38	64	47	33
43	0.9947	1.813	63	0.9920	1.781	52	88	63	46
55	0.9967	1.853	80	0.9951	1.825	67	111	81	57
61	0.9973	1.864	88	0.9960	1.839	73	123	90	64

TABLE 3.1

To compare the efficiency of each of the iterative methods it is necessary to consider the computational complexity for each iteration as well as the number of iterations required for convergence. One way of estimating this computational complexity is to count the number of arithmetic operations performed per iteration (Evans & Biggins, 1982). For m^2 internal points on the mesh these are given in Table 3.2.

Method	Computational Complexity		Execution Time/Solution for $h^{-1}=61$	
	Adds.	Mults.	Problem 1	Problem 2
S.O.R.	$5m^2$	$2m^2$	370	200
S.L.O.R.	$5m^2$	$m(3m-1)$	367	195
S.2L.O.R.	$m(6m-1)$	$m(10m-5)/2$	272	152
4pt. S.E.B.O.R.	$11m^2/2$	$5m^2/2$	303	134
9pt. S.E.B.O.R.	$53m^2/9$	$34m^2/9$	202	111

TABLE 3.2

The results shown in Table 3.1 can be combined with the number of operations per iteration required by each method given in Table 3.2 to give the total number of arithmetic operations required by each of the point and block S.O.R. methods together with the times in computer mill units to obtain the solution using optimal ω for $h^{-1}=61$.

Problem 2: The solution of the generalized Dirichlet problem

$$\frac{\partial}{\partial x}\left(e^{10(x+y)}\frac{\partial u}{\partial x}\right)+\frac{\partial}{\partial y}\left(e^{10(x+y)}\frac{\partial u}{\partial y}\right)=0\ ,$$

in the unit square with the boundary condition $u\equiv 0$ on all boundaries.

Using finite differences this equation was replaced by

$$(2e^{10(x+y)}(e^{5h}+e^{-5h}))U_{i,j}=e^{10(x+y)}e^{5h}U_{i+1,j}+e^{10(x+y)}e^{-5h}U_{i-1,j}+$$
$$e^{10(x+y)}e^{-5h}U_{i,j-1}+e^{10(x+y)}e^{5h}U_{i,j+1}\ . \qquad (3.3.29)$$

By multiplying the system (3.3.29) i.e. $AU=b$, by the diagonal matrix D_1, where the diagonal elements of D_1 are of the form,

$$e^{-10(x+y)}\ ,$$

we obtain the system of equations,

$$2(e^{5h}+e^{-5h})U_{i,j} = e^{5h}U_{i+1,j}+e^{-5h}U_{i-1,j}+e^{5h}U_{i,j+1}+e^{-5h}U_{i,j-1} , \tag{3.3.30}$$

which leads to an unsymmetric matrix. Further, by choosing the diagonal matrix $T=D_1^{-\frac{1}{2}}$ and applying a similarity transformation we obtain

$$\begin{aligned} T(D_1A)T^{-1} &= D_1^{-\frac{1}{2}}(D_1A)D_1^{\frac{1}{2}} \\ &= D_1^{\frac{1}{2}}AD_1^{\frac{1}{2}} , \end{aligned} \tag{3.3.31}$$

which gives the system of equations,

$$2(e^{5h}+e^{-5h})U_{i,j} = U_{i+1,j}+U_{i-1,j}+U_{i,j+1}+U_{i,j-1} , \tag{3.3.32}$$

and the corresponding coefficient matrix is now a symmetric, positive definite matrix.

We choose the labelling of the mesh points and groups of 4 points and 9 points as before in a red-black ordering. For this case, the initial guess was chosen to be $U_{i,j}^{(0)}\equiv 1$ in the interior of the region, with zero values on the boundary. The iteration was continued until the criteria,

$$\max_{i,j} |U_{i,j}^{(n)}| = \max_{i,j} |e_{i,j}^{(n)}| \leq 5 * 10^{-6},$$

was satisfied.

In this case as in Problem 1, the submatrices $A_{k,k}$, $k=1,\ldots,\frac{m^2}{4}$, of the 4 point groups are all the same, i.e.,

$$A_{k,k} = \begin{bmatrix} \alpha & -1 & -1 & 0 \\ -1 & \alpha & 0 & -1 \\ -1 & 0 & \alpha & -1 \\ 0 & -1 & -1 & \alpha \end{bmatrix}$$

can be easily inverted beforehand and we have,

$$A_{k,k}^{-1} = \frac{1}{\alpha(\alpha^2-4)} \begin{bmatrix} \alpha^2-2 & \alpha & \alpha & 2 \\ \alpha & \alpha^2-2 & 2 & \alpha \\ \alpha & 2 & \alpha^2-2 & \alpha \\ 2 & \alpha & \alpha & \alpha^2-2 \end{bmatrix} \tag{3.3.33}$$

where $\alpha=2.*(e^{5h}+e^{-5h})$.

Once again, we calculate initially the values of $U_{i,j}^{(n+1)}$ and $U_{i+1,j+1}^{(n+1)}$. Then, using these values we calculate the $U_{i,j+1}^{(n+1)}$ and $U_{i+1,j}^{(n+1)}$ using equation (3.3.32) as in Problem 1.

The inverse of the submatrices $A_{k,k}$ as given in (3.3.21) for $k=1,\ldots,$

$m^2/9$, is more complex and for reasons of brevity is given in Evans & Sojoodi Haghighi,(1982). Once again we first calculate the values $U_{i,j+1}^{(n+1)}$, $U_{i+1,j+2}^{(n+1)}$ and $U_{i+2,j+1}^{(n+1)}$ by multiplying the rows 2,4,6 and 8 of the inverse matrix $A_{k,k}$ by the right hand side vector and then using these values we calculate the remaining 5 points in a similar way as in Problem 1. The results are shown in Table 3.3 whilst the computational complexity and time information is recorded in Table 3.2.

method / h^{-1} mesh spacing	4 point explicit block SOR			9 point explicit block SOR			SOR	SLOR	S2LOR
	ρ(B)	ω	n	ρ(B)	ω	n	n	n	n
19	0.9105	1.425	16	0.8731	1.352	14	23	16	12
31	0.9646	1.591	26	0.9483	1.528	21	36	26	19
43	0.9814	1.683	35	0.9725	1.629	29	49	35	26
55	0.9886	1.743	44	0.9830	1.695	37	62	44	33
61	0.9907	1.765	48	0.9861	1.720	40	69	48	36

cont.

TABLE 3.3

A variety of group S.O.R. methods have been investigated to obtain some impressive results on two typical problems and confirm the results given in Parter, (1981). During the investigation it became clear that the most efficient way by which the system of equations is partitioned to give the best result depends on the nature of the problem in hand. It can be seen from Table 3.2 that if the resulting blocks on the diagonal of the coefficient matrix are the same (e.g. Problems 1 and 2) then comparing the number of iterations, the amount of arithmetic operations required per iteration and the total time incurred to obtain the solution shows that the group of nine points appears to be the most efficient algorithm.

However, the most important conclusion we can derive from these experiments is that contrary to the tradition of choosing the 2-line S.O.R. method as the most efficient iterative method, the smaller sized blocks, i.e. 9-point and 4-point seem to work better under many circumstances.

4. PRECONDITIONED ITERATIVE METHODS

This section generalises the preconditioning techniques, introduced by Evans, (1968), and defines sparse and compact preconditioned iterative methods for the numerical solution of the sparse linear system Au=b. The difference between the methods is shown to depend on whether a conditioning matrix R consists of components derived from a *splitting* or *factorization* of A. Some theoretical results for the iterative schemes are given when A has particular properties such as consistent ordering, irreducibility, diagonal dominance, positive definiteness, etc., when derived from the finite difference discretisation of a 2nd order self-adjoint elliptic partial differential equation. Finally, the application of both forms of preconditioning to the Conjugate Gradient method is presented and computational results compared.

4.1 Introduction

Here we are primarily concerned with the use of iterative methods for the solution of N linear algebraic equations of the form,

$$Au = b \ , \tag{4.1.1}$$

where the matrix A is large, non-singular with non-vanishing diagonal elements. We note that if we premultiply (4.1.1) by A^{-1}, then we obtain immediately its solution $u=A^{-1}b$. However, there are certain difficulties for computing A^{-1}, especially when (4.1.1) is derived from the approximation of elliptic partial differential equations, so instead we consider the case where the above system is premultiplied by a non-singular matrix R^{-1}, where R^{-1} is an approximate inverse of A. Thus, we transform the original system (4.1.1) into the following preconditioned form,

$$R^{-1}Au = R^{-1}b. \tag{4.1.2}$$

The matrix R will be referred to as the conditioning matrix and according to our previous observation we first require this matrix to approach A so that the spectral condition number of $R^{-1}A$ is a great deal less than the spectral condition number of A where the spectral condition number of a non-singular matrix M is defined to be $\kappa(M)=||M|| \ ||M^{-1}||$.

After we form system (4.1.2), we can define the following general iterative scheme,

$$u^{(n+1)} = u^{(n)} + \tau R^{-1}(b-Au^{(n)}) \ , \tag{4.1.3}$$

where τ is a real parameter (Forsythe & Wasow, 1960). It can be easily seen that the constructed iterative scheme (4.1.3) is completely consistent with (4.1.2) if and only if R is non-singular and $\tau \neq 0$ (see Young, [1971)).

Let us now consider the possible forms which the matrix R can possess within the contents of the above discussion. In order to simplify our considerations we let the form of R be closely associated with (a) the splitting or (b) approximate factorisation of the matrix A.

a) Therefore, if we assume that A has the splitting

$$A = D - C_L - C_U \ , \tag{4.1.4}$$

where D is the diagonal matrix with the same diagonal elements as A and C_L, C_U are strictly lower and upper triangular matrices, respectively, then a suitable form of R is

$$R = (D - \omega C_L) D^{-1} (D - \omega C_U) \ . \tag{4.1.5}$$

Evidently, for the different values of the parameters τ and ω we can easily construct Table 4.1.1 where

$$L = D^{-1} C_L \quad \text{and} \quad U = D^{-1} C_U \ . \tag{4.1.6}$$

ω	τ	R	Iterative Method
0	1	D	Jacobi (J)
0	τ_0	D	Jacobi Overrelaxation (JOR)
0	τ_0	I	Simultaneous Displacement (SD)
1	1	D(I-L)	Gauss-Seidel (GS)
ω	ω	D(I-ωL)	Successive Overrelaxation (SOR)
ω	$\omega(2-\omega)$	D(I-ωL)(I-ωU)	Symmetric SOR (SSOR), Young, 1971.
ω	ω	D(I-ωL)(I-ωU)	Extrapolated Modified Aitken (EMA), Evans, 1963.
ω	τ_0	D(I-ωL)(I-ωU)	Preconditioned Simultaneous Displacement (SPSD), Evans, 1968.

TABLE 4.1.1: Sparse Preconditioning Methods

From our previous considerations we conclude that the iterative schemes in Table 4.1.1 can be based on the idea of improving the "condition" of the original system by using different types of conditioning matrices, Missirlis (1978). Moreover, if we observe carefully the J and JOR schemes (see Table 4.1.1), then we see that although both methods have identical conditioning matrices, the optimum value $\tau_0 = \frac{2}{\bar{a}+\bar{b}}$, where, $\bar{a},\bar{b}$ are the minimum and maximum eigenvalues of the corresponding matrix $R^{-1}A$, respectively for the parameter τ assures an improvement in the rate of convergence of JOR over the J method. From this observation, it follows that given the conditioning matrix R, the most effective iterative scheme is obtained if τ attains its optimum value τ_O. However, this does not seem to be the case for the SSOR and EMA methods (see Table 4.1.1). It is therefore clearly realised that a reconstruction of the iterative schemes illustrated in Table 4.1.1 is essential.

b) Alternatively, we can consider the factorisation of A into easily invertible matrix factors. Methods for such factorisations are well known and are given in Table 4.1.2 below as well as in many standard reference books.

However, the direct factorisation process of a sparse matrix A results in a large number of the zero entries of the matrix being replaced or 'filled in' by non-zero values. Severe disadvantages arise from these entries for not only are they a source of rounding errors in any computation in which they are involved but they greatly increase the amount of work and computer storage involved.

Exact Method	R conditioning matrix	Iterative Method
Triangular Factorisation $A = LU$	$L_s U_s$	Stone, 1968 Lipitakis & Evans, 1979
Choleski Square Root $A = QQ^T$	$Q_s Q_s^T$	Dupont, 1968 Meyerink & Van der Vorst, 1977 Gustafsson, 1977
Root Free Choleski $A = LDL^T$	$L_s D_s L_s^T$	Kershaw, 1978
Normalised Symmetric Factorisation $A = DTT^T D$	$D_s T_s T_s^T D_s$	Varga, 1960 Evans & Lipitakis, 1980

TABLE 4.1.2: Compact Preconditioning Methods

Thus, in recent years, research has been directed towards a number of indirect or iterative methods in which the strategy is to reduce the fill-in terms of the factors L, U, etc. to a minimum whilst still retaining the property that $A \simeq L_s U_s$, etc. where L_s, U_s, etc. denotes the corresponding sparse matrix factors. Of course, the strategies by which each method retains its sparseness varies differently and the reader is referred to the literature for further details. Here only the strategy for the normalised method will be discussed.

4.2 An Elliptic 2D Problem

We will be concerned first with the application of the method of finite differences to obtain an approximate solution for the 'self-adjoint' two dimensional problem defined by,

$$\frac{\partial}{\partial x}(C_1(x,y)\frac{\partial U}{\partial x}) + \frac{\partial}{\partial y}(C_2(x,y)\frac{\partial U}{\partial y}) + P(x,y)U(x,y) + Q(x,y) = 0$$

$$(x,y) \in R, \qquad (4.2.1)$$

subject to the general boundary conditions,

$$\alpha U(x,y) + \beta\frac{\partial U}{\partial x}(x,y) = \gamma(x,y), \quad (x,y) \in \Gamma, \qquad (4.2.1a)$$

where the region R denotes a rectangular plane region with a boundary Γ. The functions C_1, C_2, P, Q are assumed to be 'sufficiently smooth' functions and satisfy in $\bar{R}=R+\Gamma$ the conditions,

$$C_1(x,y) > 0, \quad C_2(x,y) > 0, \quad P(x,y) \leq 0. \qquad (4.2.1b)$$

To approximate the solution to the problem (4.2.1) a network of straight lines of spaces h_x and h_y, parallel to each of the coordinate axes is superimposed over the region R. Then, a discrete averaged centred finite difference approximation for the x-derivative of the form,

$$\frac{\partial}{\partial x}(C_1(x,y)\frac{\partial U}{\partial x}) \cong C_1(x+\frac{h_x}{2},y)\left[\frac{u(x+h_x,y)-u(x,y)}{h_x^2}\right] - C_1(x-\frac{h_x}{2},y)\left[\frac{u(x,y)-u(x-h_x,y)}{h_x^2}\right], \qquad (4.2.2)$$

with an analogous expression for $\frac{\partial}{\partial y}(C_2(x,y)\frac{\partial U}{\partial y})$ can be obtained, provided that the (x,y) is a regular mesh point. In the case of an irregular mesh point, the Taylor series expansion of U(x,y) about such a point may still be used. Substitution of the above finite difference approximations for the derivatives in (4.2.1), assuming that a columnwise ordering is considered, $h_x=h_y=h$ and $u_{i,j}$ denotes u(ih,jh), leads to the following linear five-point finite difference equation at the point (ih,jh)

$$a_{i,j}u_{i,j} - b_{i,j}u_{i,j+1} - b_{i,j-1}u_{i,j-1} - c_{i,j}u_{i+1,j} - c_{i-1,j}u_{i-1,j} = s_{i,j}$$

$$(4.2.2a)$$

where $a_{i,j}, b_{i,j}, b_{i,j-1}, c_{i,j}, c_{i-1,j} > 0$ and $a_{i,j} \geq b_{i,j} + b_{i,j-1} + c_{i,j} + c_{i-1,j}$ for all i,j with equality at interior points if P=0.

Equation (4.2.2a) applied at each regular mesh point yields a set of inhomogeneous, linear, simultaneous, symmetric difference equations, which can be expressed in matrix notation as,

$$Au = b \, , \tag{4.2.3}$$

where u and b consist of the unknown approximate solution $u_{i,j}$ and the known boundary value plus the quantity $h^2Q(x,y)$ respectively.

The coefficient matrix A is a sparse, structured, real quindiagonal (N×N) matrix (N is the number of unknown mesh points) with the following properties: (i) symmetric, (ii) positive definite, (iii) diagonally dominant, (iv) irreducible, (v) has positive diagonal and non-positive off-diagonal elements and has the following form,

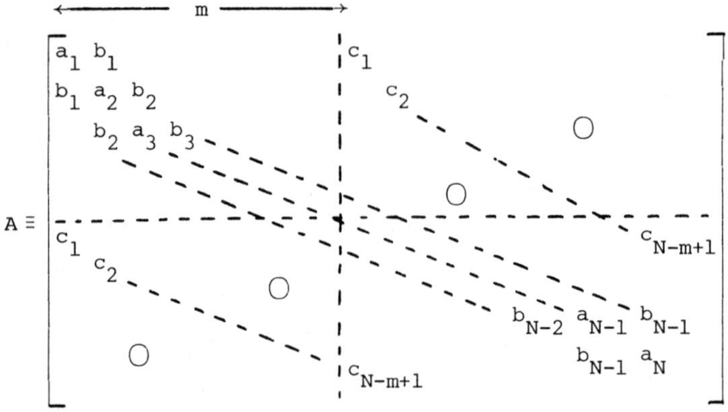

$$\equiv \left[\begin{array}{c|c} A_{11} & A_{12} \\ \hline A_{21} & A_{22} \end{array}\right] \tag{4.2.4}$$

4.3 The Sparse Preconditioned Simultaneous Displacement Method (SPSD)

In a previous work (Evans & Missirlis, 1979) the authors considered various forms of the conditioning matrix R and developed new ADI iterative schemes for the solution of (4.1.1). Here, we consider the iterative method which is produced by (4.1.3) when R is given by (4.1.5) i.e. we consider the iterative scheme defined by,

$$u^{(n+1)} = u^{(n)} + \tau(I-\omega U)^{-1}(I-\omega L)^{-1}D^{-1}(b-Au^{(n)}) \, , \tag{4.3.1}$$

where $$L = D^{-1}C_L, \qquad U = D^{-1}C_U , \tag{4.3.2}$$

and ω is a real parameter. The iterative process(4.3.1) can be regarded as belonging to a generalised class referred to as the sparse Preconditioned Simultaneous Displacement method (SPSD method). At this point we can safely state that if we let τ and ω obtain their optimum values τ_o and ω_o respectively, then the SPSD method will possess smaller spectral radius, than any other iterative scheme which employs the same conditioning matrix R. Therefore, there may be a possibility of SPSD to produce a better rate of convergence than SSOR. However any comparison must take into account the computational work involved per iteration and in Missirlis (1978) it is shown that one SPSD iteration can be computed as two half iterations such that a reduction scheme similar to Niethammer's scheme (1964) can be effectively applied.

Next we present another form for the computation of one SPSD iteration in terms of intermediate vector approximations as follows:

$$\begin{aligned} u^{(n+\frac{1}{2})} &= (1-\tau)u^{(n)}+\omega Lu^{(n+\frac{1}{2})}+(\tau-\omega)Lu^{(n)}+\tau(Uu^{(n)}+b) \\ u^{(n+1)} &= u^{(n+\frac{1}{2})}+\omega Uu^{(n+1)}-\omega Uu^{(n)} . \end{aligned} \tag{4.3.3}$$

By expressing one SPSD iteration as in (4.3.3) we can exploit the fact that when ω is fixed it is not necessary to recompute the vector $Uu^{(n)}$ in the second half iteration. Thus, if sufficient computer storage is available to accommodate an extra N-vector, the computational work involved for the SPSD method after n iterations is reduced such that to be approximately the same as with the SOR method.

An alternative form of SPSD is given by,

$$u^{(n+1)} = \mathcal{D}_{\tau,\omega}u^{(n)} + \delta_{\tau,\omega} , \tag{4.3.4}$$

where,

$$\mathcal{D}_{\tau,\omega} = I-\tau\mathcal{B}_{\omega} = I-\tau(I-\omega U)^{-1}(I-\omega L)^{-1}D^{-1}A \tag{4.3.5}$$

and

$$\delta_{\tau,\omega} = \tau(I-\omega U)^{-1}(I-\omega L)^{-1}D^{-1}b. \tag{4.3.6}$$

Since the conditioning matrix of SPSD, given by (4.3.5) is non-singular the method is completely consistent with (4.1.1) for $\tau \neq 0$ and A being non-singular.

4.4 Theoretical Analysis for Sparse Preconditioning

Now a study of the convergence properties of the SPSD method is given under the assumption that A is real and symmetric with positive diagonal elements.

Lemma 4.1: Let A be a real symmetric matrix with positive diagonal elements, then B_ω defined by (4.3.5) has real eigenvalues. Moreover, B_ω has positive eigenvalues if and only if A is positive definite.

Proof: Evidently, B_ω is similar to the matrix

$$\overline{B}_\omega = D^{-\frac{1}{2}}(D-\omega C_U)B_\omega(D-\omega C_U)^{-1}D^{\frac{1}{2}} = [D^{\frac{1}{2}}(D-\omega C_L)^{-1}]A[D^{\frac{1}{2}}(D-\omega C_L)^{-1}]^T, \quad (4.4.1)$$

which implies that $\overline{B}_\omega$ has real eigenvalues and hence the same is true for B_ω. Moreover, if we assume that A is positive definite, then,

$$\overline{B}_\omega = [D^{\frac{1}{2}}(D-\omega C_L)^{-1}A^{\frac{1}{2}}][D^{\frac{1}{2}}(D-\omega C_L)^{-1}A^{\frac{1}{2}}]^T \quad (4.4.2)$$

and since $D^{\frac{1}{2}}(D-\omega C_L)A^{\frac{1}{2}}$ is non-singular, it follows (see Theorem 2-2.6 in Young, 1971) that $\overline{B}_\omega$ is positive definite. Since B_ω is similar to $\overline{B}_\omega$, it follows that its eigenvalues are also positive. Alternatively, if we assume that B_ω is positive definite, then the same is true for $\overline{B}_\omega$ and from (4.4.1) we can easily see that A must be positive definite thus completing the proof of the Lemma.

In the following we attempt to determine the range for the parameters τ and ω such that SPSD converges. From (4.3.5) and the proof of Lemma 4.1 it follows that $\lambda>0$ is an eigenvalue of B_ω if and only if A is positive definite and

$$d = 1-\tau\lambda , \quad (4.4.3)$$

is an eigenvalue of $D_{\tau,\omega}$. On the other hand $S(D_{\tau,\omega})$, the spectral radius of $D_{\tau,\omega}$, is less than unity if and only if $|d|<1$, or equivalently if

$$0 < \tau < 2/\Lambda(B_\omega) , \quad (4.4.4)$$

where $\Lambda(B_\omega)$ denotes the maximum eigenvalue of B_ω. Also if we happened to know an eigenvector v associated with the maximum eigenvalue of B_ω, then we would be able to determine $\Lambda(B_\omega)$ exactly from the expression,

$$\Lambda(B_\omega) = \max_{v\neq 0} \frac{(v,B_\omega v)}{(v,v)} , \quad (4.4.5)$$

but this is an unrealistic expectation, thus we will attempt to determine an upper bound of $\Lambda(B_\omega)$ using (4.4.5) and (4.3.5).

Theorem 4.2: Let A be a real symmetric positive definite matrix, then for any ω in the range $0<\omega<2$, we have, $\Lambda(B_\omega) \leq \dfrac{1}{\omega(2-\omega)}$. (4.4.6)

Proof: Let us assume that λ is an eigenvalue of B_ω and $v \neq 0$ is an associated eigenvector, then $B_\omega v = \lambda v$ which by taking inner products of both sides and substituting B_ω by (4.3.5) yields the following expression for λ

$$\lambda = \frac{(v,Av)}{(v,D(I-\omega L)(I-\omega U)v)} \quad . \tag{4.4.7}$$

However,

$$D(I-\omega L)(I-\omega U) = \omega(2-\omega)A + D[(1-\omega)I + \omega L][(1-\omega)I + \omega U]$$

which when substituted in (4.4.7) gives,

$$\lambda = \frac{(v,Av)}{(v,\omega(2-\omega)Av)+(v,D[(1-\omega)I+\omega L][(1-\omega)I+\omega U]v)} \quad . \tag{4.4.8}$$

It is readily seen that the second term in the denominator of (4.4.9) is always non-negative thus (4.4.9) yields $\lambda \leq (v,Av)/(v,\omega(2-\omega)Av)$ and the proof of the theorem is complete since if A is positive definite, then by Lemma 4.1, $\lambda > 0$.

Next, the following theorem gives sufficient and necessary conditions for the SPSD method to converge.

Theorem 4.3: Let A be a real symmetric and positive definite matrix, then the SPSD method converges if and only if,

$$0 < \omega < 2 \quad \text{and} \quad 0 < \tau < 2\omega(2-\omega) \leq 2. \tag{4.4.10}$$

Proof: By combining the estimated bound on $\Lambda(B_\omega)$, given by (4.4.6), with the inequalities (4.4.4) it is readily seen that (4.4.10) are satisfied, thus completing the proof of the theorem.

Theorem 4.4: Let A be a real symmetric matrix with positive diagonal elements. If the SPSD method converges, then τ and ω satisfy the relationships (4.4.10) and A is positive definite.

Proof: If A is not positive definite, then there exists $v \neq 0$ and $\alpha \leq 0$ such that $Av = \alpha v$. From (4.4.1) and if we let $w = D^{-\frac{1}{2}}(D-\omega C_U)v$ then we have,

$$(w,\bar{B}_\omega w)/(w,w) = \alpha(v,v)/(w,w) \quad . \tag{4.4.11}$$

By (4.3.5) we also have that $(w,\bar{D}_{\tau,\omega}w)/(w,w) = 1-\tau\alpha(v,v)/(w,w)$ which implies that $S(\bar{D}_{\tau,\omega}) < 1$ provided $0 < \tau$ where $\bar{D}_{\tau,\omega} = D^{-\frac{1}{2}}(D-\omega C_U)D_{\tau,\omega}(D-\omega C_U)^{-1}D$. However, we have seen that $S(D_{\tau,\omega}) < 1$ when (4.4.4) is satisfied. This contradiction proves that A is positive definite and the proof of Theorem 4.4 is complete (see also Theorem 4.3).

The problem of determining "good" estimates τ_1,ω_1 for τ and ω respectively such that $S(D_{\tau,\omega})$ to attain the value $S(D_{\tau_1,\omega_1})$ which is near to the actual minimum value $S(D_{\tau_0,\omega_0})$ is of paramount importance. The development of the analysis will be under the condition that A is a real symmetric and positive definite matrix. Here we recall Evans (1968) that we will attempt to determine a "good" choice of ω_1 of ω by minimising $P(B_\omega)$ the P-condition number given by,

$$P(B_\omega) = \Lambda(B_\omega)/\lambda(B_\omega) \quad , \tag{4.4.12}$$

where $\Lambda(B_\omega),\lambda(B_\omega)$ denote the maximum and minimum eigenvalues of B_ω, respectively. On the other hand, τ will be determined by the expression,

$$\tau = \tau_1 = 2/(\Lambda(B_{\omega_1}) + \lambda(B_{\omega_1})) \quad , \tag{4.4.13}$$

in order for $S(D_{\tau,\omega})$ to attain its minimum value.

By Theorem 4.2 we have an upper bound for $\Lambda(B_\omega)$ and the following theorem determines a lower bound for $\lambda(B_\omega)$.

Theorem 4.5: Let A be a real symmetric and positive definite matrix and $\bar{\beta}$,M and m be numbers such that (Young, (1977)),

$$-2\sqrt{\bar{\beta}}\leq m\leq m(B),\ M(B)\leq M\leq \min(1,2\sqrt{\bar{\beta}}) \text{ and } S(LU)\leq\bar{\beta} \quad , \tag{4.4.14}$$

where M(B), m(B) denote the maximum and minimum eigenvalues of B, respectively. Then for any ω in the range $0<\omega<2$ a lower bound on $\lambda(B_\omega)$ is given by,

$$\lambda(B_\omega) \geq \begin{cases} (1-M)/(1-\omega M+\omega^2\bar{\beta}) = \phi_1(\omega) \ , & \text{if } \bar{\beta}\geq\frac{1}{4} \text{ or if } \bar{\beta}<\frac{1}{4} \text{ and } \omega\leq\omega^* \\ (1-m)/(1-\omega m+\omega^2\bar{\beta}) = \phi_2(\omega) \ , & \text{if } \bar{\beta}<\frac{1}{4} \text{ and } \omega>\omega^* \end{cases} \tag{4.4.15}$$

where for $\bar{\beta}<\frac{1}{4}$ we define ω^* by,

$$\omega^* = 2/(1+(1-4\bar{\beta})^{\frac{1}{2}}) \quad . \tag{4.4.16}$$

Proof: See Evans(1983).

Then, by combining Theorems 4.2 and 4.5 we can form an upper bound for $P(B_\omega)$ which can be studied as a function of ω.

Theorem 4.6: Let A be a real symmetric and positive definite matrix and $\bar{\beta}$,M and m satisfying (4.4.14). Then an upper bound on $P(B_\omega)$ is given by,

$$P(B_\omega) \leq \begin{cases} \dfrac{1-\omega M+\omega^2\bar{\beta}}{\omega(2-\omega)(1-M)} = \theta_1(\omega) \ , & \text{if } \bar{\beta}\geq\frac{1}{4} \text{ or if } \bar{\beta}<\frac{1}{4} \text{ and } \omega\leq\omega^* \\ \dfrac{1-\omega m+\omega^2\bar{\beta}}{\omega(2-\omega)(1-m)} = \theta_2(\omega) \ , & \text{if } \bar{\beta}<\frac{1}{4} \text{ and } \omega>\omega^* \end{cases} \tag{4.4.17}$$

where for $\bar{\beta}<\frac{1}{4}$, ω^* is defined by (4.4.16). Moreover, the bound on $P(B_\omega)$ is minimised if we let $\omega=\omega_1$ where,

$$\omega_1 = \begin{cases} 2/(1+(1-2M+4\bar{\beta})^{\frac{1}{2}}) = \omega_M , & \text{if } M\leq 4\bar{\beta} , \\ 2/(1+(1-4\bar{\beta})^{\frac{1}{2}}) = \omega^* , & \text{if } M\geq 4\bar{\beta} , \end{cases} \qquad (4.4.18)$$

and its corresponding value is given by,

$$P(B_{\omega_1}) \leq \begin{cases} [1+(1-2M+4\bar{\beta})^{\frac{1}{2}}/(1-M)]/2 = [(2-M\omega_M)/(\omega_M-M\omega_M)]/2, & \text{if } M\leq 4\bar{\beta} \\ [1+1/(1-4\bar{\beta})^{\frac{1}{2}}]/2 = 1/(2-\omega^*) , & \text{if } M\geq 4\bar{\beta}. \end{cases} \qquad (4.4.19)$$

On the other hand, if we let,

$$\tau = \tau_1 = 2\omega_1(2-\omega_1)/(1+1/P(B_{\omega_1})) , \qquad (4.4.20)$$

then $S(D_{\tau,\omega})$ attains its minimum value and is given by,

$$S(D_{\tau_1,\omega_1}) \leq \begin{cases} [2(1-\omega_M)+M\omega_M]/[2(1+\omega_M)-3M\omega_M], & \text{if } M\leq 4\bar{\beta}, \\ 1/[2/(\omega^*-1)-1] & , \text{if } M\geq 4\bar{\beta}. \end{cases} \qquad (4.4.21)$$

Proof: see Evans (1983).

For the value of τ, given by (4.4.20) we can find the minimum value of the bound for $S(D_{\tau,\omega})$, given by (4.4.21), from the formula, Evans, (1973),

$$S(D_{\tau_1,\omega_1}) = \frac{P(B_{\omega_1})-1}{P(B_{\omega_1})+1} . \qquad (4.4.22)$$

By considering the additional cases where $\bar{\beta}\geq 1/4$, and $\bar{\beta}\leq 1/4$, (4.4.19) yields the following modified bound on $P(B_{\omega_1})$,

$$P(B_{\omega_1}) \leq \begin{cases} [1+1/(1-M)^{\frac{1}{2}}]/2 , & \text{if } \bar{\beta}\leq M/4, \\ [1+(2/(1-M))^{\frac{1}{2}}]/2, & \text{if } M/4\leq\bar{\beta}\leq 1/4 \\ [1+\gamma^{-1}(2/(1-M))^{\frac{1}{2}}]/2, & \text{if } 1/4\leq\bar{\beta} \end{cases} \qquad (4.4.23)$$

where,

$$\gamma = [1+2(\bar{\beta}-1/4)/(1-M)]^{-\frac{1}{2}} , \qquad (4.4.24)$$

From the above theorem we see that good estimates for ω and τ to be used in the SPSD method are given by ω_1 and τ_1, respectively in terms of an upper bound M for S(B) and an upper bound $\bar{\beta}$ for S(LU).

4.5 The SPSD Conjugate Gradient Method (SPSD-CG Method)

In this section, we consider the application of the Conjugate

Gradient method (CG method) (Hestenes & Stiefel, (1952)) as one form of accelerating the SPD method.

Next, we present the CG method based on

$$u^{(n+1)} = Gu^{(n)} + k\ , \tag{4.5.1a}$$

where I-G is non-singular, $k=(I-G)A^{-1}b$ and G has the form,

$$G = I-(QQ^T)^{-1}A\ , \tag{4.5.1b}$$

with Q being a non-singular matrix. In this case, we define the CG method as the non-stationary second degree method (Young, 1975),

$$u^{(n+1)} = \rho_{n+1}[\gamma_{n+1}(Gu^{(n)}+k)+(1-\gamma_{n+1})u^{(n)}]+(1-\rho_{n+1})u^{(n-1)}\ , \tag{4.5.2}$$

where,

$$\gamma_{n+1} = (\hat{r}^{(n)},\hat{r}^{(n)})/(\hat{s}^{(n)},A\hat{s}^{(n)}),\ \hat{r}^{(n)} = Q^{-1}r^{(n)},\ r^{(n)} = b-Au^{(n)} \tag{4.5.3}$$

$$\hat{s}^{(n)} = Gu^{(n)}+k-u^{(n)} \text{ and } \rho_{n+1} = \left[1 - \frac{\gamma_{n+1}}{\gamma_n}\cdot\frac{(\hat{r}^{(n)},\hat{r}^{(n)})}{(\hat{r}^{(n-1)},\hat{r}^{(n-1)})}\cdot\frac{1}{\rho_n}\right]^{-1}$$

$n=1,2,\ldots$

The advantages of the CG method over the alternative accelerated techniques are:

i) It is not necessary to know the maximum and minimum eigenvalues of G and,

ii) The CG method appears to be better than the SI method applied to (4.5.1a) in the sence of minimising the $A^{\frac{1}{2}}$-norm of the error vector.

In fact, it can be proved that (Young, 1975),

$$||u^{(n)}-\bar{u}||_{A^{\frac{1}{2}}} \leq ||\tilde{u}^{(n)}-\bar{u}||_{A^{\frac{1}{2}}}\ , \tag{4.5.4}$$

where $\bar{u}$ is the exact solution of (4.1.1), $u^{(n)}$, $\tilde{u}^{(n)}$ are the approximate solutions obtained by the CG and SI methods with respect to (4.5.1a), respectively.

Let us now assume without loss of generality (see Young (1971)) that the matrix A in (4.1.1) has the splitting A=I-L-U, where $L=U^T$ and L,U are the strictly lower and upper triangular parts of A, respectively. We note that if we let,

$$Q = \tau^{-\frac{1}{2}}(I-\omega L)\ , \tag{4.5.5}$$

the iteration matrix of the SPSD method has the form (4.5.1b). By (4.5.3) and (4.5.5) we have,

$$\hat{r}^{(n)} = \sqrt{\tau}(I-\omega L)^{-1}r^{(n)} \text{ and } \hat{s}^{(n)} = (I-\omega U)^{-1}(I-\omega L)^{-1}r^{(n)} , \tag{4.5.6}$$

since the SPSD method is given by,

$$u^{(n+1)} = u^{(n)}+\tau(I-\omega U)^{-1}(I-\omega L)^{-1}r^{(n)} . \tag{4.5.7}$$

Therefore by (4.5.2) and (4.5.7) we readily see that the SPSD-CG scheme is defined by,

$$u^{(n+1)} = \rho_{n+1}(u^{(n)}+\hat{\gamma}_{n+1}(I-\omega U)^{-1}(I-\omega L)^{-1}r^{(n)})+ (1-\rho_{n+1})u^{(n-1)} , \tag{4.5.8}$$

where the iteration parameters $\hat{\gamma}_{n+1}=\gamma_{n+1}\tau$ and ρ_{n+1} are determined by (4.5.3) as follows,

$$\hat{\gamma}_{n+1} = (r^{(n)},\tilde{s}^{(n)})/(\tilde{s}^{(n)},A\tilde{s}^{(n)}),\ \tilde{s}^{(n)} = (I-\omega U)^{-1}(I-\omega L)^{-1}r^{(n)} , \tag{4.5.9}$$

and,

$$\rho_{n+1} = \left[1 - \frac{\gamma_{n+1}}{\gamma_n} \cdot \frac{(r^{(n)},r^{(n)})}{(r^{(n-1)},r^{(n-1)})} \cdot \frac{1}{\rho_n}\right]^{-1} . \tag{4.5.10}$$

Evidently, by (4.5.4) it follows that with the SPSD-CG method we will have an order of magnitude improvement in the rate of convergence as compared to SOR.

4.6 The Compact Preconditioning Simultaneous Displacement Method (CPSD)

It is well known (Cuthill & Varga, 1954), that for a symmetric and positive definite matrix A there exists a unique factorization of the form,

$$A = DT'TD , \tag{4.6.1}$$

where D is a unique, positive diagonal matrix, T is a unique, real, upper triangular matrix with unit diagonal elements and T' denotes the transpose of T. With A given by (4.2.4) and D and T in (4.6.1) are of the form:

$$D \equiv \text{diag}\{d_1,d_2,\ldots,d_N\} \equiv \left[\begin{array}{c|c} D_{11} & O \\ \hline O & D_{22} \end{array}\right] \tag{4.6.2}$$

and

$$T \equiv \left[\begin{array}{ccccc|ccccc}
1 & e_1 & & & & t_{11} & & & & \\
 & 1 & e_2 & & & t_{21} & t_{12} & & & O \\
 & & \ddots & \ddots & & \vdots & t_{22} & \ddots & & \\
 & O & & 1 & e_{m-2} & t_{m-2,1} & \vdots & \ddots & \ddots & t_{1,N-m+1} \\
 & & & & 1 & e_{m-1}+t_{m-1,1} & & & \ddots & t_{2,N-m+1} \\
\hline
 & & & & & 1 & e_m+t_{m-1,2} & \ddots & & \vdots \\
 & & O & & & & 1 & \ddots & \ddots & t_{m-2,N-m+1} \\
 & & & & & & O & \ddots & 1 & e_{N-1}+t_{m-1,N-m+1} \\
 & & & & & & & & & 1
\end{array}\right]$$

(the first m columns are marked $\longleftarrow m \longrightarrow$)

$$\equiv \left[\begin{array}{c|c} T_{11} & T_{12} \\ \hline O & T_{22} \end{array}\right] \qquad (4.6.3)$$

where it can be seen that a considerable amount of 'fill-in' has occurred during the factorisation process.

Then, the linear system (4.2.3) can be solved directly by re-writing it in the form,

$$DT'TD.u = b \ , \qquad (4.6.4)$$

or equivalently,

$$T'T(Du) = D^{-1}b \ , \qquad (4.6.5)$$

and by introducing the auxiliary vectors y and g where,

$$y = Du \ ; \qquad g = D^{-1}b \ , \qquad (4.6.6)$$

the problem is reduced to solving,

$$T'Ty = g \ ,$$

where y is obtained in terms of an auxiliary vector h, i.e.,

$$T.y = h \text{ and } T'h = g \ , \qquad (4.6.7)$$

and the final solution u is obtained from $u=D^{-1}y$.

Now, the elements of the D and T matrices can be obtained from the matrix identity,

$$\left[\begin{array}{c|c} D_{11} & O \\ \hline O & D_{22} \end{array}\right]\left[\begin{array}{c|c} T'_{11} & O \\ \hline O & T'_{22} \end{array}\right]\left[\begin{array}{c|c} T_{11} & O \\ \hline O & T_{22} \end{array}\right]\left[\begin{array}{c|c} D_{11} & O \\ \hline O & D_{22} \end{array}\right] \equiv$$

$$\left[\begin{array}{c|c} A_{11} & A_{12} \\ \hline A'_{12} & A_{22} \end{array}\right] \qquad (4.6.8)$$

and a recursive procedure using the relations,

$$A_{11} = D_{11}T'_{11}T_{11}D_{11}; \qquad A_{12} = D_{11}T'_{11}T_{12}D_{22};$$

and $$A_{22} = D_{22}T'_{12}T_{12}D_{22}+D_{22}T'_{22}T_{22}D_{22} \qquad (4.6.9)$$

can be obtained.

This normalized algorithm (Benson and Evans, 1979) gives the elements of D and T as,

$$d_1 = \sqrt{a_1}\ ; \quad d_i = \{a_i - (\frac{b_{i-1}}{d_{i-1}})^2\}^{\frac{1}{2}}; \quad e_{i-1} = \frac{b_{i-1}}{d_{i-1}d_i}$$

for i=2,3,...,m-1. (4.6.10)

Then,

for j=1,2,3,...,N-m+1

$$x_1 = \frac{c_j}{d_j}\ ; \quad v = \frac{b_{m+j-2}}{d_{M+j-2}}\ , \qquad (4.6.11)$$

whilst for j≤m-2

$$x_i = -e_{i+j-2}x_{i-1}\ , \quad \text{for } i=2,3,\ldots,m-j. \qquad (4.6.12)$$

Then, for all j>1

$$x_i = -e_{i+j-2}x_{i-1} - \sum_{k=1}^{i-1} x_k t_{k-i+m,i+j-m}\ , \qquad (4.6.13)$$

for either i=(m-j+1),(m-j+2),...,(m-1) and all j≤m-1

or i=2,3,...,(m-1) for j>m-1.

Then,

$$d_{m+j-1} = \{a_{m+j-1} - \sum_{k=1}^{m-2} x_k^2-(x_{m-1}+v)^2\}^{\frac{1}{2}}\ ; \quad e_{m+j-2} = \frac{v}{d_{m+j-1}} \qquad (4.6.14)$$

and $$t_{i,j} = \frac{x_i}{d_{m+j-1}}\ , \quad i=1,2,\ldots,m-1. \qquad (4.6.15)$$

The normalized algorithm when applied to many of the common P.D.E.'s of Mathematical Physics, involving two space dimensions can be shown to possess the property that the elements $t_{i,j}$ of the upper triangular matrix T satisfy the following theorem (Lipitakis, 1978).

Theorem 4.7: Let A be an (N×N) matrix of bandwidth m, as is given by (4.2.3) with the properties §2(i)-(v), and consider the factorization A=DT'TD, where D,T are given by (4.6.2), (4.6.3) respectively. Let $t_{i,j}$, $i \in [1,m-1]$, $j \in [1,N-m+1]$ be the elements of T-matrix and r be the number of terms retained in the bandwidth m, and $1 \leq r \leq 4$, $r \ll m$. Then, the elements $t_{i,j}$ are monotonically decreased in absolute value for $i \in [1,m-r]$ (i.e. the sequence $|t_{1,j}|, |t_{2,j}|, \ldots, |t_{m-r,j}|$ $j \in [1,N-m+1]$ decreases monotonically).

Proof: See Evans (1983).

Consider now an approximate normalised factorization of the form,

$$A \simeq D_s^r T_s^r T_s^{r'} D_s^r \quad , \tag{4.6.16}$$

where D_s^r, T_s^r are the sparse forms of the D,T matrices and $T_s^{r'}$ denotes again the transpose of T_s^r (Lipitakis, 1978).

As before D_s^r, T_s^r have the following forms,

$$D_s^r \equiv \text{diag}\{\tilde{d}_1, \tilde{d}_2, \ldots, \tilde{d}_{m-1}, \tilde{d}_m, \ldots, \tilde{d}_N\}, \tag{4.6.17}$$

and

$T_s^r \equiv$

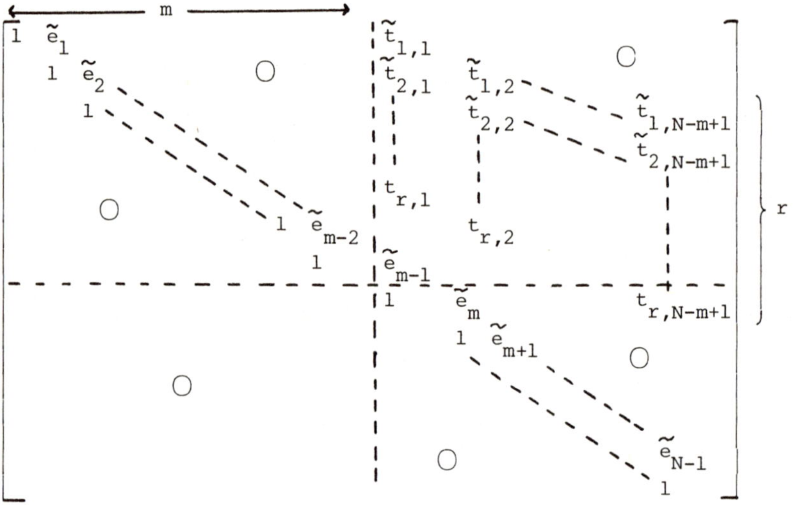

(4.6.18)

where only the r-outermost off diagonal entries of the 'fill-in' terms have now been retained.

In a similar manner as before the algorithmic procedure for determining the elements of the D_s^r and T_s^r factors can be expressed in the compact form:-

$$\tilde{d}_1 = \sqrt{a_1} \; ; \quad \tilde{d}_i = \{a_i - (\frac{b_{i-1}}{\tilde{d}_{i-1}})^2\}^{\frac{1}{2}} \; ; \quad \tilde{e}_{i-1} = \frac{b_{i-1}}{\tilde{d}_{i-1}\tilde{d}_i}$$

$$\text{for } i=2,3,\ldots,m-1. \tag{4.6.19}$$

Then <u>for j=1,2,3,...,N-m+1</u>, we have

$$\tilde{x}_1 = \frac{c_j}{\tilde{d}_j} \; ; \quad \tilde{v} = \frac{b_{m+j-2}}{\tilde{d}_{m+j-2}} \tag{4.6.20}$$

and <u>for j≤r-1</u>

$$\tilde{x}_i = -\tilde{e}_{i+j-2}\tilde{x}_{i-1} \;, \quad \text{for } i=2,3,\ldots,r+1-j \;. \tag{4.6.21}$$

Then, <u>for j>1 and r>1</u>

$$\tilde{x}_i = -\tilde{e}_{i+j-2}\tilde{x}_{i-1} - \sum_{k=1}^{i-1} \tilde{x}_k \tilde{t}_{k-i+r+1,i+j-r-1} \;, \tag{4.6.22}$$

and for either $i=(r-j+2),(r-j+3),\ldots,r$ and all $j \le r$

or $i=2,3,\ldots,r$ for $j>r$.

Then, we have,

$$\tilde{d}_{m+j-1} = \{a_{m+j-1} - \sum_{k=1}^{r} (\tilde{x}_k^2 - \tilde{v}^2)\}^{\frac{1}{2}} \;, \tag{4.6.23}$$

$$\tilde{e}_{m+j-2} = \tilde{v}/\tilde{d}_{m+j-1} \;, \tag{4.6.24}$$

and,

$$\tilde{t}_{i,j} = \tilde{x}_i/\tilde{d}_{m+j-1} \;, \quad \text{for } i=1,2,3,\ldots,r. \tag{4.6.25}$$

Since by definition, we require the conditioning matrix R to approximate closely to A then it seems pertinent to consider the alternative form of R as,

$$R_r = D_s^r T_s^{r'} T_s^r D_s^r \;. \tag{4.6.26}$$

Thus, the compact Preconditioned Simultaneous Displacement CPSD method is defined to be,

$$u^{(n+1)} = u^{(n)} + \tau(D_s^r T_s^{r'} T_s^r D_s^r)^{-1}(b-Au^{(n)}) \;, \tag{4.6.27}$$

or equivalently in the form,

$$(T_s^{r'} T_s^r) D_s^r \delta u^{(n+1)} = \tau(D_s^r)^{-1} r^{(n)} \;, \tag{4.6.28}$$

where,

$$\delta u^{(n+1)} = u^{(n+1)} - u^{(n)} \;, \quad r^{(n)} = b-Au^{(n)} \;. \tag{4.6.29}$$

By introducing the auxiliary vectors v and g where,

$$v = D_s^r u \;; \quad g = (D_s^r)^{-1} r = (D_s^r)^{-1} b - (D_s^r)^{-1} A (D_s^r)^{-1} v \;, \tag{4.6.30}$$

the C.P.S.D. method reduces to solving the normalised systems for each iteration, i.e.,

$$T_s^{r'} T_s^r \delta v^{(n+1)} = \tau g^{(n)} \;, \tag{4.6.31}$$

and δv is obtained in terms of an auxiliary vector h, where,

$$T_s^r \delta v = h \quad \text{and} \quad T_s^{r'} h = \tau g \;, \tag{4.6.32}$$

i.e. $h_1 = \tau g_1$;

$$h_i = \tau g_i - \tilde{e}_{i-1} h_{i-1}, \quad i=2,3,\ldots,m-1; \tag{4.6.33}$$

and

$$h_i = \tau g_i - \tilde{e}_{i-1} h_{i-1} - \sum_{k=i-m+1}^{i-m+r} \tilde{t}_{k-i+m,i-m+1} h_k$$

$$i=m,m+1,\ldots,N. \tag{4.6.35}$$

A back substitution process yields δv in terms of h, the components being given by the equations,

$$\delta v_N = h_N, \tag{4.6.36}$$

and

$$\delta v_i = h_i - \tilde{e}_i \delta v_{i+1} - \sum_{j=p}^{q} \tilde{t}_{i-j+m,j-m+1} \delta v_j, \tag{4.6.37}$$

the summation p,q are given in Evans and Lipitakis (1980).

4.7 The Compact Preconditioned Conjugate Gradient Method

In this section, we introduce the Normalized Compact Preconditioned Conjugate Gradient method for solving the given linear system (4.2.3). As with the preconditioned Conjugate Gradient method(Evans, 1973), the method is very efficient, making no assumptions about the structure of the coefficient matrix A, requires no estimation of iterative parameters and is easy to program. Furthermore the new method because of the normalization and the choice of the optimum fill-in parameter r, compares favourably with other similar preconditioned methods recently developed, Chandra et al (1975), Meyerink & Van der Vorst (1977), Gustaffson (1977), and Kershaw (1978).

Instead of considering the linear system (4.2.3) we now consider the linear system

$$(D_s^r T_s^{r'} T_s^r D_s^r)^{-1} Au = (D_s^r T_s^{r'} T_s^r D_s^r)^{-1} b, \tag{4.7.1}$$

where $D_s^r T_s^{r'} T_s^r D_s^r$ is the known approximate factorization of A, as given in Section 6, with D_s^r, T_s^r defined as in (4.6.25), (4.6.26) respectively.

Once the factorization has been computed, the system (4.7.1) is solved by the C.G. method assuming that $(D_s^r T_s^{r'} T_s^r D_s^r)^{-1}$ is symmetric and positive definite.

Let u_0 be an arbitrary initial approximation of the solution u form the residual,

$$r_0 = b - Au_0. \tag{4.7.2}$$

We proceed to solve,

$$(D_s^r T_s^{r'} T_s^r D_s^r) r_0^+ = r_0, \tag{4.7.3}$$

and set,

and set $$\sigma_0 = r_0^+ \; . \qquad (4.7.4)$$

Then, for i=0,1,2,..., calculate the vectors $u_{i+1}, r_{i+1}, \sigma_{i+1}$ and the scalar quantities a_i, β_{i+1} as follows:

$$a_i = \frac{(r_i, r_i^+)}{(\sigma_i, A\sigma_i)} \; , \qquad (4.7.5)$$

$$u_{i+1} = u_i + a_i\sigma_i \; , \quad \text{and} \quad r_{i+1} = r_i - a_i A\sigma_i \; . \qquad (4.7.6)$$

Then, we solve, $$(D_s^r T_s^{r'} T_s^r D_s^r) r_{i+1}^+ = r_{i+1} \; , \qquad (4.7.7)$$

and evaluate,

$$\beta_{i+1} = \frac{(r_{i+1}, r_{i+1}^+)}{(r_i, r_i^+)} \; , \text{ and } \sigma_{i+1} = r_{i+1}^+ + \beta_{i+1}\sigma_i \; . \qquad (4.7.8)$$

A normalized form of the iterative scheme (4.7.2)-(4.7.8). Henceforth called the N.I.C.G. method, Lipitakis (1978) is given below:

Form, $$r_0 = b - Au_0 \; , \quad \text{set } \tilde{r}_0 = (D_s^r)^{-1} r_0 \; , \qquad (4.7.9)$$

solve, $$(T_s^{r'} T_s^r) . r_0^* = \tilde{r}_0 \; , \text{ and set } \tilde{\sigma}_0 = r_0^* \; . \qquad (4.7.10)$$

Then, for i=0,1,2,3,... calculate the vectors $u_{i+1}, \tilde{r}_{i+1}, \tilde{\sigma}_{i+1}$ and the scalar quantities $\tilde{a}_i, \tilde{\beta}_{i+1}$ as follows:

$$\tilde{a}_i = \frac{(\tilde{r}_i, r_i^*)}{(\tilde{\sigma}_i, A\tilde{\sigma}_i)} \; , \qquad (4.7.11)$$

$$u_{i+1} = u_i + \tilde{a}_i\tilde{\sigma}_i \; , \quad \text{and} \quad \tilde{r}_{i+1} = \tilde{r}_i - \tilde{a}_i A\tilde{\sigma}_i \; . \qquad (4.7.12)$$

Then, solve

$$(T_s^{r'} T_s^r) r_{i+1}^* = \tilde{r}_{i+1} \; , \qquad (4.7.13)$$

and evaluate,

$$\tilde{\beta}_{i+1} = \frac{(\tilde{r}_{i+1}, r_{i+1}^*)}{(\tilde{r}_i, r_i^*)} \quad \text{and } \tilde{\sigma}_{i+1} = r_{i+1}^* + \tilde{\beta}_{i+1}\tilde{\sigma}_i \; , \qquad (4.7.14)$$

to complete the computational process.

4.8 Computational Results

We now present some computational results for numerical experiments carried out on the following model problem. Given a continuous function g(x,y) defined on the boundary Γ of the unit square $R: 0 \leq x \leq 1,\ 0 \leq y \leq 1$, find a function U(x,y) continuous in the closed square and satisfying in the interior R the Laplace equation $\partial^2 U/\partial x^2 + \partial^2 U/\partial y^2 = 0$. On the boundary we require $U(x,y)=g(x,y)$.

By considering the five-point discrete analogue and using the natural ordering it can be shown that (Young, 1971)

$$S(B) = \cos \pi h , \tag{4.8.1}$$

where h is the mesh size. In addition, the matrix A is real, symmetric, positive definite and consistently ordered. Moreover, it is known that (Ehrlich, 1963) $S(LU)=\frac{1}{4}\cos^2 \frac{\pi h}{2(1-h)}$ hence we can let $\bar{\beta}$ given by,

$$S(LU) \lesssim \bar{\beta} = \frac{1}{4} \cos^{2\pi} \frac{h}{2} . \tag{4.8.2}$$

By noting that $2\sqrt{\bar{\beta}}=\cos \frac{\pi h}{2} \gtrsim 4\bar{\beta}= \cos^2 \frac{\pi h}{2} \gtrsim \cos\pi h=M$ we can apply Theorem 4.6 to obtain successively,

$$\omega_1 = 2/(1+\sqrt{3} \sin(\pi h/2)) \text{ and } P(B_{\omega_1}) = (1+\sqrt{3}/(2\sin(\pi h/2)))/2. \tag{4.8.3}$$

Therefore, for sufficiently small h we have,

$$RR(D_{\tau_1,\omega_1}) \sim P(B_{\omega_1})/2 \sim (1 + \frac{\sqrt{3}}{\pi} h^{-1})/4 \tag{4.8.4}$$

where for any convergent iteration method (4.5.1a), the quantity $R(G)=\log S(G)$ is the rate of convergence whereas the quantity $RR(G)=1/R(G)$ is the reciprocal rate of convergence.

In order to obtain information to confirm the theories in practice of the iterative procedures presented in this paper, we considered the numerical solution of the Laplace equation in the unit square with zero boundary values. The starting vector $u^{(0)}$ used was the vector with all its components equal to unity. The number of iterations for the different mesh sizes of the numerical experiments as well as the values of the parameters are presented in Table 4.8.1.

TABLE 4.8.1: Comparison of Sparse Preconditioning Displacement-Conjugate Gradient Method with S.O.R.

h^{-1}	ω_0	$\bar{\rho}=\tau_0$	PSD-CG	ω_b	S.O.R.
20	1.7640	0.6993	14	1.7295	61
40	1.8750	0.4264	20	1.8547	121
60	1.9157	0.3031	25	1.9005	181

For all cases, the optimum values τ_0 and ω_0 were used and the convergence criteria was chosen to be 10^{-6}. From Table 4.8.1 we note that the number of iterations required for convergence behaves approximately as $h^{-\frac{1}{2}}$ which is a substantial improvement over the behaviour h^{-1} of S.O.R. However if we take into account the computational work involved as the other criteria then the SPSD-CG method involves considerably more computational effort (Missirlis, (1978)).

Also, for comparison purposes in Table 4.8.2 we present the results for the CG method and NICG methods for 4 values of the fill-in parameter r as presented in Section 4.7.

TABLE 4.8.2: Comparison of Compact Preconditioning Methods, N.I.C.G. for various values of r, the fill-in parameter with the C.G. method

h^{-1}	C.G.	N.I.C.G.			
		r=1	2	3	4
20	55	17	13	10	9
40	101	28	20	17	14
60	160	40	28	23	20

A summary of the storage and number of multiplications per iteration required for the methods is given in Table 4.8.3 from which we can establish that an upper bound on the arithmetic work per iteration for the N.I.C.G. method is O(20N) multiplications when r=4 and O(14N) multiplications when r=1.

TABLE 4.8.3: Two dimensional case. Storage requirements and number of multiplications for N×N ($N=n^2$) systems, where n=m-1.

Equation	Vectors Stored	Storage Required	No.of mults. per iter.	No. of mults. for the sol. of $(T'_sT_s)r_i^*=\tilde{r}_i$
Laplace:				
C.G.	$r,u,\sigma A\sigma$	4N	6N+2	0
N.I.C.G.	$\tilde{r},u,\tilde{\sigma},A\tilde{\sigma},a,b,s,\nu$ plus r-vectors in T_s matrix	(8+r)N-rn	(2r+8)N-rn-r	(2r+2)N-rn-r-2
General:				
C.G.	$r,u,\sigma,A\sigma,a,b,c$	7N-n	10N-4n+2	0
N.I.C.G.	$\tilde{r},u,\tilde{\sigma},A\tilde{\sigma},a,b,c,s,\nu$ plus r-vectors in T_s matrix	(9+r)N-n	(2r+12)N-(r+4)n-r	(2r+2)N-rn-r-2

5. COMPARISON BETWEEN ITERATIVE AND DIRECT METHODS

An iterative method has the significant advantage in that it works with the original matrix and involves no creation of additional non-zeros (fill-ins). In most cases the matrix A is not stored but generated since it suffices to have code that can generate the non-zeros when they are required. These methods are therefore well suited to very large problems involving networks and grids.

For instance, the 5-point finite difference approximation to Laplace's equation in a 2 or 3 space dimension with r unknowns in the x,y (and z) directions involves r^2 and r^3 unknowns. Using direct methods the maximum semi-bandwidth is r and r^2 and the number of multiplications is approximately $\frac{1}{4}r^4$ and $\frac{1}{7}r^7$. When S.O.R. is used with optimal parameter ω, the number of multiplications per iteration is about $2r^2$ and $2r^3$, the maximal eigenvalues are about $(1-4\pi^2/r^2)$ at $\omega=1$ and $(1-2\pi/r)$ at the optimal ω so that the number of iterations needed to gain 1 decimal place is about $0.06r^2$ at $\omega=1$ and $0.37r$ at the optimum ω. This corresponds to $0.77r^3$ and $0.77r^4$ multiplications for 1 decimal place in 2 and 3 dimensions with the optimum ω. Thus, direct and iterative methods are likely to be fairly competitive in two dimensions but iterative methods are clearly superior in three dimensions.

In a more general situation however, much depends on the problem. Well conditioning favours iterative methods, all of which then require a small number of iterations as was shown for conjugate gradients in section 4, but the new preconditioning techniques (Evans, 1983) are powerful enough to cope with ill-conditioning problems too. Large numbers of variables favour iterative methods, unless the structure results in little fill-in for a direct method (for example, if $\underline{A}$ decomposes to block triangular form or is a band matrix with small band-width), since the cost of a direct method is dependent on the structure only.

Where several sets of equations with the same matrix, i.e. mildly non-linear problems or many right hand sides, are to be solved, the factorization may be retained so that second and subsequent sets may be solved much more rapidly. For example, the operation counts for Laplace's equation in a square and a cube reduce to only $\frac{4}{3}r^3$ and $\frac{4}{5}r^5$ respectively. Direct methods are therefore superior to iterative methods for "many-off" cases.

6. REFERENCES

Duff, I.S. (1982). Research Directions in Sparse Matrix Computations. A.E.R.E. R. 10547. H.M.S.O.

Wilkinson, J.H. (1965). The Algebraic Eigenvalue Problem. Oxford University Press.

Isaacson, E. and Keller, H.B. (1966). Analysis of Numerical Methods. Wiley.

Rose, D.J. (1970). Symmetric Elimination on Sparse Positive Definite Systems and the Potential Network Problem. Ph.D. Thesis. Harvard University, Cambridge, Massachusetts.

Sherman, A.H. (1975). On the Efficient Solution of Sparse Systems of Linear and Non-Linear Equations. Ph.D. Thesis, Yale University.

Harary, F. (1969). Graph Theory. Addison-Wesley.

Rose, D.J. (1972). A Graph-Theoretic Study of the Numerical Solution of Sparse Positive Definite Systems of Linear Equations. In 'Graph Theory and Computing', Author Read, R.C., Acad. Press.

Markowitz, H.M. (1957). The Elimination Form of the Inverse and Its Application to Linear Programming. Management Science, 3, 255-269.

Duff, I.S. and Reid, J.K. (1974). A Comparison of Sparsity Orderings for Obtaining Pivotal Sequences in Gaussian Elimination. J.I.M.A. 14, 281-291.

Zlatev, Z. (1978). On Some Pivotal Strategies in Gaussian Elimination by Sparse Technique. Kobenhavns University, Matematisk Institut.

Duff, I.S. (1977). MA28 - a Set of Fortran Subroutines for Sparse Unsymmetric Matrices. A.E.R.E., R.8730, Harwell, Oxon.

Duff, I.S. and Reid, J.K. (1977). Some Design Features of a Sparse Matrix Code. A.E.R.E. R. CSS 48, Harwell, Oxon.

Zlatev, Z., Barker, V.A. and Thomson, P.G. (1978). SSLEST: A Fortran IV Subroutine for Solving Sparse Systems of Linear Equations (User's Guide). Inst. for Numerical Analysis. Technical University of Denmark, ISSN 0105-4988, Report No. 78-01.

Reid, J.K. (1971). A Note on the Stability of Gaussian Elimination. J.I.M.A., 8, 374-375.

George, J.A. (1973). Nested Dissection of a Regular Finite Element Mesh. S.I.N.U.M., 10, 345-363.

Woo, P.T., Roberts, S.J. and Gustavson, F.G. (1973). Application of Sparse Matrix Techniques to Reservoir Simulation, Paper SPE 4544, Society of Petroleum Engineers of AIME.

Birkhoff, G. and George, J.A. (1973). Elimination by Nested Dissection. In J.F. Traub, Complexity of Sequential and Parallel Numerical Algorithms. 221-269. Acad. Press.

Duff, I.S., Erisman, A.M. and Reid, J.K. (1974). On George's Nested Dissection Method. A.E.R.E. Report, Harwell. England.

Rose, D.J. and Whitten, G.F., (1974). Automatic Nested Dissection. In Proc. ACM Nat.Conf. 82-88.

Varga, R.S. (1962). Matrix Iterative Analysis. Prentice Hall.

Young, D.M. (1971). Iterative Solution of Large Linear Systems. Academic Press.

Strang, G. and Fix. G. (1973). An Analysis of the Finite Element Method. Prentice Hall.

Forsythe, G.E. and Wasow, W.R. (1960). Finite Difference Methods for Partial Differential Equations. J. Wiley.

Ames, W.F. (1969). Numerical Methods for Partial Differential Equations. Nelson.

Evans, D.J. and Biggins, M.J. (1982). The Solution of Elliptic Partial Differential Equations by a New Block Overrelaxation Technique, Int.Jour.Comp.Math. 10, 269-282.

Parter, S.V. (1981). Block Iterative Methods. In Elliptic Problem Solvers, ed. M.H. Schultz, 375-382. Acad. Press.

Evans, D.J. (1968). The Use of Preconditioning in Iterative Methods for Solving Linear Equations with Symmetric Positive Definite Matrices. J.I.M.A. 4, 295-314.

Young, D.M. (1977). On the Accelerated SSOR Method for Solving Large Linear Systems. Adv. in Math. 23, 215-271.

Evans, D.J. (1983). Preconditioning Methods: Theory and Applications. Gordon and Breach. New York.

Missirlis, N.M. (1978). Preconditioned Iterative Methods for Solving Elliptic Partial Differential Equations. Ph.D. Thesis, University of Technology, Loughborough.

Evans, D.J. (1963). The Extrapolated Modified Aitken Iteration Method for Solving Elliptic Difference Equations. Comp.J. 6, 193-201.

Stone, H.L. (1968). Iterative Solution of Implicit Approximations of Multi-dimensional Partial Differential Equations, SIAM J. Numer.Anal. 5, 530-558.

Lipitakis, E.A. and Evans, D.J. (1979). A Sparse LU Factorization Procedure for the Solution of Parabolic Differential Equations, In Numerical Methods in Thermal Problems, edit. R.W. Lewis & K. Morgan, 954-966. Pineridge Press, Swansea, U.K.

Dupont T., Kendall R and Rachford, H.H. Jr. (1968). An Approximate Factorisation Procedure for Solving Self-Adjoint Elliptic Difference Equations, S.I.A.M. J.Numer.Anal. 5, 559-573.

Meyerink, J.A., Van der Vorst, H.A. (1977). An Iterative Solution Method of Linear Systems of which the Coefficient Matrix is a Symmetric M-matrix. Math. of Comp. 31, 148-162.

Chandra, R., Eisenstat, S.C. and Schultz, M.H. (1975). Conjugate Gradient Methods for Partial Differential Equations. In Adv. in Comp. Meth. for P.D.E.'s A.I.C.A. 60-64.

Gustafsson, I. (1977). A Class of First Order Factorization Methods, Rept. 77.04R, Comp.Sc. Chalmers Univ. of Technology and Univ. of Göteborg.

Kershaw, D.S. (1978). The Incomplete Cholesky-Conjugate Gradient Method for the Iterative Solution of Systems of Linear Equations, Jour. Comp.Phys. 26, 43-65.

Evans, D.J. and Lipitakis, E.A. (1980). A Normalised Implicit Conjugate Gradient Method for the Solution of Large Sparse Systems of Linear Equations. Comp.Meth. in Appl.Mech.Eng. 23, 1-20.

Evans, D.J. and Missirlis N.M. (1979). The Modified Alternating Direction Preconditioning Method for the Numerical Solution of Elliptic Self-Adjoint Partial Differential Equations. B.I.T. 19, 172-185.

Niethammer, W. (1964). Relaxation bei Komplexen matrizen. Math.Zeit. 86, 34-40.

Evans, D.J. (1973). Comparison of the Convergence Rates of Iterative Methods for Solving Linear Equations with Preconditioning. Greek Math.Soc., Carathéodory Symp. 106-135.

Hestenes, M.R. and Stiefel, E. (1952). Methods of Conjugate Gradients for Solving Linear Systems. NBS J. of Res. 49, 409-436.

Young, D.M. (1975). Notes on the Conjugate Gradient Method, unpub. Univ. of Texas, Austin.

Cuthill, E.H. and Varga, R.S. (1959). A Method of Normalized Block Iteration. J.A.C.M. 6, 236-244.

Benson, A. and Evans, D.J. (1979). A Normalized Algorithm for the Solution of Symmetric General Five-Term Linear Systems of Semi-Band-width m. Comp. & Apl. Maths. 5, 299-304.

Lipitakis, E.A. (1978). Computational and Algorithmic Techniques for the Solution of Elliptic and Parabolic P.D.E.'s in Two and Three Space Dimensions. Ph.D. Thesis, University of Technology, Loughborough.

Evans, D.J. (1973). The Analysis and Application of Sparse Matrix Algorithms in the Finite Element Method. The Mathematics of Finite Elements and Applications. In Proc. edit. J.R. Whiteman, 427-447. Acad. Press, London.

Ehrlich, L.W. (1963). The Block Symmetric Successive Overrelaxation Method. Ph.D. Thesis. Univ. of Texas, Austin.

4

ITERATIVE METHODS FOR SPARSE LINEAR SYSTEMS: SOME RECENT DEVELOPMENTS

M.N. Missirlis
Department of Applied Mathematics, University of Athens, Panepistimiopolis 621, Athens, Greece.

Abstract. This chapter describes the preconditioning procedure for the construction of first order iterative methods. Such a procedure possesses the advantage of establishing á priori criteria which ensure the efficient performance of the produced iterative scheme. Next, it surveys some new extrapolated iterative methods which are produced using the above procedure. The study of their convergence always reveals an improved convergence rate as compared with the classical iterative schemes. Particular attention is given to the Extrapolated Successive Overrelaxation (ESOR) method.

1. INTRODUCTION

The use of finite difference/element methods for the discretisation of boundary value problems involving partial differential equations often leads to a sparse linear algebraic system of equations which has the form

$$Au = b \ , \tag{1.1}$$

where A is a known non-singular N×N real matrix and u,b N×1 real vectors with b known and u the vector of unknowns. For such problems the order of the coefficient matrix A varies from 10^3 to 10^6. Our aim in this chapter will be the study of various iterative methods for solving numerically (1.1) since there is no need of storing A and the accumulation of rounding errors is less serious than for most direct methods. In particular, we adopt the preconditioning approach [2] for the construction of a linear stationary first order iterative method which consists of the following steps:

(i) Define a non-singular real matrix of order N such that: (a) R is a "good" approximation of A, (b) "easily" invertible and (c) the spectral condition number of $R^{-1}A$, $\kappa(R^{-1}A)$ is as small as possible ($\kappa(R^{-1}A)<\kappa(A)$).

(ii) Premultiply (1.1) by R^{-1} to obtain the preconditioned system

$$\mathcal{R}u = r \ , \tag{1.2}$$

where $\mathcal{R} = R^{-1}A$ and $r = R^{-1}b$. (1.3)

(iii) For an arbitrary vector $u^{(0)}$ generate the sequence $\{u^{(n)}\}$, n=0,1,2, ... defined by the extrapolated method

$$u^{(n+1)} = u^{(n)}+\tau(r-\mathcal{R}u^{(n)}) ,$$

or equivalently, by

$$u^{(n+1)} = G_\tau u^{(n)} + k_\tau, \quad n=0,1,2,... \tag{1.4}$$

where

$$G_\tau = I-\tau\mathcal{R} , \; k_\tau = \tau r , \tag{1.5}$$

and $\tau(\neq 0)$ is a real parameter.

(iv) Determine the extrapolation parameter τ such that the sequence $\{u^{(n)}\}$, n=0,1,2,... converges as fast as possible to the unique solution of (1.1).

Remark. In the ideal but prohibitive case where R=A it follows that $u^{(n+1)}=(1-\tau)u^{(n)}+\tau A^{-1}b$. For the optimal choice $\tau_0=1$ the solution of (1.1) is obtained immediately and (1.4) degenerates into the basic (unextrapolated) iterative method

$$u^{(n+1)} = Gu^{(n)} + k, \quad n=0,1,2,... \tag{1.6}$$

where

$$G = I-\mathcal{R} \text{ and } k=r. \tag{1.7}$$

Evidently, requirements (i) and (ii) of the preconditioning procedure were motivated by the above remark. In other words, by examining the form of R we should be able to conjecture whether an iterative method will be more(less)efficient than another according to whether R is a better(worse) approximation to A. Let us postpone for the moment any further discussion on the form of R and focus our attention on the requirements (iii) and (iv). It is well known [12],[13] that (1.4) converges for an arbitrary starting vector $u^{(0)}$ to the unique solution of (1.1) if and only if

$$S(G_\tau) < 1 , \tag{1.8}$$

where

$$S(G_\tau) = \max_{\lambda \in \sigma(G_\tau)} |\lambda| , \tag{1.9}$$

and $\sigma(G_\tau)=\{\lambda:\det(\lambda I-G_\tau)=0\}$ denotes the spectrum of G_τ. When (1.8) is satisfied we can choose τ so that $S(G_\tau)$ is less than S(G) thus obtaining an iterative scheme with higher rate of convergence. Our problem therefore is to select τ from an interval which will be determined by (1.8), such that $S(G_\tau)$ is minimum. Suppose now that the eigenvalues μ of $\mathcal{R}$ are real and

$$\mu \in [\underline{a},\bar{a}] \equiv I_1 .$$

Because of (1.5), the eigenvalues of G_τ and those of $\mathcal{R}$ satisfy the relationship

$$\lambda = 1-\tau\mu \quad , \tag{1.10}$$

hence (1.9) becomes

$$S(G_\tau) = \max_{\mu \subset I_1} |\lambda| \quad . \tag{1.11}$$

From (1.8) it follows that the necessary and sufficient conditions for convergence are either

$$\underline{a} > 0 \text{ and } 0<\tau<2/\bar{a} \quad , \tag{1.12a}$$

or

$$\bar{a} < 0 \text{ and } 2/\underline{a}<\tau<0 \quad . \tag{1.12b}$$

Clearly, (1.12) indicate that (1.4) is convergent if and only if all the eigenvalues μ of R are positive (negative) and τ lies in (1.12a),((1.12b)). It should be mentioned here that the extrapolated method (1.4) is convergent while the basic method may diverge, i.e. it is not required $S(G)<1$.

Next, we present an alternative approach for the determination of the optimum value τ_0 of τ which minimises $S(G_\tau)$ to the one which is known [5]. A similar treatment, in case μ is complex, leads in a straightforward manner to the development of the convergence analysis of (1.4) [8].

By (1.10) we have

$$|\lambda|^2 = \tau^2\mu^2-2\tau\mu+1 \equiv \lambda(\mu) \quad . \tag{1.13}$$

In view of (1.11), we first have to study the behaviour of $\lambda(\mu)$ as a function of μ for fixed τ. Since $\text{sign}(\partial|\lambda|^2/\partial\mu)=\text{sign}(\tau\mu-1)$ we have to distinguish the following cases: (i) $0<\tau\leq 1/\bar{a}$, (ii) $1/\underline{a}\leq\tau$ and (iii) $1/\bar{a}\leq\tau\leq 1/\underline{a}$, where we assume $\underline{a}>0$. If τ lies in (i) then $\lambda(\mu)$ is a decreasing function of μ, thus

$$\max_{\mu \in I_1} |\lambda|^2 = \lambda(\underline{a}) \quad . \tag{1.14}$$

Similarly, if τ lies in (ii), then

$$\max_{\mu \in I_1} |\lambda|^2 = \lambda(\bar{a}) \quad . \tag{1.15}$$

In case that (iii) is valid we have

$$\max_{\mu \in I_1} |\lambda|^2 = \max\{\lambda(\bar{a}),\lambda(\underline{a})\},$$

or

$$\max_{\mu \in I_1}|\lambda|^2 = \begin{cases} \lambda(\underline{a}) \ , & \text{if } 1/\bar{a}\leq\tau\leq\tau_0 \\ \lambda(\bar{a}) \ , & \text{if } \tau_0\leq\tau\leq 1/\underline{a} \end{cases} \tag{1.16}$$

where

$$\tau_0 = \frac{2}{\bar{a}+\underline{a}} \quad . \tag{1.17}$$

Summarising our results of (i),(ii) and (iii) we obtain

$$\max_{\mu \in I_1} |\lambda|^2 = \begin{cases} \lambda(\underline{a}) \;, & \text{if } 0<\tau\leq\tau_0 \\ \lambda(\bar{a}) \;, & \text{if } \tau_0\leq\tau<+\infty . \end{cases} \tag{1.18}$$

Finally, studying the behaviour of $\max_{\mu \in I_1}|\lambda|^2$ with respect to τ we easily conclude that its minimum is attained at $\tau=\tau_0$ and its corresponding value is given by

$$S(G_{\tau_0}) = \frac{P(R)-1}{P(R)+1} \;, \tag{1.19}$$

where

$$P(R) = \frac{\bar{a}}{\underline{a}} \geq 1 \;. \tag{1.20}$$

The above quantity $P(R)$ is the P-condition number of the matrix R and its minimum value is attained when $\bar{a}=\underline{a}=a$ in which case $P=1, \tau_0=1/a$ and

$$S(G_{\tau_0}) = 0 \;. \tag{1.21}$$

Thus a further reduction on $S(G_{\tau_0})$ will be obtained by the minimisation of $P(R)$. Similar results hold when $\bar{a}<0$, where by analogous treatment we easily find that

$$S(G_{\tau_0}) = \frac{1-P(R)}{1+P(R)} \;, \tag{1.22}$$

with

$$P(R) = \frac{\underline{a}}{\bar{a}} \leq 1. \tag{1.23}$$

Evidently, in this case (1.22) is an increasing function of $P(R)$ and $S(G_{\tau_0})$ will be further reduced by maximising $P(R)$. Let us now assume that instead of (1.4) we use (1.6) in our preconditioning procedure. Then, if λ are the eigenvalues of G we have

$$\hat{\lambda} = 1-\mu \;,$$

thus

$$S(G) = \max_{\mu \in I_1} |\hat{\lambda}| \;, \tag{1.24}$$

and

$$|\hat{\lambda}|^2 = \mu^2-2\mu+1 \equiv \hat{\lambda}(\mu) \;. \tag{1.25}$$

For convergence $S(G)<1$ which because of (1.25) yields

$$0 < \mu < 2 \;,$$

since $\mu \in I_1$, it follows that the necessary and sufficient conditions for convergence of (1.6) are

$$0 < \underline{a} \quad \text{and } \bar{a} < 2 \;. \tag{1.26}$$

On the other hand, we note that $\text{sign}(\partial|\hat{\lambda}|^2/\partial\mu)=\text{sign}(\mu-1)$ which indicates that we have to distinguish two cases as to whether (i) $0<\mu\leq 1$ or (ii) $1\leq\mu<2$. Suppose that (i) holds, then $0<\underline{a}$ and $\bar{a}\leq 1$, whereas $|\hat{\lambda}|^2$ is a decreasing function of μ, hence $\max_{\mu \in I_1}|\hat{\lambda}|^2=\lambda(\underline{a})$ or

$$S(G) = 1-\underline{a} \ . \tag{1.27}$$

Similarly, if (ii) is valid, then $1 \leq \underline{a}$ and $\bar{a}<2$, whereas $\max_{\mu \in I_1} |\hat{\lambda}|^2 = \lambda(\bar{a})$ or

$$S(G) = \bar{a}-1 \ . \tag{1.28}$$

Remark. There must exist an additional condition ($\bar{a}<2$) on the eigenvalues of R for convergence of (1.6) as compared with (1.4). Moreover, it is a simple matter to verify that

$$S(G_{\tau_O}) < S(G) \ , \tag{1.29}$$

implying that the extrapolated method will always yield a faster rate of convergence. The question which therefore arises is: Do all the classical first order iterative methods satisfy the requirements of the preconditioning procedure? The answer to this question brings forward new iterative schemes which obviously will be the extrapolated versions of the classical ones. In Section 2 we define four iterative methods which were recently discovered and present their matrix and component formulae, whereas in Section 3 we establish some general convergence theorems when the coefficient matrix A possesses particular properties such as irreducibility and weak diagonal dominance, being positive definite, consistently ordered, L-matrix and M-matrix. Finally, Section 4 develops the convergence analysis of all the iterative schemes considered, when A is a consistently ordered 2-cyclic matrix and the Jacobi iteration matrix B possesses only real eigenvalues.

2. EXTRAPOLATED LINEAR FIRST ORDER ITERATIVE METHODS

Our aim in this section will be the presentation of the extrapolated versions of the conventional first order methods. Let us assume for the remainder of this section that $A=(a_{i,j})$, $b=(b_i)$ and $u=(u_i)$, then (1.1) can be written as

$$\sum_{j=1}^{N} a_{i,j}u_j = b_i \ , \ i=1(1)N \ . \tag{2.1}$$

Throughout this chapter we let the coefficient matrix A possess the splitting

$$A = D - C_L - C_U \ , \tag{2.2}$$

where D is a diagonal matrix with the same non-vanishing diagonal elements as A and C_L,C_U are the strictly lower and upper triangular part of A, respectively. For the simple choice

$$R = D \ ,$$

(1.4) yields

$$u^{(n+1)} = B_\tau u^{(n)} + c \ , \tag{2.3}$$

where

$$B_\tau = I-\tau D^{-1}A \text{ and } c = D^{-1}b \ . \tag{2.4}$$

Using the matrix notation (2.2), we can write (2.3) as

$$u^{(n+1)} = [(1-\tau)I+\tau B]u^{(n)} + \tau c \ , \tag{2.5}$$

where

$$B = L+U \ , \ L = D^{-1}C_L \quad \text{and } U = D^{-1}C_U \ . \tag{2.6}$$

As is well known [12],[13], (2.3) or (2.5) is the Extrapolated Jacobi (EJ) (or sometimes referred to as the Jacobi Overrelaxation (JOR)) method. When $\tau=1$ we note that EJ degenerates into the Jacobi (J) method. For actual implementation we need the component form of (2.5) which is given by the formula

$$u_i^{(n+1)} = (1-\tau)u_i^{(n)} + \tau \sum_{j=1}^{N} b_{i,j}u_j^{(n)} + \tau c_i \ , \ i=1(1)N \ , \tag{2.7}$$

where

$$b_{i,j} = \begin{cases} -\dfrac{a_{i,j}}{a_{i,i}} \ , & \text{if } i\neq j \ , \\ 0 \ , & \text{if } i=j \ , \end{cases} \quad \text{and } c_i = \frac{b_i}{a_{i,i}} \ . \tag{2.8}$$

Further by letting

$$R = D - C_L \ , \tag{2.9}$$

which is a better approximation to A than its previous form, (1.4) yields

$$u^{(n+1)} = L_{\tau,1}\, u^{(n)} + \tau(I-L)^{-1}c \ , \tag{2.10}$$

where

$$L_{\tau,1} = I-\tau(I-L)^{-1}D^{-1}A \ . \tag{2.11}$$

By expressing $L_{\tau,1}$ in terms of L and U, (2.10) becomes

$$u^{(n+1)} = (1-\tau)u^{(n)}+Lu^{(n+1)}+(\tau-1)Lu^{(n)}+\tau Uu^{(n)}+\tau c. \tag{2.12}$$

We will refer to (2.12) as the Extrapolated Gauss-Seidel (EGS) method [7] since for $\tau=1$ it coincides with the GS method.

The form (2.12) of EGS yields the following formula for the computation of the components of $u^{(n+1)}$

$$u_i^{(n+1)} = (1-\tau)u_i^{(n)} + \sum_{j=1}^{i-1} b_{i,j}u_j^{(n+1)} + (\tau-1)\sum_{j=1}^{i-1} b_{i,j}u_j^{(n)} + $$
$$+ \tau \sum_{j=i+1}^{N} b_{i,j}u_j^{(n)} + \tau c_i \ , \ i=1(1)N. \tag{2.13}$$

Clearly, the above iteration scheme would be difficult to discover, if one followed the classical approach of producing first the "component" form of the method and then its corresponding "matrix" form. For the same reason it took some time for the discovery of the Extrapolated Successive Overrelaxation (ESOR) [11],[4],[6],[10] since the introduction of SOR. ESOR is obtained by letting

$$R = D - \omega C_L \ , \tag{2.14}$$

where ω is a real parameter whose role, at this early stage, can be thought of as perturbing the previous form (2.9) of R such as to become a better approximation to A. Thus (1.4) yields

$$u^{(n+1)} = \mathcal{L}_{\tau,\omega} u^{(n)} + \tau(I-\omega L)^{-1} c, \tag{2.15}$$

where

$$\mathcal{L}_{\tau,\omega} = I - \tau(I-\omega L)^{-1} D^{-1} A \ . \tag{2.16}$$

In order to derive the component equation of ESOR we write (2.15) as

$$u^{(n+1)} = (1-\tau)u^{(n)} + \omega L u^{(n+1)} + (\tau-\omega) L u^{(n)} + \tau U u^{(n)} + \tau c \tag{2.17}$$

which in turn yields the following iteration formula

$$u_i^{(n+1)} = (1-\tau)u_i^{(n)} + \sum_{j=1}^{i-1} b_{i,j} u_j^{(n+1)} + (\tau-\omega) \sum_{j=1}^{i-1} b_{i,j} u_j^{(n)} + \tau \sum_{j=i+1}^{N} b_{i,j} u_j^{(n)} + \\ + \tau c_i \ , \quad i=1(1)N. \tag{2.18}$$

It is easy to see that for $\tau=\omega$, ESOR yields the well known SOR method. According to our analysis ESOR is in fact the extrapolated version of the method which is produced from (2.15) when $\tau=1$, i.e., the scheme

$$u^{(n+1)} = \mathcal{L}_{1,\omega} u^{(n)} + (I-\omega L)^{-1} c \ , \tag{2.19}$$

where

$$\mathcal{L}_{1,\omega} = I - (I-\omega L)^{-1} D^{-1} A \ . \tag{2.20}$$

By expressing (2.19) in terms of L and U yields

$$u^{(n+1)} = \omega L u^{(n+1)} + (1-\omega) L u^{(n)} + U u^{(n)} + c$$

which in turn produces the following component iteration formula

$$u_i^{(n+1)} = \omega \sum_{j=1}^{i-1} b_{i,j} u_j^{(n+1)} + (1-\omega) \sum_{j=1}^{i-1} b_{i,j} u_j^{(n)} + \sum_{j=i+1}^{N} b_{i,j} u_j^{(n)} + \\ + c_i, \quad i=1(1)N. \tag{2.21}$$

In the sequel, the above method will be referred to as the Semi-Successive Overrelaxation (SESOR) method for as we shall see later its rate of convergence is approximately half that of SOR when A is a consistently ordered matrix [7]. Finally, we briefly mention that by letting

$$R = (D-\omega C_L) D^{-1} (D-\omega C_U) \ , \tag{2.22}$$

we have the so called Preconditioned Simultaneous Displacement (PSD) method [1],[2] defined by

$$u^{(n+1)} = \mathcal{D}_{\tau,\omega}u^{(n)} + \tau(I-\omega U)^{-1}(I-\omega L)^{-1}c\,, \tag{2.23}$$

where

$$\mathcal{D}_{\tau,\omega} = I-\tau(I-\omega U)^{-1}(I-\omega L)^{-1}D^{-1}A\,. \tag{2.24}$$

For actual computation we use the following two-level form of (2.23)

$$u^{(n+\frac{1}{2})} = (1-\tau)u^{(n)} + \omega Lu^{(n+\frac{1}{2})} + (\tau-\omega)Lu^{(n)} + \tau Uu^{(n)} + \tau c \tag{2.25}$$

and

$$u^{(n+1)} = u^{(n+\frac{1}{2})} + \omega Uu^{(n+1)} - \omega Uu^{(n)}\,, \tag{2.26}$$

or in component form

$$u_i^{(n+\frac{1}{2})} = (1-\tau)u_i^{(n)} + \omega\sum_{j=1}^{i-1} b_{i,j}u_j^{(n+\frac{1}{2})} + (\tau-\omega)\sum_{j=1}^{i-1} b_{i,j}u_j^{(n)} + \tau\sum_{j=i+1}^{N} b_{i,j}u_j^{(n)} + \tau c_i\,,\quad i=1(1)N\,. \tag{2.27}$$

and

$$u_i^{(n+1)} = u_i^{(n+\frac{1}{2})} + \omega\sum_{j=i+1}^{N} b_{i,j}u_j^{(n+1)} - \omega\sum_{j=i+1}^{N} b_{i,j}u_j^{(n)}\,,\quad i=N(-1)1. \tag{2.28}$$

Clearly, PSD is the extrapolated form (note that (2.25) or (2.27) is an ESOR iteration) of the scheme which is obtained by (2.23) when $\tau=1$ (Preconditioned Jacobi method [3]), whereas for $\tau=\omega(2-\omega)$, the well known Symmetric SOR (SSOR) method is derived. We conclude this section by noting (a) in all the iterative schemes presented, requirements (i), (ii) and (iii) of the preconditioning procedure are satisfied, (b) since in PSD the matrix R has the expression $D-\omega(C_L+C_U)+\omega^2 DLU$ which best approximates A, as compared with the previous forms of R, it is conjectured that PSD will attain the fastest convergence rate under certain conditions. In the sequel we will not deal with PSD as the convergence analysis of this method has been presented elsewhere [2], and (c) instead of C_L in the form of R one may use C_U. This modification will cause the computation of $u_i^{(n+1)}$ in the reverse order which has no effect on the convergence rate.

3. SOME GENERAL CONVERGENCE THEOREMS

This section establishes some convergence theorems for the iterative methods considered in Section 2 when A has particular properties.

Theorem 3.1. Let A be an irreducible matrix with weak diagonal dominance. Then, (a) The GS method converges and EGS converges for $0<\tau\leq 1$, (b) SESOR converges for $0\leq\omega\leq 1$, whereas ESOR converges when $0<\omega\leq 1$ and $0<\tau\leq\omega\leq 1$.

Proof. Let us assume that $S(\mathcal{L}_{\tau,\omega})\geq 1$; then for some eigenvalue λ of $\mathcal{L}_{\tau,\omega}$ we have $|\lambda|\geq 1$. Furthermore, $\det(\mathcal{L}_{\tau,\omega}-\lambda I)=\det(Q)=0$, where $Q=I-[(\tau-\omega+\omega\lambda)/$

$(\lambda+\tau-1)]L-[\tau/(\lambda+\tau-1)]U$. By letting $\lambda^{-1}=qe^{i\theta}$, where q and θ are real and $0<q\leq 1$, we can easily prove that $|\tau/(\lambda+\tau-1)|\leq|(\tau+\omega(\lambda-1))/(\lambda+\tau-1)|\leq 1$. This shows that Q has weak diagonal dominance. However, Q is also irreducible, hence $\det(A)\neq 0$. Consequently $S(\mathcal{L}_{\tau,\omega})<1$. □

Theroem 3.2. Let A be a positive definite matrix. Then $||\mathcal{L}_{\tau,\omega}||_{A^{\frac{1}{2}}}<1$, if the matrix $M=\tau^{-1}[(2-\tau)D+(\tau-\omega)(C_L+C_U)]$ is positive definite or, equivalently, if

$$0 < \tau < 2 \text{ and } \tau+(2-\tau)/\mu_{min} < \omega < \tau+(2-\tau)/\mu_{max} , \qquad (3.1)$$

where μ_{min}, μ_{max} denote the minimum and maximum eigenvalues of B, respectively.

Proof. From Theorem 3-5.3 of [13] it follows that $||\mathcal{L}_{\tau,\omega}||_{A^{\frac{1}{2}}}<1$ if M is positive definite. Evidently, M is positive definite if and only if $D^{-\frac{1}{2}}MD^{-\frac{1}{2}}=\tau^{-1}[(2-\tau)I+(\tau-\omega)\tilde{B}]=\tilde{M}$, where $\tilde{B}=D^{\frac{1}{2}}BD^{-\frac{1}{2}}$, is positive definite. Clearly $\tilde{M}$ has positive diagonal elements if and only if $0<\tau<2$. Moreover, $\tilde{M}$ is positive definite if and only if its eigenvalues are positive, hence

$$2-\tau+(\tau-\omega)/\mu_i > 0 , \qquad (3.2)$$

where $\{\mu_i\}_{i=1}^N$ are the eigenvalues of B, or equivalently

$$(\omega-\tau)\mu_{max} < 2-\tau \quad \text{and} \quad (\omega-\tau)\mu_{min} < 2-\tau , \qquad (3.3)$$

which if combined yield (3.1) (note $\mu_{max}\leq 0\leq\mu_{min}$ since $trB=0$). □

Corollary 3.3. Let A be a positive definite matrix. Then:
(a) $||\mathcal{L}_{1,1}||_{A^{\frac{1}{2}}}<1$, (b) $||\mathcal{L}_{\tau,1}||_{A^{\frac{1}{2}}}<1$ if the matrix $M=\tau^{-1}[(2-\tau)D+(\tau-\omega)(C_L+C_U)]$ is positive definite or, equivalently, if $0<\tau<(2-\mu_{min})/(1-\mu_{min})$ and (c) $||\mathcal{L}_{1,\omega}||_{A^{\frac{1}{2}}}<1$ if the matrix $M=D+(1-\omega)(C_L+C_U)$ is positive definite or, equivalently, if $1+1/\mu_{min}<\omega<1+1/\mu_{max}$.

Proof. Lt follows immediately from Theorem 3.2. □

Theorem 3.4. If A is an L-matrix of order N and if $0\leq\omega\leq\tau\leq 1$ $(\tau\neq 0)$. Then:
(a) $S(\mathcal{L}_{\tau,\omega})<1$ if and only if $S(B)<1$. (b) $S(\mathcal{L}_{\tau,\omega})<1$ and $S(B)<1$ if and only if A is an M-matrix; if $S(B)<1$, then $S(\mathcal{L}_{\tau,\omega})\leq 1-\tau+\tau S(B)$, (c) If $S(B)\geq 1$ and $S(\mathcal{L}_{\tau,\omega})\geq 1$, then $S(\mathcal{L}_{\tau,\omega})\geq 1-\tau+\tau S(B)\geq 1$.

Proof. Since $\mathcal{L}_{\tau,\omega}$ is a non-negative matrix, $\bar{\lambda}=S(\mathcal{L}_{\tau,\omega})$ is an eigenvalue of $\mathcal{L}_{\tau,\omega}$ and for some $v\neq 0$ we have $\mathcal{L}_{\tau,\omega}v=\bar{\lambda}v$ and $(\bar{\beta}L+U)v=[(\bar{\lambda}-1+\tau)/\tau]v$, where $\bar{\beta}=[(\tau+\omega(\bar{\lambda}-1))/\tau]$, which implies that $(\bar{\lambda}-1+\tau)/\tau\leq S(\bar{\beta}L+U)$. This relationship yields the following results:

(i) if $\bar{\lambda}\leq 1$, then $\bar{\lambda}\leq\tau\bar{\mu}+1-\tau$.

(ii) if $\bar{\lambda}\geq 1$, then $\bar{\mu}\geq 1$ which implies

(iii) if $\bar{\mu}<1$, then $\bar{\lambda}<1$. Since $B\geq 0$, it follows that for some $v\neq 0$ we have

$B v=\bar{\mu} v$ and $Q v=(1-\tau+\tau\bar{\mu}) v$, where $Q=(I-\alpha L)^{-1}[(1-\tau) I+(\tau-\omega) L+\tau U]$ and $\alpha=\omega/(1-\tau+\tau\bar{\mu})$. Therefore, $1-\tau+\tau\bar{\mu} \leqslant S(Q)$. If $\bar{\mu} \geqslant 1$ and since $\alpha \leqslant \omega$, we have $(I-\alpha L)^{-1} \leqslant (I-\omega L)^{-1}$ which implies that $Q \leqslant L_{\tau,\omega}$, then $1-\tau+\tau\bar{\mu} \leqslant S(Q) \leqslant S(L_{\tau,\omega})=\bar{\lambda}$. Thus, we have shown that: (iv) if $\bar{\mu} \geqslant 1$, then $\bar{\lambda} \geqslant 1-\tau+\tau\bar{\mu} \geqslant 1$. This implies: (v) if $\bar{\lambda}<1$, then $\bar{\mu}<1$. By (iii) and (v) we have (a). By (iii) and (v) we have (a). By (i) and Theorem 2-7.2 of [13] we have (b), whereas by (iv) we have (c). □

Corollary 3.5. If A is an L-matrix of order N and if $0 \leqslant \omega \leqslant 1$. Then, (a) $S(L_{1,\omega})<1$ if and only if $S(B)<1$, (b) $S(L_{1,\omega})<1$ and $S(B)<1$ if and only if A is an M-matrix; if $S(B)<1$ then $S(L_{1,\omega}) \leqslant S(B)$, (c) if $S(B) \geqslant 1$ and $S(L_{1,\omega}) \geqslant 1$, then $S(L_{1,\omega}) \geqslant S(B) \geqslant 1$.

Proof. It follows from Theorem 3.4. □

Theorem 3.6. If A is an M-matrix, then $S(L_{\tau,\omega})<1$ if

$$1 \leqslant \tau \leqslant \omega<\frac{2}{1+S(B)} \quad . \tag{3.4}$$

Proof. If $1 \leqslant \tau \leqslant \omega$, then $T_{\tau,\omega}=(I-\omega L)^{-1}[(1-\tau) I+(\omega-\tau) L+\tau U]$ is non-negative and $|L_{\tau,\omega}| \leqslant T_{\tau,\omega}$. If $\bar{\gamma}=S(T_{\tau,\omega})$, then there exists $v \neq 0$ such that $T_{\tau,\omega} v=\bar{\gamma} v$, implying that $[\tau U+(\omega-\tau+\bar{\gamma}\omega) L] v=(\bar{\gamma}+1-\tau) v$. If $\bar{\gamma} \geqslant 1$, then $(\omega-\tau+\bar{\gamma}\omega)/\tau \geqslant 1$, hence $\bar{\gamma}+1-\tau \leqslant(\omega-\tau+\bar{\gamma}\omega)\bar{\mu}$ or $\omega \geqslant(1+\bar{\gamma})/(1+\bar{\gamma}\bar{\mu}) \geqslant 2/(1+\bar{\mu})$. Therefore, if (3.5) holds, then $\bar{\gamma}<0$. By Theorem 2-1.16 of [13] it follows that $S(L_{\tau,\omega}) \leqslant \bar{\gamma} \leqslant 1$. □

Corollary 3.7. If A is an M-matrix, then $S(L_{\tau,1})<1$ if

$$1 \leqslant \omega<\frac{2}{1+S(B)} \; .$$

Proof. It follows from Theorem 3.6. □

The above material was first presented in [7] except some modifications on Theorem 3.2 and its Corollary.

4. CONVERGENCE ANALYSIS FOR CONSISTENTLY ORDERED MATRICES

This section develops the convergence theory for the extrapolated iterative methods when A is a consistently ordered (2-cyclic) matrix and the Jacobi iteration matrix B possesses only real eigenvalues.

Theorem 4.1. If A is a consistently ordered matrix with non-vanishing diagonal elements such that B has real eigenvalues $\{\mu_i\}_{i=1}^N$ with $\underline{\mu}=\min_i|\mu_i| \neq 0$ and $\bar{\mu}=\max_i|\mu_i|$, then EGS converges if and only if

$$\bar{\mu}<1 \quad \text{and } 0<\tau<2/(1-\underline{\mu}^2) \tag{4.1a}$$

or

$$\underline{\mu}>1 \quad \text{and } 2/(1-\bar{\mu}^2)<\tau<0 \; . \tag{4.1b}$$

Proof. If μ, λ are eigenvalues of B and the matrix $\Lambda_1=(I-L)^{-1}D^{-1}A$, respectively, then it is easy to verify (see Theorem 5-2.2 of [13]) that

$$\lambda = 1-\mu^2 . \tag{4.2}$$

According to (1.12) let us first assume $\lambda>0$, then (4.2) yields $\mu^2<1$ or $\bar{\mu}<1$. Moreover, since

$$\underline{\mu}^2 \leqslant \mu^2 \leqslant \bar{\mu}^2 \tag{4.3}$$

it follows that

$$\bar{\lambda} = \max_{\mu^2}\{\lambda\} = 1-\underline{\mu}^2 \quad \text{and} \quad \underline{\lambda} = \min_{\mu^2}\{\lambda\} = 1-\bar{\mu}^2 \tag{4.4}$$

and because of (1.12a), (4.1a) holds. Similarly, we can easily prove that (4.1b) holds by using (1.12b). □

Corollary 4.2. Under the hypothesis of Theorem 4.1 and if $\underline{\mu}=0$, EGS converges if and only if

$$\bar{\mu} < 1 \text{ and } 0<\tau<2 . \tag{4.5}$$

Proof. Since $\mu=0$ is an eigenvalue of B it follows from (4.2) that $\lambda=1>0$, thus (4.1a) holds only with $\underline{\mu}=0$. □

Corollary 4.3. Under the hypothesis of Theorem 4.1 and if $\underline{\mu}=\bar{\mu}=\mu$, EGS converges if and only if (a) $\mu<1$ and $0<\tau<2/(1-\mu^2)$ or (b) $\mu>1$ and $-2/(\mu^2-1)<\tau<0$.

Proof. It follows from Theorem 4.1. □

Theorem 4.4. Let A be a consistently ordered matrix with non-vanishing diagonal elements such that the matrix B has real eigenvalues $\{\mu_i\}_{i=1}^N$ with $\underline{\mu}=\min_i|\mu_i|$ and $\bar{\mu}=\max_i|\mu_i|$. Then $S(L_{\tau,1})$ is minimised for

$$\tau_0 = 2/[2-(\bar{\mu}^2+\underline{\mu}^2)] \tag{4.6}$$

and its corresponding value is given by the expression

$$S(L_{\tau,1}) = \begin{cases} (\bar{\mu}^2-\underline{\mu}^2)/[2-(\bar{\mu}^2+\underline{\mu}^2)] & , \text{ if } \bar{\mu}<1 \\ (\bar{\mu}^2-\underline{\mu}^2)/[(\bar{\mu}^2+\underline{\mu}^2)-2] & , \text{ if } \underline{\mu}>1 . \end{cases} \tag{4.7}$$

Proof. Substituting the expressions (4.4) into (1.17), (1.19) and (1.22) we easily find (4.6) and (4.7), respectively, ($\bar{a}=\bar{\lambda}$ and $\underline{a}=\underline{\lambda}$). □

Corollary 4.5. Under the hypothesis of Theorem 4.4 and if $\underline{\mu}=0$, then $S(L_{\tau,1})$ is minimised for

$$\tau_0 = 2/(2-\bar{\mu}^2) \tag{4.8}$$

and its corresponding value is given by

$$S(L_{\tau_0,1}) = \bar{\mu}^2/(2-\bar{\mu}^2). \tag{4.9}$$

Proof. It follows immediately by letting $\underline{\mu}=0$ in (4.6) and (4.7). □

Corollary 4.6. Under the hypothesis of Theorem 4.4 and if

$$\underline{\mu} = \bar{\mu} = \mu , \tag{4.10}$$

then for

$$\tau_0 = 1/(1-\mu^2) , \tag{4.11}$$

we have

$$S(L_{\tau_0,1}) = 0 . \tag{4.12}$$

Proof. It follows from Theorem 4.4. □

The following theorem shows a comparison of rates of convergence of the EGS and GS methods.

Theorem 4.7. If $\bar{\mu}<1$ then $(R(.)=-\log S(.))$

$$\lim_{\bar{\mu}\to 1^-}\left[\frac{R(L_{\tau_0,1})}{R(L_{1,1})}\right] = \begin{cases} 2/(1-\underline{\mu}^2), & \text{if } \underline{\mu}\neq 0 \\ 2 & \text{, if } \underline{\mu}=0 \end{cases} \tag{4.13}$$

where $L_{1,1}$ is the iteration matrix of the GS method.

Proof. Evidently $R(L_{\tau_0,1})=-\log S(L_{\tau_0,1}) \sim 2/\kappa(\Lambda_1)$ for $\kappa(\Lambda_1)=(1-\underline{\mu}^2)/(1-\bar{\mu}^2)$ $>>1$, that is when $\bar{\mu}\to 1^-$. Also $R(L_{1,1})=-\log S(L_{1,1})=-\log\bar{\mu}^2 \sim 1-\bar{\mu}^2$ and (4.13) is easily derived. □

From Theorem 4.7 we conclude that EGS converges asymptotically at least twice as fast as GS, whereas its convergence rate becomes very rapid as $\underline{\mu}$ goes away from zero (Corollary 4.6).

In the remainder of this section we (i) establish the convergence conditions of ESOR and SESOR and (ii) determine the optimum values of the involved parameters in both methods in the sense of minimising their spectral radius [6].

Theorem 4.8. Let A be a consistently ordered matrix with non-vanishing diagonal elements. If μ is an eigenvalue of B and λ satisfies

$$(1-\lambda)^2 = \mu^2(1-\lambda\omega) \quad , \tag{4.14}$$

then λ is an eigenvalue of the matrix

$$\Lambda_\omega = (I-\omega L)^{-1}D^{-1}A \; . \tag{4.15}$$

Conversely, if λ is an eigenvalue of Λ_ω and if μ satisfies (4.14), then μ is an eigenvalue of B.

Proof. The proof of this theorem is analogous to that of Theorem 5-2.2 in [13] and is therefore omitted. □

A sufficient and necessary condition for ESOR to converge is $S(L_{\tau,\omega})<1$. Thus, if we let $\lambda=a+ib$, where a,b are real numbers and $i=\sqrt{-1}$, be an eigenvalue of Λ_ω, then ESOR converges if and only if $|1-\tau(a+ib)|<1$ or

$$\tau^2(a^2+b^2) < 2\tau a \tag{4.16}$$

which implies that $\qquad \tau a > 0 \quad .$ (4.17)

Therefore, (4.16) is equivalent to either

$$a>0,\ 0<\tau<2a/(a^2+b^2) \quad \text{or } a<0 \text{ and } 2a/(a^2+b^2)<\tau<0 \tag{4.18}$$

(when b=0, then (4.18) yields (1.12)).

Theorem 4.9. If A is a consistently ordered matrix with non-vanishing diagonal elements such that the matrix B has real eigenvalues $\{\mu_i\}_{i=1}^N$ with $\underline{\mu}=\min_i|\mu_i|\neq 0$ and $\bar{\mu}=\max_i|\mu_i|$, then ESOR converges if and only if $\bar{\mu}=S(B)<1$ and the parameters ω,τ lie in any of the corresponding domains given by Table 4.1.

ω-Domain*	τ-Domain*	Condition
$-\infty<\omega\leq\omega'(\bar{\mu})$	$0<\tau<2/\lambda_+(\bar{\mu})$	-
$\omega'(\bar{\mu})\leq\omega\leq 2$	$0<\tau<h(\omega,\underline{\mu})$	-
$2\leq\omega<2/\bar{\mu}^2$	$0<\tau<h(\omega,\bar{\mu})$	-
$2/\underline{\mu}^2<\omega\leq\omega''(\bar{\mu})$	$h(\omega,\underline{\mu})<\tau<0$	$1-\underline{\mu}^2<(1-\bar{\mu}^2)^{\frac{1}{2}}$
$\omega''(\bar{\mu})\leq\omega<+\infty$	$2/\lambda_-(\bar{\mu})<\tau<0$	"

*$h(\omega,\mu)$ and $\lambda_\pm(\mu)$ are given by (4.25) and (4.20) respectively, whereas $\omega'(\mu)$ and $\omega''(\mu)$ are defined by (4.22).

TABLE 4.1

Proof. If μ,λ are eigenvalues of B and Λ_ω, respectively, then

$$\lambda^2-(2-\omega\mu^2)\lambda+1-\mu^2 = 0 \ . \tag{4.19}$$

The roots of (4.19) are given by

$$\lambda_\pm(\mu) = (2-\omega\mu^2\pm\Delta^{\frac{1}{2}})/2 \ , \tag{4.20}$$

where

$$\Delta \equiv \Delta(\mu) = \mu^2[\omega^2\mu^2-4(\omega-1)] \ . \tag{4.21}$$

The kind of $\lambda_\pm(\mu)$ (real or complex) depends upon the sign of Δ which in turn is determined according to the position of ω with respect to the quantities

$$\omega'(\mu) = 2/[1+(1-\mu^2)^{\frac{1}{2}}] \text{ and } \omega''(\mu) = 2/[1-(1-\mu^2)^{\frac{1}{2}}] \tag{4.22}$$

which obviously are the roots of Δ. In the sequel we examine the sign of Δ by distinguishing the following three basic cases. Case I: All $\lambda_\pm(\mu)$ are complex, Case II: All $\lambda_\pm(\mu)$ are real and Case III: Some $\lambda_\pm(\mu)$ are complex and others are real.

Case I: In this case $\lambda\pm(\mu)$ are complex conjugate pairs and $\Delta(\mu)<0$ for all μ^2 such that $0<\underline{\mu}^2\leq\mu^2\leq\bar{\mu}^2$, hence $\omega'(\bar{\mu})\leq\omega\leq\omega''(\bar{\mu})$. Evidently, $|\lambda_\pm(\mu)|^2 = 1-\mu^2$ and $Re\lambda_\pm(\mu)=(2-\omega\mu^2)/2$. Next, we also note that $Re\lambda_\pm(\mu)>0$ implies $\omega<2/\bar{\mu}^2$, whereas in order for $Re\lambda_\pm(\mu)<0$ we must have $2/\underline{\mu}^2<\omega''(\bar{\mu})$ or

$$1-\underline{\mu}^2<(1-\bar{\mu}^2)^{\frac{1}{2}} \tag{4.23}$$

in which case $\omega>2/\underline{\mu}^2$. We therefore split the above interval of ω into the following: (i) $\omega'(\bar{\mu})\leq\omega<2/\bar{\mu}^2$, (ii) $2/\underline{\mu}^2<\omega\leq\omega''(\bar{\mu})$ and examine each sub-case separately. When ω lies in the interval given by (i), $Re\lambda_\pm(\mu)>0$, and from (4.18) it follows that τ must lie in the range

$$0<\tau< \min_{\underline{\mu}^2 \leq \mu^2 \leq \bar{\mu}^2} h(\omega,\mu) \;, \tag{4.24}$$

where

$$h(\omega,\mu) = (2-\omega\mu^2)/(1-\mu^2) \;. \tag{4.25}$$

However, $\text{sign}[\partial h(\omega,\mu)/\partial\mu^2]=\text{sign}(2-\omega)$, hence for this subcase the ranges for ω and τ such that ESOR converges are

$$\omega'(\bar{\mu})\leq\omega\leq 2 \quad \text{and} \quad 0<\tau<h(\omega,\underline{\mu}) \;, \tag{4.26}$$

or

$$2\leq\omega<2/\bar{\mu}^2 \quad \text{and} \quad 0<\tau<h(\omega,\bar{\mu}) \;. \tag{4.27}$$

Alternatively, if (ii) is valid, then (4.23) must hold and the range of τ is (see (4.18))

$$h(\omega,\underline{\mu})<\tau<0 \;. \tag{4.28}$$

Case II: In this case we assume that all $\lambda_{\pm}(\mu)$ are real. If they are all positive, then $-\infty<\omega\leq\omega'(\underline{\mu})$ and (4.18) yields (b=0) $0<\tau<2/\lambda_{+}(\mu)$. But, $\lambda_{+}(\mu)$ is an increasing function of μ^2, hence $0<\tau<2/\lambda_{+}(\bar{\mu})$. Similarly, if all $\lambda_{\pm}(\mu)$ are negative we find $\omega''(\underline{\mu})\leq\omega<+\infty$ and $2/\lambda_{-}(\bar{\mu})<\tau<0$.

Case III: Since some of $\lambda_{\pm}(\mu)$ are complex and the others are real, ω lies in either of the following ranges (i) $\omega'(\underline{\mu})\leq\omega\leq\omega'(\bar{\mu})$ or (ii) $\omega''(\bar{\mu})\leq\omega\leq\omega''(\underline{\mu})$. Let us first suppose that ω lies in (i). Then, the real $\lambda_{\pm}(\mu)$ are positive and $0<\tau<2/\lambda_{+}(\mu)$ (Case II), whereas for the complex $\lambda_{\pm}(\mu)$, we have that their real parts are positive and $0<\tau<h(\omega,\mu)$ (Case I). However, it is readily verified that $0<2/\lambda_{+}(\mu)<h(\omega,\mu)$ implying that when ω lies in (i) the interval for τ is $0<\tau<2/\lambda_{+}(\bar{\mu})$. Alternatively, if ω lies in (ii), the real $\lambda_{\pm}(\mu)$ are negative ifand only if $2/\underline{\mu}^2<\omega$, hence the range becomes $\max\{2/\underline{\mu}^2,\omega''(\bar{\mu})\}\leq\omega\leq\omega''(\underline{\mu})$, where equality on the left holds only when $2/\underline{\mu}^2<\omega''(\bar{\mu})$. Evidently for this range we have $2/\lambda_{-}(\mu)<\tau<0$ (Case II). For the complex $\lambda_{\pm}(\mu)$, the range of τ is $h(\omega,\mu)<\tau<0$ (Case I). But $h(\omega,\mu)< 2/\lambda_{-}(\mu)<0$, therefore the necessary and sufficient conditions in this case are

$$\max\{2/\underline{\mu}^2,\omega''(\bar{\mu})\}\leq\omega\leq\omega''(\underline{\mu}) \quad \text{and} \quad 2/\lambda_{-}(\bar{\mu})<\tau<0.$$

Summarising the results of the above Cases I, II and III we easily conclude that the region of convergence for ESOR is described by the ranges given in Table 4.1. □

Corollary 4.10. Under the hypothesis of Theorem 4.9 and if $\underline{\mu}=0$, ESOR converges if and only if $\bar{\mu}<1$ and either

$$-\infty<\omega\leq 1 \quad \text{and} \quad 0<\tau<2/\lambda_{+}(\bar{\mu}) \;, \tag{4.29}$$

or

$$1\leq\omega\leq 2 \quad \text{and} \quad 0<\tau<2 \;, \tag{4.30}$$

or

$$2\leq\omega<2/\bar{\mu}^2 \quad \text{and} \quad 0<\tau<h(\omega,\bar{\mu}) \;. \tag{4.31}$$

Proof. Following a similar treatment as in the proof of Theorem 4.9 we distinguish the following two basic cases.

Case I: Assume that all $\lambda_{\pm}(\mu)$ are complex, then their real parts are positive since when $\mu=0$, then $\mathrm{Re}\lambda_{\pm}(0)=1>0$. Thus the convergence ranges of ω and τ are either (Case I of Theorem 4.9) $\omega'(\bar{\mu})\leq\omega\leq 2$ and $0<\tau<2$ or $2\leq\omega<2/\bar{\mu}^2$ and $0<\tau<h(\omega,\bar{\mu})$.

Case II: Suppose now that some $\lambda_{\pm}(\mu)$ are complex and the others are real. Reasoning as in Case I, ω lies in the range $-\infty<\omega\leq\omega'(\bar{\mu})$, since $2/\bar{\mu}^2<\omega''(\bar{\mu})$. Evidently, the real $\lambda_{\pm}(\mu)$ are positive hence $0<\tau<2/\lambda_{+}(\bar{\mu})$, whereas the complex $\lambda_{\pm}(\mu)$ have positive real parts and $0<\tau<2$. Next, since $2\leq 2/\lambda_{+}(\bar{\mu})$ if and only if $\omega\geq 1$, it follows that the convergence intervals for this case are given by (4.29) and (4.30). □

Corollary 4.11. Under the hypothesis of Theorem 4.9 and if $\underline{\mu}=0$ SOR converges if and only if $\bar{\mu}<1$ and $0<\omega<2$.

Proof. It easily follows from Corollary 4.10. □

Corollary 4.12. Under the hypothesis of Theorem 4.9 and if $0<\underline{\mu}=\bar{\mu}=\mu<1$, ESOR converges if and only if $\mu<1$ and the parameters ω,τ lie in any of the corresponding domains given by Table 4.2.

ω-Domain	τ-Domain
$-\infty<\omega\leq\omega'(\mu)$	$0<\tau<2/\lambda_{+}(\mu)$
$\omega'(\mu)\leq\omega<2/\mu^2$	$0<\tau<h(\omega,\mu)$
$2/\mu^2<\omega\leq\omega''(\mu)$	$h(\omega,\mu)<\tau<0$
$\omega''(\mu)\leq\omega<+\infty$	$2/\lambda_{-}(\mu)<\tau<0$

TABLE 4.2

Proof. It is easily derived by following a similar approach to the proof of Theorem 4.9. □

Theorem 4.13. If A is a consistently ordered matrix with non-vanishing diagonal elements such that the matrix B has real eigenvalues, then SESOR converges if and only if

$$\bar{\mu}<1 \text{ and } \omega_1<\omega<\omega_2\ , \tag{4.32}$$

where

$$\omega_1 = \frac{\bar{\mu}^2-1}{2\bar{\mu}^2} \qquad \text{and} \qquad \omega_2 = \frac{1+\bar{\mu}^2}{\bar{\mu}^2}\ . \tag{4.33}$$

Proof. Let ν denote the eigenvalues of $L_{1,\omega}$, then from (2.20) we have $\nu=1-\lambda$ where λ is an eigenvalue of Λ_{ω}. Expressing (4.19) in terms of ν we find

In addition, we note that

$$\mathrm{sign}(\partial g/\partial\omega) = \mathrm{sign}(\tau) \ . \tag{4.42}$$

For the case where λ is real $S(L_{\tau,\underline{\omega}})$ is minimised at (see (1.17))

$$\tau = \tau_0 = 2/(\bar{\lambda}+\underline{\lambda}) \ , \tag{4.43}$$

where

$$\bar{\lambda} = \begin{cases} \max\limits_{\underline{\mu}^2 \le \mu^2 \le \bar{\mu}^2} \{\lambda\} & , \text{ if } \lambda > 0 \\ \min\limits_{\underline{\mu}^2 \le \mu^2 \le \bar{\mu}^2} \{\lambda\} & , \text{ if } \lambda < 0 \end{cases} \ , \quad \underline{\lambda} = \begin{cases} \min\limits_{\underline{\mu}^2 \le \mu^2 \le \bar{\mu}^2} \{\lambda\} & , \text{ if } \lambda > 0 \\ \max\limits_{\underline{\mu}^2 \le \mu^2 \le \bar{\mu}^2} \{\lambda\} & , \text{ if } \lambda < 0 \end{cases} \tag{4.44}$$

and its corresponding value is given by the expression (see (1.19) and (1.22))

$$S(L_{\tau_0,\omega}) = |\kappa(\Lambda_\omega)-1|/[\kappa(\Lambda_\omega)+1] \ , \ \kappa(\Lambda_\omega)=\bar{\lambda}/\underline{\lambda} \ . \tag{4.45}$$

A simple study of the behaviour of $\lambda_\pm(\mu)$ as a function of μ^2 reveals that

$$\mathrm{sign}(\partial\lambda_+(\mu)/\partial\mu^2) = +1 \text{ and } \mathrm{sign}(\partial\lambda_-(\mu)/\partial\mu^2) = -1. \tag{4.46}$$

<u>Theorem 4.14</u>. Let A be a consistently ordered matrix with non-vanishing diagonal elements such that the matrix B has real eigenvalues $\{\mu_i\}_{i=1}^N$ with $\underline{\mu}=\min\limits_i|\mu_i|\neq 0$, $\bar{\mu}=\max\limits_i|\mu_i|<1$. Then, the expressions of $S(L_{\tau_0,\omega})$ for the different ranges of ω are presented in Table 4.4. Moreover, if $1-\underline{\mu}^2<(1-\bar{\mu}^2)^{\frac{1}{2}}$, $S(L_{\tau,\omega})$ is minimised for

$$\omega_0 = \omega'(\bar{\mu}) = 2/[1+(1-\bar{\mu}^2)^{\frac{1}{2}}], \ \tau_0 = (2-\omega_0\underline{\mu}^2)/[2(1-\underline{\mu}^2)] \tag{4.47}$$

and its corresponding value is given by the expression

$$S(L_{\tau_0,\omega_0}) = [\underline{\mu}(\bar{\mu}^2-\underline{\mu}^2)^{\frac{1}{2}}]/[(1-\underline{\mu}^2)^{\frac{1}{2}}(1+(1-\bar{\mu}^2)^{\frac{1}{2}})] \ . \tag{4.48}$$

<u>Proof</u>. Following a similar treatment as in Theorem 4.9 we distinguish again three basic cases. Next, for each case we determine the expression of $S(L_{\tau_0,\omega})$ via (4.40) or (4.45) for the different ranges of ω.

<u>Case I</u>: Let us assume that all $\lambda_\pm(\mu)$ are complex, then ω lies in the range $\omega'(\bar{\mu})\le\omega\le\omega''(\bar{\mu})$. According to Table 4.3 we consider three subcases.

(i) $-\infty<\tau<0$. This implies $\mathrm{Re}\lambda_\pm(\mu)=(2-\omega\mu^2)/2<0$ or $2/\underline{\mu}^2<\omega$, hence this subcase exists if $2/\underline{\mu}^2<\omega''(\bar{\mu})$ or, equivalently if

$$1-\underline{\mu}^2<(1-\bar{\mu}^2)^{\frac{1}{2}} \ . \tag{4.49}$$

Therefore, the range of ω becomes $2/\underline{\mu}^2<\omega\le\omega''(\bar{\mu})$, whereas from (4.39) and (4.40) it follows that

$$\tau_0 = q(\omega,\underline{\mu}) \text{ and } S(L_{\tau_0,\omega}) = \{-\underline{\Delta}/[4(1-\underline{\mu}^2)]\}^{\frac{1}{2}} \ , \tag{4.50}$$

where $\underline{\Delta}=\Delta(\underline{\mu})$.

$$\nu^2-\mu^2\omega\nu+\mu^2(\omega-1) = 0 \ . \tag{4.34}$$

Therefore, $S(L_{1,\omega})<1$ if and only if the roots of (4.34) are less than one in modulus or, equivalently, (Lemma 6-2.1 in [13]) if $|\mu^2(\omega-1)|<1$ and $|\mu^2\omega|<1+\mu^2(\omega-1)$ which in turn yield (4.32). □

In the sequel our aim will be the determination of the optimum values τ_0,ω_0 of τ,ω respectively such that $S(L_{\tau,\omega})$ is minimised. Let ξ be an eigenvalue of $L_{\tau,\omega}$, then because of (2.16) we have the following relationship

$$\xi = 1-\tau\lambda \ , \tag{4.35}$$

where λ is an eigenvalue of Λ_ω. Next, we first minimise the expression

$$\max_{\underline{\mu}^2\leq\mu^2\leq\bar{\mu}^2}|\xi| \tag{4.36}$$

with respect to τ for the different ranges of ω. Secondly, we find the value of ω for which the above expression attains its minimum value. From (4.35) we obtain

$$|\xi|^2 = \tau^2(a^2+b^2)-2\tau a+1 \tag{4.37}$$

or in terms of μ and ω, (4.37) yields

$$|\xi|^2 \equiv g(\tau,\omega,\mu^2) = \tau^2(1-\mu^2)-\tau(2-\omega\mu^2)+1 \ . \tag{4.38}$$

Evidently, $|\xi|^2$ is minimised if we let

$$\tau_0 \equiv q(\omega,\mu) = (2-\omega\mu^2)/[2(1-\mu^2)] \tag{4.39}$$

and its corresponding value is given by the expression

$$\min_\tau |\xi|^2 \equiv f(\omega,\mu^2) = -\Delta/[4(1-\mu^2)] \ . \tag{4.40}$$

Next, we examine the behaviour of $g(\tau,\omega,\mu^2)$ as a function of μ^2. Since

$$\text{sign}(\partial g/\partial\mu^2) = \sin(\tau(\omega-\tau)) \ , \tag{4.41}$$

we distinguish six cases in order to determine the sign$(\tau(\omega-\tau))$ which are presented in Table 4.3 together with the maximum value of $g(\tau,\omega,\mu^2)$ with respect to μ^2 for each case.

ω-Domain	τ-Domain	$\tau(\omega-\tau)$	$\max\{g(\tau,\omega,\mu^2)\}$
$-\infty<\omega<0$	$-\infty<\tau<\omega$	<0	$g(\tau,\omega,\underline{\mu}^2)$
	$\omega\leq\tau<0$	≥0	$g(\tau,\omega,\bar{\mu}^2)$
	$0<\tau<+\infty$	<0	$g(\tau,\omega,\underline{\mu}^2)$
$0\leq\omega<+\infty$	$-\infty<\tau<0$	<0	$g(\tau,\omega,\underline{\mu}^2)$
	$0<\tau\leq\omega$	≥0	$g(\tau,\omega,\bar{\mu}^2)$
	$\omega<\tau<+\infty$	<0	$g(\tau,\omega,\underline{\mu}^2)$

TABLE 4.3

(ii) $0<\tau\leq\omega$. Similarly, in this case $\omega'(\bar{\mu})\leq\omega<2/\bar{\mu}^2$. Moreover, since $\tau_o\leq\omega$ or $2/(2-\bar{\mu}^2)\leq\omega$, it follows that the range of ω becomes $2/(2-\bar{\mu}^2)\leq\omega<2/\bar{\mu}^2$, whereas

$$\tau_o = q(\omega,\bar{\mu}) \text{ and } S(L_{\tau_o,\omega}) = \{-\bar{\Delta}/[4(1-\bar{\mu}^2)]\}^{\frac{1}{2}} , \qquad (4.51)$$

where $\bar{\Delta}=\Delta(\bar{\mu})$.

(iii) $0<\omega<\tau<+\infty$. Again $\omega'(\bar{\mu})\leq\omega<2/\bar{\mu}^2$ and since $\omega<\tau_o$, it follows that $\omega<2/(2-\underline{\mu}^2)$. Therefore, for this case to exist $\omega'(\bar{\mu})<2/(2-\underline{\mu}^2)$ or (4.49) must hold. Evidently, the range of ω is $\omega'(\bar{\mu})\leq\omega<2/(2-\underline{\mu}^2)$ while (4.50) holds for this subcase also.

Case II: Here we consider the case where all $\lambda_\pm(\mu)$ are real. If they are positive, then $-\infty<\omega\leq\omega'(\underline{\mu})$ and $S(L_{\tau_o,\omega})$ is given by (4.45) with $\bar{\lambda}=\lambda_+(\bar{\mu})$ and $\underline{\lambda}=\lambda_-(\bar{\mu})$ thus

$$S(L_{\tau_o,\omega}) = \bar{\Delta}^{\frac{1}{2}}/(2-\omega\bar{\mu}^2) \qquad (4.52)$$

where τ_o is determined by (4.43).

Alternatively, if all $\lambda_\pm(\mu)$ are negative $\omega''(\underline{\mu})\leq\omega<+\infty$ and

$$S(L_{\tau_o,\omega}) = \bar{\Delta}^{\frac{1}{2}}/(\omega\bar{\mu}^2-2) . \qquad (4.53)$$

Case III: Assume now that some of $\lambda_\pm(\mu)$ are complex and the others are real. This implies that ω lies in either of the following intervals: (i) $\omega'(\underline{\mu})\leq\omega\leq\omega'(\bar{\mu})$ or (ii) $\omega''(\bar{\mu})\leq\omega\leq\omega''(\underline{\mu})$. If ω lies in (i), then all real $\lambda_\pm(\mu)$ are positive and all complex $\lambda_\pm(\mu)$ have positive real parts. Therefore, in the real case (see Case II) $S(L_{\tau_o,\omega})=\Delta^{\frac{1}{2}}/(2-\omega\mu^2)$, whereas in the complex case $S(L_{\tau_o,\omega})=\{-\Delta/[4(1-\mu^2)]\}^{\frac{1}{2}}$. However, since $(2-\omega\mu^2)^2\leq4(1-\mu^2)$ it follows that $S(L_{\tau_o,\omega})$ is given by (4.52). Similarly, when ω lies in (ii) we find that $S(L_{\tau_o,\omega})$ is given by (4.53).

Summarising our results so far we can construct Table 4.4 where the expressions of $S(L_{\tau_o,\omega})$ and the corresponding ranges of ω are presented.

Case	ω-Domain	Condition	$S(L_{\tau_o,\omega})$
1	$-\infty<\omega\leq\omega'(\bar{\mu})$	-	$\bar{\Delta}^{\frac{1}{2}}/(2-\omega\bar{\mu}^2)$
2	$\omega'(\bar{\mu})\leq\omega<2/(2-\underline{\mu}^2)$	*	$\{-\underline{\Delta}/[4(1-\underline{\mu}^2)]\}^{\frac{1}{2}}$
3	$2/(2-\bar{\mu}^2)\leq\omega<2/\bar{\mu}^2$	-	$\{-\bar{\Delta}/[4(1-\bar{\mu}^2)]\}^{\frac{1}{2}}$
4	$2/\underline{\mu}^2<\omega\leq\omega''(\bar{\mu})$	*	$\{-\underline{\Delta}/[4(1-\underline{\mu}^2)]\}^{\frac{1}{2}}$
5	$\omega''(\bar{\mu})\leq\omega<+\infty$	*	$\bar{\Delta}^{\frac{1}{2}}/(\omega\bar{\mu}^2-2)$

* implies that condition (4.49) holds.

TABLE 4.4

By studying the behaviour of $S(L_{\tau_0,\omega})$ as a function of ω, for all the cases of Table 4.4 we can easily find that it is minimised either at $\omega_0=\omega'(\bar{\mu})$ or at $\omega_0=\omega''(\bar{\mu})$. A simple comparison of the two candidate minimum values of $S(L_{\tau_0,\omega})$ reveals that the one for $\omega_0=\omega'(\bar{\mu})$ is the smallest. □

Corollary 4.15. Under the hypothesis of Theorem 4.14 and if $\underline{\mu}=0$, then $S(L_{\tau,\omega})$ is minimised at $\tau_0=\omega_0=\omega'(\bar{\mu})$ and its corresponding value is

$$S(L_{\omega_0,\omega_0}) = \omega_0-1 = [1-(1-\bar{\mu}^2)^{\frac{1}{2}}]/[1+(1-\bar{\mu}^2)^{\frac{1}{2}}] . \tag{4.54}$$

Proof. Since we always have $\mathrm{Re}\lambda_{\pm}(\mu)>0$ and $\tau>0$ it follows (see (4.41)) that $\mathrm{sign}(\partial g/\partial\mu^2)=\mathrm{sign}(\omega-\tau)$ and $g(\tau,\omega,\mu^2)$ is an increasing function of ω (see (4.42)). In the sequel we distinguish two cases.

Case I: Suppose that all $\lambda_{\pm}(\mu)$ are complex, then ω lies in the range $\omega'(\bar{\mu})\leq\omega\leq 2/\bar{\mu}^2$ and

$$\max_{0\leq\mu^2\leq\bar{\mu}^2} |\xi|^2 = \begin{cases} g(\tau,\omega,0) = (\tau-1)^2 & , \text{ if } \omega\leq\tau \\ g(\tau,\omega,\bar{\mu}^2) & , \text{ if } \omega\geq\tau . \end{cases} \tag{4.55}$$

Case II: In this case we assume that some of $\lambda_{\pm}(\mu)$ are complex and the others are real. Then, $S(L_{\tau_0,\omega})$ is given by (4.52) (see Case III of Theorem 4.14) and ω lies in the range $-\infty<\omega\leq\omega'(\bar{\mu})$. The results of Cases I and II are summarised in Table 4.5 (note that $\min_{\tau\leq\omega}(\tau-1)=\omega-1$).

Case	ω-Domain	$S(L_{\tau_0,\omega})$
1	$-\infty<\omega\leq\omega'(\bar{\mu})$	$\bar{\Delta}^{\frac{1}{2}}/(2-\omega\bar{\mu}^2)$
2	$\omega'(\bar{\mu})\leq\omega\leq\tau$	$\omega-1$
3	$\omega'(\bar{\mu})\leq\tau\leq\omega<2/\bar{\mu}^2$	$\{-\bar{\Delta}/[4(1-\bar{\mu}^2)]\}^{\frac{1}{2}}$

TABLE 4.5

Studying the behaviour of $S(L_{\tau_0,\omega})$ as a function of ω we easily arrive at the conclusion that its minimum is attained at $\tau_0=\omega_0=\omega'(\bar{\mu})$ (Case 2). □

Corollary 4.16. Under the hypothesis of Theorem 4.14 and if $\underline{\mu}=0$ $S(L_{\omega,\omega})$ is minimised at $\omega_0=\omega'(\bar{\mu})$ and its corresponding value is given by (4.54).

Proof. It follows immediately from Corollary (4.15). □

Corollary 4.17. Under the hypothesis of Theorem 4.14 and if

$$0<\underline{\mu} = \bar{\mu} = \mu<1 , \tag{4.56}$$

then for either $\omega_0=\omega'(\mu)$ and $\tau_0=1/(1-\mu^2)^{\frac{1}{2}}$ or $\omega_0=\omega''(\mu)$ and $\tau_0=-1/(1-\mu^2)$, we have

$$S(L_{\tau_0,\omega_0}) = 0 . \tag{4.57}$$

Proof. Following a similar approach we can construct Table 4.6. By studying the behaviour of $S(L_{\tau_o,\omega})$ as a function of ω, we easily conclude that its minimum is attained for $\omega_o=\omega'(\mu)$ and $\omega_o=\omega''(\mu)$. Since, these values of ω_o are the roots of $\Delta(\mu)$, (4.57) is obtained. Finally, the optimum values of τ are obtained by using (4.39). □

Theorem 4.18. Let A be a consistently ordered matrix with non-vanishing diagonal elements such that the matrix B has real eigenvalues $\{\mu_i\}_{i=1}^N$, with $\underline{\mu}=\min_i|\mu_i|\neq 0$ and $\bar{\mu}=\max_i|\mu_i|<1$. Then, $S(L_{1,\omega})$ is minimised for

$$\omega_o = \omega'(\bar{\mu}) = 2/[1+(1-\bar{\mu}^2)^{\frac{1}{2}}] \tag{4.58}$$

and its corresponding value is given by the expression

$$S(L_{1,\omega}) = \bar{\mu}(\omega_o-1)^{\frac{1}{2}} = \bar{\mu}^2[1-(1-\bar{\mu}^2)^{\frac{1}{2}}]/[1+(1-\bar{\mu}^2)^{\frac{1}{2}}] . \tag{4.59}$$

Proof. By developing an analogous treatment to Theorem 4.14 we construct Table 4.6.

Case	ω-Domain	$S(L_{1,\omega})$
1	$-\infty<\omega\leq\omega'(\underline{\mu})$	$(\omega\bar{\mu}^2+\bar{\Delta}^{\frac{1}{2}})/2$
2	$\omega'(\underline{\mu})\leq\omega\leq\omega'(\bar{\mu})$	$\max\{\mu(1-\omega)^{\frac{1}{2}},(\omega\mu^2+\Delta^{\frac{1}{2}})/2\}$
3	$\omega'(\bar{\mu})\leq\omega<+\infty$	$\bar{\mu}(\omega-1)^{\frac{1}{2}}$

TABLE 4.6

However, since $\text{sign}[(\omega\mu^2+\Delta^{\frac{1}{2}})^2/4-\mu^2(1-\omega)]=+1$ it follows that for any ω in the range $\omega_1<\omega<\omega_2$ (see Theorem 4.13) we have

$$S(L_{1,\omega}) = \begin{cases} (\omega\bar{\mu}^2+\bar{\Delta}^{\frac{1}{2}})/2, & \text{if } \omega_1<\omega\leq\omega'(\bar{\mu}) \\ \bar{\mu}(\omega-1)^{\frac{1}{2}}, & \text{if } \omega'(\bar{\mu})\leq\omega<\omega_2 . \end{cases} \tag{4.60}$$

A simple study of the behaviour of $S(L_{1,\omega})$ as a function of ω reveals that for $\omega_1<\omega\leq\omega'(\bar{\mu})$ decreases for increasing ω, whereas for $\omega'(\bar{\mu})\leq\omega<\omega_2$ it increases for increasing ω. Thus $\omega_o=\omega'(\bar{\mu})$ and from (4.60), (4.59) follows. □

Before we close this section the following comments, concerned with the derived results, can be made. Table 4.1 presents all the necessary and sufficient conditions for ESOR to converge. Although the same problem was also tackled independently in [4] and [10] our approach differs from those in the aforementioned references in that it is more straightforward. As a result we extended the convergence region found in [10] (see Theorem 2A) by adding the first and the last two conditions of Table 4.1. On the other hand, Corollary 4.10 shows that ESOR converges for a wider range of parameters τ and ω than SOR whose convergence conditions are obtained as a by-product of the whole analysis (Corollary 4.11). By modifying slightly the simple approach followed in Theorem 4.9 we were able to determine the optimum values for the parameters τ and ω not only for ESOR but for SESOR also. The basic conclusions from our analysis are: (i) clearly ESOR attains a faster rate of convergence than SOR when A is a consistently ordered matrix and B possesses real eigenvalues such that $\underline{\mu}\neq 0$, $\bar{\mu}<1$ and $1-\underline{\mu}^2<(1-\bar{\mu}^2)^{\frac{1}{2}}$ and (ii) as $\bar{\mu}\to 1^-$ the convergence rate of SESOR is approximately half that of SOR since from (4.59) it follows that $R(L_{1,\omega_0})=R(B)+R(L_{\omega_0,\omega_0})/2$, where L_{ω_0,ω_0} denotes the SOR iteration matrix. It is noted that the optimum value for the spectral radius does not depend upon $\underline{\mu}$, whereas the optimum value of ω_0 is identical for both SOR and SESOR methods. ESOR's superiority is due to the fact that in this method one can fully exploit the spectrum of the eigenvalues of B to achieve the best possible results, while such a possibility in SOR is precluded. This is emphasised in Corollary 4.17, where under the special condition (4.56) an exceptionally fast rate of convergence is obtained by employing the ESOR method. Further, we note that the basic criterion of using ESOR is to check whether the value of $\underline{\mu}$ is away from zero. This requirement is derived from the fact that if $\underline{\mu}\to 0^+$ (Corollary 4.15) we have that $S(L_{\tau_0,\omega_0})\to S(L_{\omega_0,\omega_0})$, whereas if $\underline{\mu}\to\bar{\mu}$, (Corollary 4.17) we have that $S(L_{\tau_0,\omega_0})\to 0^+$. The quantity $\underline{\mu}$ is necessarily zero whenever the matrix A is consistently ordered of odd order, whereas when A is of even order we clearly need a formula for its estimation. Of course, the determination of $\underline{\mu}$ is the additional work required in the ESOR method (as compared with SOR) and may incur some extra computational effort. On the other hand, condition (4.49) yields a lower bound for $\underline{\mu}$ $(1-(1-\bar{\mu}^2)^{\frac{1}{2}}<\underline{\mu}^2)$, where if we let $\bar{\mu}=1-\varepsilon$, $\varepsilon<<1$, then $\underline{\mu}\geq 1-(\varepsilon/2)^{\frac{1}{2}}$. To realise the information that this result offers, let us take $\varepsilon=0.1$, then $\underline{\mu}\geq 0.77639$. This corresponds to the situation where all the eigenvalues of B are scattered inside the intervals (-1, -0.77639) and (0.77639,1). Although, such cases do not often arise in practical problems, this should not obscure the overall performance evaluation of ESOR since the method is a generalised version of SOR and as such it is expected to exhibit its real power in more general problems.

A simple numerical experiment carried out in [10] shows that for a particular class of non-cyclic matrices ESOR converges twice as fast as SOR.

In [9] a similar theory to the present section is developed when B is a skew-symmetric matrix. It is shown that the optimum value of ω for ESOR and SESOR are identical and less than unity, a fact which justifies the replacement of "over-relaxation" with the term "under-relaxation" (ESUR, SESUR and SUR). However, the main result of the above work is that ESUR possesses a faster rate of convergence than SUR under the only condition $\underline{\mu}\neq 0$. In particular, it is shown that $S(L_{\tau,\omega})$ is minimised at

$$\omega_0 = 2/[1+(1+\bar{\mu}^2)^{\frac{1}{2}}] , \quad \tau_0 = (2+\omega_0\underline{\mu}^2)/[2(1+\underline{\mu}^2)]$$

and its corresponding value is given by the expression

$$S(L_{\tau_0,\omega_0}) = [\underline{\mu}(\bar{\mu}^2-\underline{\mu}^2)^{\frac{1}{2}}]/[(1+\underline{\mu}^2)^{\frac{1}{2}}(1+(1+\bar{\mu}^2)^{\frac{1}{2}}] .$$

REFERENCES

[1] EVANS D.J. and MISSIRLIS N.M., "The preconditioned simultaneous displacement method (PSD method)", MACS 22, (1980), 256-263.

[2] EVANS D.J. and MISSIRLIS N.M., "Preconditioned iterative methods for the numerical solution of partial differential equations", pp.115-178 in "Preconditioning Methods, Analysis and Applications", edit. D.J. Evans, Gordon & Breach, Pub. 1983.

[3] EVANS D.J. and MISSIRLIS N.M., "On the preconditioned Jacobi method for solving large linear systems", Computing 29, 167-173, 1982.

[4] HADJIDIMOS A., "Accelerated overrelaxation method", Math.Comp. 32, (1978), 149-157.

[5] ISAACSON E. and KELLER H.B., "Analysis of numerical methods", John Wiley and Sons, New York, 1966.

[6] MISSIRLIS N.M. and EVANS D.J., "The Extrapolated Successive Over-relaxation (ESOR) method for consistently ordered matrices", Intern. J. of Math. and Math. Sciences, (to appear).

[7] MISSIRLIS N.M. and EVANS D.J., "On the convergence of some generalised preconditioned iterative methods", SIAM J.Numer. Anal. 18, (1981), 591-596.

[8] MISSIRLIS N.M., "Analysis of the first order iterative method for linear systems with complex eigenvalues", 7th Congress of Balcan Mathematicians, Athens, December 1983.

[9] MISSIRLIS N.M., "On solving a certain class of non-symmetric linear systems (submitted).

[10] NIETHAMMER, W., "On different splittings and the associated iteration method", SIAM J.Numer.Anal. 16, (1979), 186-200.

[11] SISLER M., "Über ein zweiparametriges Iterations verfahren", Apl.Mat. 18, (1973), 325-332.

[12] VARGA R.S., "Matrix Iterative Analysis", Prentice Hall, Englewood Cliffs, New Jersey, (1962).

[13] YOUNG D.M., "Iterative solution of large linear systems", Academic Press, New York, (1971).

5 ON THE SOLUTION OF SPARSE NON-LINEAR EQUATIONS AND SOME APPLICATIONS

R. P. Tewarson
Department of Applied Mathematics and Statistics
State University of New York at Stony Brook
Long Island, New York 11794

Abstract. Large and sparse systems of non-linear algebraic equations arise in realistic and comprehensive modeling in a variety of application areas. We will discuss some of the mathematical techniques and ideas that have led to efficient algorithms for the numerical solution of such systems. Structural information about the model, which is available in many applications, can be used in designing a solution algorithm to minimize the overall run-time and storage. Many situations arise where subsets of the system are much easier to solve than the rest of the non-linear system. Highly efficient algorithms, exhibiting excellent performance in a variety of applications, have resulted from the developmental work with methods in which the subsystems are solved more frequently than the rest of the system. Also sparse low rank updates have been found to be very useful in practice. We shall describe applications of the above mentioned techniques and concepts in the area of energy-economy modeling, fluid flow networks, and the numerical solution of differential equations.
(This research has been supported in part by the National Institutes of Health Grant No. AM17583)

I INTRODUCTION

Our day to day living invariably involves some degree of interaction with large complex systems; for example, we drive over bridges and highway networks, have to deal with bureaucracies, depend on major corporations, are affected by global weather patterns. In many instances the system may not itself be physically large but its complexity forces a large number of its components to be considered as separate entities.

In order to explain, predict and generally get a better understanding of such large systems, mathematical models that are reasonably comprehensive and realistic are used. Such models usually involve the numerical solution of large systems of non-linear differential and/or integral equations, because analytical solutions are not possible. The systems of non-linear algebraic equations resulting from the use of numerical methods tend to be of very large size, because, in order to incur low truncation errors, it is necessary to divide the region of interest into a large number of small subregions. The determination of the parameters of the models and their validation, often on the basis of inadequate and inaccurate data, requires the repeated solution of the associated algebraic equations which are often underdetermined.

The solution of a large system of non-linear algebraic equations is generally a very difficult problem and some knowledge of efficient algorithms for this purpose is very desirable. Because a lack of familiarity with suitable algorithms, often requires that good models must either be entirely discarded or at least significantly compromised.

We will describe some solution algorithms and the theory on which they are based. These algorithms make use of qualitative and quantitative information about the model under consideration. Remarkable success has been achieved with this approach in many important application areas. It involves selecting a small subset of the variables and equations and expressing these variables as functions of the rest of the variables by using the remaining equations. This reduces the problem to the solution of the small selected system of equations in the few chosen variables. Obviously, this approach is used recursively.

II LARGE SYSTEMS AND THEIR SUBSYSTEMS

Large systems are broken down into a number of small components for conceptual and modeling purposes. In most cases, it is physically impossible for each component to interact with all the others, as the following examples show: in a large company the head office interacts with each branch office but the latter interact with only a few others; in an ecosystem each species interacts with a common environment but only a few other species; in hierarchical systems, each component reports to and is instructed by the next level of command.

In each of the cases mentioned above, it is easy to ascertain the component which basically determines the whole system. The variables and equations associated with these components are called critical, basic, important, sensitive, global or leaders, bosses, trouble makers, etc. The remaining variables and equations are labelled noncritical, nonbasic, unimportant, dependent, local or workers, followers, etc.

In some systems it is not easy to determine the basic components. In such cases, graph theoretic methods that include the model structure or numerical methods using the domain of attraction and convergence rate can be used to determine a basic set of components. However, such methods are not yet fully developed and much work still remains to be done in this direction.

The determination of basic variables and equations is very closely related to the following practical questions. Where to take a finer grid and/or make more accurate measurements when studying the global interaction of air currents, temperature and pressure for weather forecasting? What are the critical finite elements so that disection methods can be used in designing bridges and other structures? How to find the points of attachment for partitioning and tearing electrical and other networks? What are the critical areas in diffusion and transport problems (water purification, renal modeling, nuclear reactor simulation, etc.)

The separation of the variables and equations into the 'basic' and 'nonbasic' categories has been called decomposition, splitting, partitioning, tearing, disection and/or modification. See [3-6, 10-13, 18, 20, 21, 32, 34, 35, 43, 48] for additional details and applications.

III ALGORITHMS FOR SOLVING SPARSE NON-LINEAR EQUATIONS

We will need the so-called 'structure matrix' of the non-linear system of algebraic equations associated with a given model. It is defined as the matrix whose i^{th} row and j^{th} column element is one if and only if the jth variable occurs in the ith equation, otherwise the elements are zeros. The structure matrix associated with a given model can be permuted to one of the desirable forms given in [43] for the Gaussian elimination. For example, by renumbering the variables and equations of the basic components of the model as the last ones, the structure matrix may turn out to be a Doubly Bordered Block Band Form [43]. In case the non-basic components of the model do not interact with each other, the corresponding structure matrix, after suitable permutations of the components, will be of Doubly Bordered Block Diagonal Form [43]. The borders correspond to the variables and equations of the basic components (head office and environment, respectively, see the examples given in Sec. I).

The equations associated with the basic components of a model are nearly always more difficult to solve than those corresponding to the non-basic components. This is due to their high sensitivity to round-off errors and large storage requirements. As a result of the large number of variables in the basic equations, the corresponding rows of the structure matrix are quite dense. On the other hand, the non-basic components involve a much smaller number of variables and, therefore, the corresponding rows of the structure matrix are quite sparse. In view of these facts, we will often use the terms 'easy' and 'difficult' for the 'non-basic' and 'basic' equations and variables, respectively.

Let the equations and variables under consideration be renumbered (permuted) and written in the form

$$f(x,y) = 0 \tag{1a}$$

and

$$g(x,y) = 0, \tag{1b}$$

where the set of equations $f(x,y) = 0$ is much easier to solve for x when y is given, than the set $g(x,y) = 0$ for x or y. Let the dimensions of f and x be n, and of g and y be m.

We assume that all possible information about the model and the partitioning and tearing algorithms have been used to write the given equations in the above form. In case it was possible to permute the original set of equations such that the structure matrix of the resulting equations (1) is Block Diagonal, Block Lower Triangular or a combination of these forms, then the equations (1) can be solved sequentially, starting with the top left hand corner. The storage required is that of the maximum diagonal block. Steward [39] and subsequently Erickson [14] and Sargent [36] have discussed this in detail. On the other hand, if the structure matrix has a Band Lower Triangular or even Band form, then the situation is not so simple. In some cases, the diagonal elements are the only one that can be chosen as pivots due to model considerations; e.g., they are the principal variables in an 'element' or 'compartment'. In these cases, the nonlinear Gauss-Seidel method can be used, and the rate of convergence is reasonable due to the large number of zeros in the upper triangular part of 'structure matrix'. Naturally, the convergence can be accelerated by the use of any one of the well-known techniques.

We are primarily concerned with cases where the 'structure matrix' is of the Bordered Block Lower Triangular, Bordered Band Lower Triangular, Bordered (Singly or Doubly) Block diagonal, Bordered (Singly or Doubly) Block Lower Triangular, or a combination of these forms. Furthermore, the use of the non-linear Gauss-Seidel and related iterative methods with acceleration techniques, still leads to slow convergence and Newton's method, due to its quadratic convergence, is preferred. This is the situation when stiff differential equations arising in comprehensive and realistic models involving the countercurrent multipliers must be solved [38, 44]. In this case the structure matrix of (1) is of Doubly Bordered Band Form. The borders correspond to the central compartment and the band is due to the interaction (flow) of the rest of the compartments with their immediate neighbors.

If (x^*,y^*) is a locally unique solution of (1), then, from the Implicit Function Theorem, under suitable conditions, it follows that corresponding to every y in a neighborhood of y^*, there exists a unique x such that $x = T(y)$ and $f(T(y),y) = 0$. Using these facts in the 'difficult' equation $g(x,y) = 0$, we get, for y in the neighborhood of y^*,

$$g(T(y),y) = 0. \tag{2}$$

Thus g is now a function of only y, and if the dimensions of g and y are much smaller than those of f and x, then the solution of (2) involves a much smaller amount of storage than that of (1). We use Newton's method to solve (2); if y^k is the kth approximation to y^* (how to make a good choice for y^0 is discussed in the last part of this section), then we have

$$y^{k+1} = y^k - z^k, \tag{3}$$

where z^k is the solution of the linear system

$$\left(\frac{dg}{dy}\right)_k z^k = g^k, \tag{4}$$

and $\left(\frac{dg}{dy}\right)_k$ and g^k denote, respectively, $\frac{dg}{dy}$ and g evaluated at $(T(y^k),y^k)$.

To solve (4), we must compute the small Jacobian $\left(\frac{dg}{dy}\right)$. There are three methods by which this can be achieved [22, 27, 41, 42, 44].

Method 1

We consider x as a function of y, since for all y in a neighborhood of y^*, $x = T(y)$. In view of this fact, from (1) we have

$$\frac{df}{dy} = \frac{\partial f}{\partial x}\frac{dx}{dy} + \frac{\partial f}{\partial y} \tag{5a}$$

and

$$\frac{dg}{dy} = \frac{\partial g}{\partial y} + \frac{\partial g}{\partial x}\frac{dx}{dy}\ . \tag{5b}$$

We know that the condition $f(T(y),y) = 0$ is satisfied for y in a neighborhood of y^*, hence $\frac{df}{dy} = 0$; and it follows from (5a) that

$$\frac{dx}{dy} = -\left(\frac{\partial f}{\partial x}\right)^{-1}\frac{\partial f}{\partial y};$$

substituting the above in (5b), we have

$$\frac{dg}{dy} = \frac{\partial g}{\partial y} - \frac{\partial g}{\partial x}\left(\frac{\partial f}{\partial x}\right)^{-1}\frac{\partial f}{\partial y}\ , \tag{6}$$

where all the partial derivatives are evaluated at the point $(T(y^k),y^k)$. In equation (6), it is not necessary to invert $\frac{\partial f}{\partial x}$, because Gaussian elimination can be directly applied to the part of the Jacobian of equations (1), evaluated at $(T(y^k),y^k)$, which corresponds to x:

$$\begin{bmatrix} \frac{\partial f}{\partial x} & \frac{\partial f}{\partial y} \\ \frac{\partial g}{\partial x} & \frac{\partial g}{\partial y} \end{bmatrix} \overset{GE}{\rightarrow} \begin{bmatrix} U & L^{-1}\frac{\partial f}{\partial y} \\ 0 & \frac{\partial g}{\partial y} - \frac{\partial g}{\partial x}\left(\frac{\partial f}{\partial x}\right)^{-1}\frac{\partial f}{\partial y} \end{bmatrix}, \tag{7}$$

where $\frac{\partial f}{\partial x} = LU$, L and U are, respectively, lower and upper triangular matrices.

If $\frac{\partial f}{\partial x}$ is a lower (Block) Triangular matrix, then only $\frac{\partial f}{\partial y}$ and $\frac{\partial g}{\partial y}$ are needed in storage, and the columns of $\frac{\partial f}{\partial x}$ and $\frac{\partial g}{\partial x}$ can be generated and eliminated one (or a few) at a time.

To determine $T(y^k)$ for a given y^k, we must solve $f(x,y^k)$ for x. Newton's method can be used for this purpose. We start with some initial approximation x^{ko} for $T(y^k)$ and then compute a sequence of iterates $x^{ko},x^{k1},x^{k2},\ldots$ by the formula

$$x^{k,p+1} = x^{kp} - z^{kp}, \tag{8}$$

where z^{kp} is the solution of the linear equations

$$(\frac{\partial f}{\partial x})z^{kp} = f(x^{kp},y^k), \qquad (9)$$

and $\frac{\partial f}{\partial x}$ is evaluated at (x^{kp},y^k). It can be shown that, under suitable conditions, the sequence $x^{ko},x^{k1},x^{k2},\ldots$ converges to $T(y^k)$.

The partial derivatives in (7) and (9) are obtained by a consistent difference approximation. For example, $\frac{\partial f}{\partial x}$ at the point (x^{kp},y^k) can be approximated as follows:

$$(\frac{\partial f}{\partial x})e_j \approx \frac{f(x^{kp}+he_j,y^k) - f(x^{kp},y^k)}{h}, \; j = 1,2,\ldots,n, \qquad (10)$$

where e_j is the jth column of the identity matrix of order n.

It is now evident that in Method 1 we use (8), (9) and (10) to compute $T(y^k)$, and then Gaussian elimination, as shown in (7), to compute $\frac{dg}{dy}$ at the point $(T(y^k),y^k)$. The computer storage requirements are those for $\frac{\partial f}{\partial y}$ and $\frac{\partial g}{\partial y}$ if $\frac{\partial f}{\partial x}$ is a Lower (Block) Triangular matrix. In the next two methods $\frac{\partial f}{\partial y}$ is not required and the storage requirements are, therefore, substantially decreased.

Method 2

We compute $\frac{dg}{dy}$ by the approximation [22, 27, 44]

$$(\frac{dg}{dy})e_j \approx \frac{g(T(y^k+he_j),y^k+he_j) - g(T(y^k),y^k)}{h}, \; j=1,2,\ldots,m \qquad (11)$$

where $T(y^k)$ and $T(y^k + he_j)$ are computed as in the previous method. It is evident that this computation requires the solution of $f(x,y) = 0$ $m + 1$ times (m being the dimension of y). Thus the decrease in storage is, to some extent, offset by an increase in the overall computing effort.

Method 3

The equation $g(T(y),y) = 0$ is solved by using a Quasi-Newton method. The matrix $(\frac{\partial g}{\partial y})^{-1}$ is used as an initial approximation for $(\frac{dg}{dy})^{-1}$ and either a rank one or rank two update is used to improve the approximate inverse. Of course, each update involves the evaluation of $T(y)$ for a given y.

Our experience in solving practical problems by each of these three methods has shown that, in the easy equations $f(x,y) = 0$, only one or two Newton iterations of the type (8) were necessary for each value of y^k to get a usable $T(y^k)$.

Having determined $\frac{dg}{dy}$ at the point $(T(y^k),y^k)$, the same value can then be used for the succeeding values of k until the rate of convergence becomes too slow. It can, however, be updated by a quasi-Newton method. We have observed that the re-evaluation of the small Jacobian $\frac{dg}{dy}$ is rarely needed and, therefore, in most cases the computation of $\frac{dg}{dy}$ at only the point $(T(y^0),y^0)$ is sufficient. Convergence conditions for the above mentioned methods and results of some numerical computations are given in [44].

It is interesting to note that the methods described in [7, 30, 40] are special cases of the first method described above.

Iterative methods; e.g., Newton's method, require a good initial approximation to the desired root of the given system of equations. We can make use of the continuation method [2, 9, 16, 17, 29, 33, 37, 46] to get this approximation as follows. Instead of (1), let us consider the system

$$f(x,ty) = 0, \tag{12a}$$

$$g(x,y) = 0 \tag{12b}$$

for $0 \leq t \leq 1$.

If $t = 1$, then (12) is the same as the original system (1). The solution of (12a) for $t = 0$ is often trivial to determine, because it is a completely 'local' solution since the 'feedback' has been set to zero. If the structure matrix has one of the bordered forms mentioned earlier, then for $t = 0$ the border vanishes and we have a Lower Triangular Form (Block Diagonal or Band), and naturally the solution is easy to find. In order to get the solution of (12) for $t = 1$, the value of t is increased from zero to one, in small steps, and the converged solution at one step is used as the initial approximation for the next step. Some very promising theoretical and practical results have been obtained by using the above-mentioned continuation methods for counter-current models [17, 25, 26, 28, 35]. In general, the growing literature on practical applications of the continuation method can be used [2, 9, 25, 26, 28, 29, 33, 37].

In principle, it is possible to use sparse quasi-Newton updates [19, 24, 31, 45, 47] on the whole Jacobian but this still requires the stor of the whole Jacobian which is generally impossible for large models.

IV SOME APPLICATIONS

We now briefly describe some applications where the use of sparse non-linear algorithms is especially advantageous.

1 ENERGY-ECONOMY MODEL.

Let us consider the following model [1, 8, 15]:

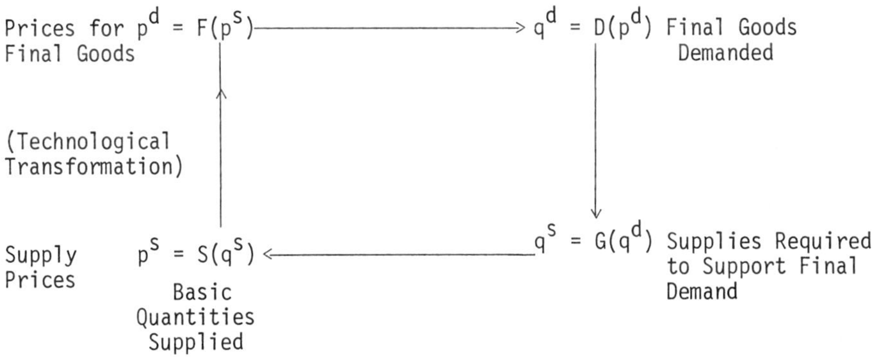

A vector of basic quantities q^s is produced by the supplier. There is a vector of supply prices $p^s = S(q^s)$ which results. These basic supplies are transformed by technological processes "up the newtork" into final goods which are demanded by the consumer. The prices for the final goods are a function $F(p^s)$ of the basic supply prices. Based on a vector of final prices, the consumer buys, or demands D of the final products. Clearly, in the marketplace, quantities are produced, prices are set, and quantities are bought. A descriptive theory of what these quantities and prices are is that they are those for which the network is in equilibrium. Equilibrium is achieved when the consumer will buy the final goods supported by the basic supplies produced at the prices implied by the basic level of production. A final function $G(q^d)$ gives the basic supplies required to support the final demand q^d.

We can write the supply and demand equations at an equilibrium point q^s as

$$S(q^s) - p^s = 0,$$
$$F(p^s) - p^d = 0,$$
$$D(p^d) - q^d = 0,$$

and

$$G(q^d) - q^s = 0,$$

where the superscripts s and d refer to the supply and demand sides of the network. The Jacobian or the structure matrix of the above equations can be exhibited as follows:

Equations	p^s	p^d	q^d	q^s
$S(q^s) - p^s = 0$	$-I_a$	0	0	$\frac{\partial S}{\partial q^s}$
$F(p^s) - p^d = 0$	$\frac{\partial F}{\partial p^s}$	$-I_b$	0	0
$D(p^d) - q^d = 0$	0	$\frac{\partial D}{\partial p^d}$	$-I_c$	0
$G(q^d) - q^s = 0$	0	0	$\frac{\partial G}{\partial q^d}$	$-I_d$

(Column heads under "Variables".)

If we perform block Gaussian elimination on the above matrix then we have

$$\begin{bmatrix} -I_a & 0 & 0 & \frac{\partial S}{\partial q^s} \\ 0 & -I_b & 0 & \frac{\partial F}{\partial p^s}\frac{\partial S}{\partial q^s} \\ 0 & 0 & -I_c & \frac{\partial D}{\partial p^d}\frac{\partial F}{\partial p^s}\frac{\partial S}{\partial q^s} \\ 0 & 0 & 0 & -I_d + \frac{\partial G}{\partial q^d}\frac{\partial D}{\partial p^d}\frac{\partial F}{\partial p^s}\frac{\partial S}{\partial q^s} \end{bmatrix},$$

and clearly the necessary small Jacobian is the bottom right hand side submatrix.

2 FLUID FLOW NETWORKS

Let us consider a simple flow network shown below. Fluid enters under pressure at the tops of tubes 1 and 3 and after flowing in the directions indicated by arrows, exits from the top of tube 2 and the bottom of tube 3. The bath, which is labelled 4, has an overflow so that its volume remains constant. The walls of the tubes are semi-permeable and therefore some part of the solvent and some of the solutes permeate and/or are pumped in or out of the tubes to the common bath.

Let us define the following:

x:normalized distance measured from the top of the bath (x = 0 at the top; x = 1 at the bottom).

For i = 1,2,3,4 and p = 1,2,3,4

$y_{ip}(x)$:axial flow $p \neq 4$,

:pressure if p = 4

$f_{ip}(x)$:transmural flux

$y(x) = (y_{ip}(x))$:vector with $y_{ip}(x)$ as components.

Then the differential equations are

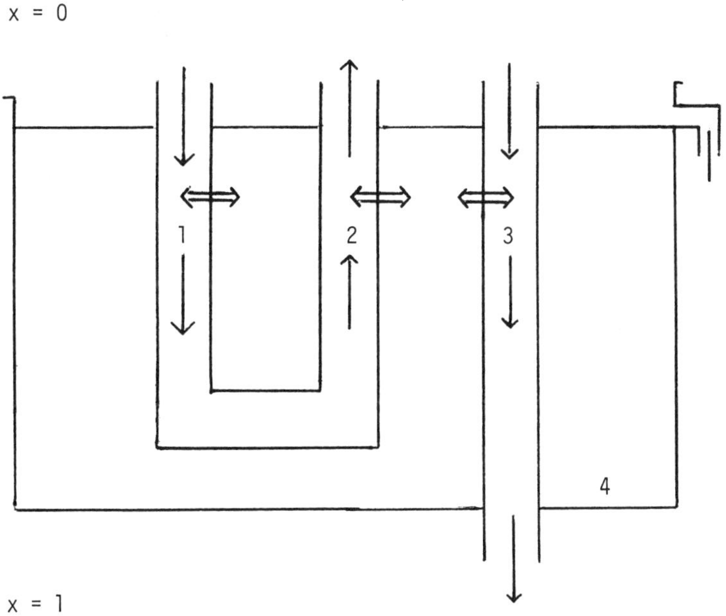

A Simple Flow Network

(→axial flows, ⟺transmural flows)

Variables

Tubes	1	2	3	4
1	X			X
2	X	X		X
3			X	
4	X	X	X	X

The X mark in position (2.1) denotes that tube 1 flows into tube 2, and the first three X marks in row 4 or column 4 denote the transmural interaction of tubes 1, 2, 3 with the bath. It turns out that each of the diagonal blocks corresponding to tubes 1, 2 and 3 are themselves block bidiagonal due to the fact that each ϕ_{ipj} is a function of y_{ipj} and $y_{ip,j-1}$. Thus the tube equations and variables lead to a block bidiagonal matrix, which is easy to solve. Therefore, the whole problem can be solved as a function of equations and variables associated with the bath.

IV GENERAL TWO POINT BOUNDARY VALUE DIFFERENTIAL EQUATIONS

In case of the fluid flow networks just described, due to equation (15), the Jacobian has X marks in all positions in the bottom row and also in the last column; therefore, general boundary value problems, the whole matrix is block bidiagonal, except there is an X in position (4,1) due to the boundary conditions. In other words, the Jacobian is a cyclic matrix and methods of Section II can be readily utilized.

$$\frac{dy_{ip}(x)}{dx} = f_{ip}(x,y(x)) \tag{13}$$

with the two-point boundary conditions

$$L(y(0),y(1)) = 0, \tag{14}$$

where L is usually a linear operator. The transmural fluxes are connected by the equations

$$\sum_{i=1}^{3} f_{ip}(x) + f_{4p}(x) = 0. \tag{15}$$

In order to solve (13) we first let

$$h = 1/n,\ x_j = hj,\ j = 0,1,2,\ldots,n,$$

$$y_{ipj} = y_{ip}(x_j) \text{ and } f_{ipj} = f_{ip}(x_j,\underline{y}(x_j)),$$

then integrate (13) to get

$$y_{ipj} - y_{ip,j-1} - \int_{x_{j-1}}^{x_j} f_{ip}(x,y(x))dx = 0.$$

If the integrand in the above equation is evaluated by the well known Trapezoidal Rule (TR) and we let

$$\phi_{ipj} = y_{ipj} - y_{ip,y-1} - (h/2)(f_{ipj} + f_{ip,j-1}),$$

then in view of the fact that TR has an $O(h^3)$ error, we have

$$\phi_{ipj} + O(h^3) = 0,\ j = 1,2,\ldots,n. \tag{16}$$

Since L is linear, some of the variables y_{ip0} and/or y_{ipn} are either given explicitly or as functions of the other variables. This leads to only as many variables y_{ipj} as there are equations in (16).

If L is non-linear then additional variables and equations are included with the set of bath variables and equations.

The structure of the Jacobian of (16) is given by the following:

1. Adler, J. R., Cazalet, E. G. et al. (1978). The DFI energy-economy modeling system. Decision Focus Report (draft), Decision Focus, Incorporated, Palo Alto, California.
2. Avila, J. H. (1974). The feasibility of continuation methods for non-linear equations, SIAM J. Num. Anal. 11, 102-120.
3. Barker, V. A. (1976). Sparse Matrix Techniques. Springer-Verlag.
4. Björck, A., Plemmons, R. J., Schneider, H. (1981). Large Scale Matrix Problems. North Holland.
5. Brameller, A., Allan, R. N., Harman, Y. M. (1976). Sparsity. Pitman.
6. Bunch, J. R. & Rose, D. J. (1975). Sparse matrix computations, Proc. of Conf. held at Argonne National Laboratory, Academic Press.
7. Buzbee, B. L. (1976). A capacitance matrix technique, in Bunch and Rose, 365-373.
8. Cazalet, Edward G. (1977). Generalized equilibrium modeling: the methodology of the SRI-GULF model. SRI Report, SRI International, Menlo Park, California.
9. Deuflhard, P., Pesch, H. J. & Rentrop, P. A. (1976). Modified continuation method for the numerical solution of non-linear two-point boundary value problems by shooting technqiues, Numer. Math. 26, 327-343.
10. Duff, I. S. (1976). A survey of sparse matrix research, Computer Science and Systems Division, AERE Harwell, England, Report no. CSS28.
11. Duff, I. S. & Stewart, G. W. (1979). Sparse matrix proceedings, SIAM, Philadelphia, PA.
12. Duff, I. S. (1981). Sparse matrices and their uses. Academic Press.
13. Duff, I. S. (1982). Research directions in sparse matrix computations. Harwell Report AERE R10547.
14. Erickson, J. (1976). A note on the decomposition of systems of sparse non-linear equations, BIT, 16, 426-465.
15. Falk, J. E., McCormick, G. P. and Soland, R. M. (1979). Methodology of the leap model, Final Report to Brookhaven National Laboratories. Inst. for Management Sci. & Engr., George Washington University, Washington, D. C. 20052.
16. Farah-Zad, P. (1977). On the numerical solution of differential equations for network of exchanging flow tubes, Ph.D. Thesis, Applied Mathematics and Statistics Dept., State Univ. of New York, Stony Brook, New York.
17. Farah-Zad, P. & Tewarson, R. P. (1979). Numerical continuation method for the solution of a set of parallel flow tubes, Comp. Bio. Med., 9, 21-27.
18. George, A. & Liu, J. W. H. (1981). Computer solution of large sparse positive definite systems. Prentice Hall.
19. Griewank, A. & Toint, P. L. (1982). Local convergence analysis for partitioned quasi-Newton updates, Numer. Math. 39, 429-448.
20. Himmalbeau, D. M. (Ed.) (1973). Decomposition of large-scale problems, American Elsevier, New York.
21. Jennings, A. (1977). Matrix computation for engineers and scientists, Wiley & Sons, New York.
22. Juang, L. L. (1976). Solution of bordered type non-linear equations, Ph.D. Thesis, Applied Mathematics and Statistics Dept., State Univ. of New York, Stony Brook, New York.
23. Kernevez, J. P. & Bunow, B. (1980). Numerical exploration of bifurcating branches of solutions to reaction-diffusion equations describing the kinetics of immobilized enzymes. In Numerical

Methods for Engineering, 2nd International Congress, E. Absi, R. Glowinski, P. Lascaux, & H. Veysseyre, Eds. Dunod, Paris, p. 65.
24. Koshy, M. & Tewarson, R. P. (1983). On the use of restricted pseudo-inverses for the unified derivation of quasi-Newton updates. Submitted for publication.
25. Kubicek, M. (1976). Dependence of solution of non-linear systems on a parameter, Algorithm 502, ACM Transactions on Math. Software 2, p. 98.
26. Lory, P. (1980). Enlarging the domain of convergence for multiple shooting by the homotopy method, Numer. Math. 35, 231-240.
27. Mejia, R., Kellogg, R. B. & Stephenson, J. L. (1977). Comparison of numerical methods for renal network flows, J. Comp. Phys. 23, 53-62.
28. Mejia, R. & Stephenson, J. L. (1980). Multiple solutions of convection diffusion equations describing the renal concentrating mechanism, Num. Method for Engr., GAMNI 2 Second Inf. Cong. DUNOD, pp. 1003-1013.
29. Menzel, R. & Schwetlick, H. (1976). Über einen ordnungsbegriff bei einbettungsalgorithmen zur losüng nightlinearer gleichungen, Computing 16, 187-199.
30. Miklosko, J. (1976). Method of shooting for pentadiagonal system of linear equations, ZAMM 56, 273-276.
31. Murtagh, B. A. and Saunders, W. A. (1982). A projected lagrangian algorithm and its implementation for sparse non-linear constraints, Math. Progra. 16 (MAR):84-117.
32. Reid, J. K. (Ed.) (1970). Large sparse sets of linear equations, Proc. of Conf. at Oxford, Academic Press, London.
33. Rheinboldt, W. C. (1975). An adpative continuation process for solving systems of nonlinear equations, Report No. TR-393, Computer Science Dept., University of Maryland, College Park, MD.
34. Rheinboldt, W. C. (1975). On the solution of large, sparse sets of non-linear equations. In Computational Mechanics (Lecture notes in Mathematics, No. 461), J. T. Oden (Ed.), Springer Verlag, New York, 169-194.
35. Rose, D. J. and Willoughby, R. A. (Ed.) (1971). Sparse matrices and their applications, Pro. of Conf. at IBM Research Center, NY, September 9-10. Plenum Press, New York.
36. Sargent, R. W. H. (1977). The decomposition of systems of procedures and algebraic equations. In Numerical Analysis, Proc., Biennial Conf. Dundee, Watson, G. A. (Ed.), Springer-Verlag, NY, 158-178.
37. Schmidt, W. F. (1978). Adaptive step size selection for use with the continuation method, Int. J. Num. Meth. Engr. 12, 677-694.
38. Stephenson, J. L., Mejia, R. & Tewarson, R. P. (1976). A model of solute and water movement in the kidney, Proc. Nat. Acad. of Sc., USA, 73, 252-256.
39. Steward, D. V. (1962). On an approach to techniques for the analysis of the structure of large systems of equations, SIAM Rev. 4, 321-342.
40. Tewarson, R. P. (1968). Solution of linear equations with coefficient matrix in band form, Nord. Tids. Inf. Bhld. (BIT) 8, 53-58.
41. Tewarson, R. P. (1976). Sparse matrix methods and mathematical models of the renal concentrating mechanism. Summer Computer Simulation Conference, Washington, D. C., 500-501 (invited paper).

42. Tewarson, R. P., Stephenson, J. L., Kydes, A. & Mejia, R. (1976). Use of sparse matrix techniques in numerical solution of differential equations for renal counterflow systems. Computers and Biomedical Research 9, 507-520.
43. Tewarson, R. P. (1973). Sparse Matrices. Academic Press, New York. (Russian edition, Moscow, 1977.)
44. Tewarson, R. P., Stephenson, J. L. & Juang, L. L. (1978). A note on solution of large sparse systems of non-linear equations, J. Math. Anal. and Appl. 63, 439-445.
45. Tewarson, R. P. (1979). A unified derivation of quasi-Newton Methods for solving non-sparse and sparse non-linear equations. Computing 21, 113-125.
46. Todd, M. J. (1983). Piecewise-linear homotopy algorithms for sparse systems of non-linear equations, SIAM J. Com. 21 (2):204-214.
47. Toint, P. (1981). A sparse quasi-Newton update derived variationally with a non-diagonally weighted Frobenius norm. Math. Comp. 37 425-434.
48. Willoughby, R. A. (Ed.) (1969). Proceedings of the symposium on sparse matrices and their applications, Yorktown Heights, New York, IBM Report RA1 (#11707).

6 SOLUTION OF SPARSE EIGENVALUE PROBLEMS

A. Jennings
Civil Engineering Department, Queens University,
Belfast BT7 1NN, N. Ireland.

Abstract. Transformation, eigenvalue search and vector iterative methods are discussed in the context of their application to sparse eigensolutions. Particular attention is paid to symmetric problems including the symmetric linear generalized eigenvalue problem which occurs in vibration frequency analysis and other applications.

1 INTRODUCTION

The selection of which of the various solution methods is most appropriate to use for a particular sparse eigenvalue problem will depend heavily on the properties of the equations and the nature of the eigensolution required. This discourse is intended to give a broad idea of which methods may be most appropriate for some of the commonest types of sparse eigenvalue problem.

A major source of sparse eigenvalue problems is in the analysis of vibrations and resonance in structural and other types of systems. Included in this category are, for instance, the response of structures to earthquakes, the prevention of flow induced instability in aircraft structures, tall buildings, heat exchangers, marine piles etc., and the analysis of wave induced resonances in harbours. However there are many other areas where sparse eigenvalue problems might arise such as in the analysis of structural buckling, the calculation of molecular orbits and the stochastic analysis of systems such as the job sequencing in time sharing computers.

2 BASIC CONCEPTS

Basic eigenvalue properties

An nxn matrix A generally has n eigenvalues $\lambda_1, \lambda_2, \ldots, \lambda_n$ and associated (right) eigenvectors $q_1, q_2, \ldots, q_n$ such that

$$Ax_i = \lambda_i x_i \tag{2.1}$$

The full set of n equations can therefore be specified as

$$AX = X\Lambda \tag{2.2}$$

where $X = [x_1 x_2 \ldots x_n]$ and $\Lambda = \lceil \lambda_1 \lambda_2 \ldots \lambda_n \rfloor$ is the diagonal matrix of eigenvalues. It can be shown that, if A is symmetric, the eigenvalues must be real and the normalised eigenvalues satisfy the orthogonality condition

$$X^T X = I \tag{2.3}$$

Table 1 shows the properties of the main classes of eigenvalue problems. A Hermitian matrix A = B + iC having an eigenvector $x_j = u_j + iv_j$ satisfies the eigenvalue equation

$$\begin{bmatrix} B & -C \\ C & B \end{bmatrix} \begin{bmatrix} u_j \\ v_j \end{bmatrix} = \lambda \begin{bmatrix} u_j \\ v_j \end{bmatrix} \tag{2.4}$$

Here the left-hand matrix is symmetric (since $B = B^T$ and $C = -C^T$) and therefore a Hermitian matrix eigenvalue problem can be solved with software for real symmetric eigenvalues and purpose built software can be constructed by methods which are extensions of the methods for symmetric matrices.

Three fundamentally different solution methods

It is possible to identify three basically different methods of solving eigenvalue problems, the principles in each being frequently combined to produce the most effective modern techniques.

Table 1 Some classes of eigenvalue problem

Matrix A	Eigenvalues, λ_i	Eigenvector properties
{Real symmetric positive definite}	Real positive	Real orthogonal
Real symmetric	Real	Real orthogonal
Real Hermitian	Real	Complex unitary
Real unsymmetric	{Real with possibility of complex pairs}	{Real or complex pairs}
Complex	Complex	Complex

For small eigenvalue problems of order 2 or 3 it is normal to convert to the determinantal equation

$$|A - \lambda I| = 0 \qquad (2.5)$$

and reduce this to the characteristic equation which may then be solved by well known root finding techniques. The reduction to characteristic equation form is particularly awkward to implement on a computer and the whole solution process is not only numerically cumbersome but may also be highly sensitive to rounding errors. It is a prime example of a method which is not suitable to numerical computation. However a variation of this method is to evaluate $|A - \lambda I|$ for n + 1 values of λ and then obtain the characteristic equation as the n-th order polynomial which fits these points. This method is easier to implement on a computer. Although the computational requirement is of order n^4 for a full matrix, it does yield possible savings when A is sparse and also, when partial eigensolutions are required, eigenvalues may be sufficiently well approximated by means of a lower order polynomial.

The Jacobi transformation method of determining eigenvalues of symmetric matrices dates from the middle of the 19^{C} but was particularly cumbersome to implement by hand because at each transformation the matrix needed to be recopied to have a proper record of the calculations. When digital computers became available in the 1950's improvements of it, in particular Givens' and Householder's methods for symmetric matrices, were heralded as great advances in the art of numerical computation. The LR and QR transformation methods were also developed for unsymmetric eigenvalue problems. These methods are particularly suitable where all the eigenvalues of dense matrices are required. However they certainly do not provide a panacea for the person with sparse eigenvalue problems, since these are generally of large order and requiring only partial eigensolution.

The power method involves repeatedly multiplying a trial vector by the matrix until it converges to the eigenvector corresponding to the eigenvalue of largest modulus. When convergence is achieved the corresponding eigenvalue is known from the magnification of the vector through each iteration. It is particularly simple to implement on a computer in such a way that advantage can be taken of sparsity in the matrix. However, in its basic form, it only yields one eigenvalue and its convergence rate could be disastrously slow.

Generalized eigenvalue problems

Very many eigenvalue problems appear in generalized form rather than standard form. The linear generalized form

$$Ax = \lambda Bx \tag{2.6}$$

is common with the quadratic generalized form

$$\lambda^2 Ax + \lambda Bx + Cx = 0 \tag{2.7}$$

not so common but certainly of significance. If such equations are obtained with dense matrices, particularly ones of small order, it is possible to transform them into standard form. For instance, if the appropriate matrices are non-singular, eq. (2.6) transforms to

$$(B^{-1}A)x = \lambda x \tag{2.8}$$

and eq. (2.7) to

$$\begin{bmatrix} -A^{-1}B & -A^{-1}C \\ I & 0 \end{bmatrix} \begin{bmatrix} y \\ x \end{bmatrix} = \lambda \begin{bmatrix} y \\ x \end{bmatrix} \tag{2.9}$$

However when the matrices are large and sparse, the sparsity is lost in the transformed matrices. Hence sparse solution techniques have had to be developed for generalized as well as standard eigenvalue problems.

3 A PROBLEM GIVING RISE TO SPARSE EIGENVALUE EQUATIONS

Vibration of a plate using finite differences

As an example of a physical problem giving rise to a sparse eigenvalue problem consider the lateral vibration of a uniform simply supported square flat plate of side a. The governing differential equation is

$$\frac{\partial^4 w}{\partial x^4} + 2\frac{\partial^4 w}{\partial x^2 \partial y^2} + \frac{\partial^4 w}{\partial y^2} = \frac{m\omega^2}{D} w \tag{3.1}$$

where m is the mass/unit area, ω the natural frequency, D the modulus of rigidity and w(x,y) the lateral deflection (Timoshenko & Young 1955;

Timoshenko & Woinowsky-Krieger 1959). The standard finite difference approximation on a square mesh of spacing h is shown in Fig. 1 where $\lambda = m\omega^2h^4/D$. Thus using the 5 x 5 mesh shown in Fig. 2 equations of the form of (2.1) can be obtained in which x is the displacement vector $\{w_1 w_2 \ldots w_{16}\}$ and A is the matrix shown in Fig. 3.

If the frequencies are written in the form

$$\omega = \frac{k}{a^2}\sqrt{\frac{D}{m}} \qquad (3.2)$$

FIG. 1 Molecule for plate vibration finite difference formula

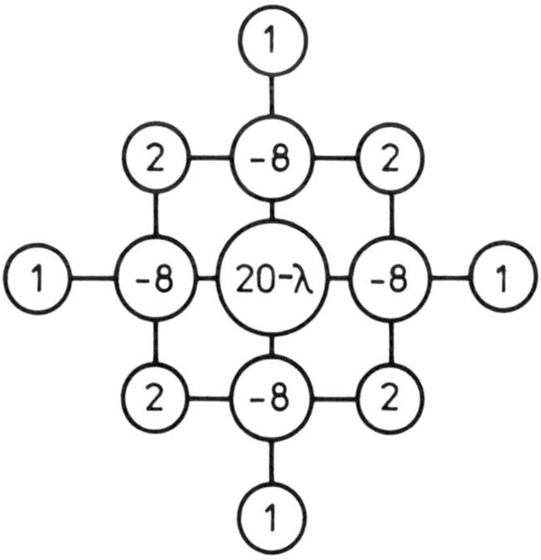

FIG. 2 5 x 5 finite difference mesh

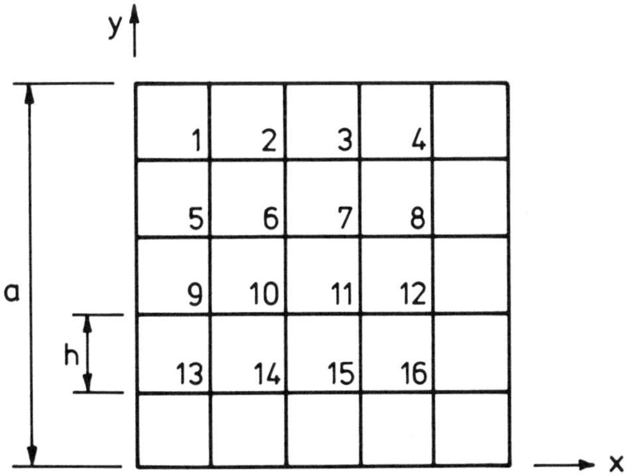

values of the constant k can be obtained from the sixteen eigenvalues of A using the relationship $k = \lambda^{\frac{1}{2}} a^2/h^2$. These are compared with the first twenty-four values obtained from classical theory in Table 2. Only the lowest frequency has been predicted with an accuracy that might be acceptable. One way to improve the accuracy is to use a finer mesh. Solving the same problem with a 21 x 21 mesh involving the eigensolution of a 400 x 400 matrix yields four frequencies with errors of less than 1% (Table 2). Therefore, in common with many other large sparse eigenvalue problems, only a partial eigensolution is required.

FIG. 3 Matrix for 5 x 5 finite difference mesh

$$\left[\begin{array}{cccc|cccc|cccc|cccc}
18 &&&&&&&&&&&&&&& \\
-8 & 19 &&&&&&&&&&&&&& \\
1 & -8 & 19 &&&&&&&&&&&&& \\
 & 1 & -8 & 18 &&&&&&&&&&&& \\
\hline
-8 & 2 & & & 19 & & \text{symmetric} &&&&&&&&& \\
2 & -8 & 2 & & -8 & 20 &&&&&&&&&& \\
 & 2 & -8 & 2 & 1 & -8 & 20 &&&&&&&&& \\
 & & 2 & -8 & & 1 & -8 & 19 &&&&&&&& \\
\hline
1 & & & & -8 & 2 & & & 19 &&&&&&& \\
 & 1 & & & 2 & -8 & 2 & & -8 & 20 &&&&&& \\
 & & 1 & & & 2 & -8 & 2 & 1 & -8 & 20 &&&&& \\
 & & & 1 & & & 2 & -8 & & 1 & -8 & 19 &&&& \\
\hline
 & & & & 1 & & & & -8 & 2 & & & 18 &&& \\
 & & & & & 1 & & & 2 & -8 & 2 & & -8 & 19 && \\
 & & & & & & 1 & & & 2 & -8 & 2 & 1 & -8 & 19 & \\
 & & & & & & & 1 & & & 2 & -8 & & 1 & -8 & 18
\end{array}\right]$$

Table 2 Vibration frequencies for a square flat plate. (i,j) = no. of half wavelengths in x and y directions respectively

Mode. nos.	(i,j)	Theoretical k value	5 x 5 mesh		21 x 21 mesh	
			k value	error	k value	error
1	(1,1)	19.7392	19.10	3.2%	19.702	0.2%
2,3	(1,2)&(2,1)	49.348	44.1	10.7%	49.04	0.6%
4	(2,2)	78.957	69.1	12.5%	78.37	0.7%
5,6	(1,3)&(3,1)	98.696	75.0	24%	97.2	1.5%
7,8	(2,3)&(3,2)	128.305	100.0	22%	126.5	1.4%
9,10	(1,4)&(4,1)	167.783	100.0	40%	163.1	2.8%
11	(3,3)	177.653	130.9	26%	174.7	1.7%
12,13	(2,4)&(4,2)	197.39	125.0	37%	192.4	2.5%
14,15	(3,4)&(4,3)	246.74	155.9	37%	240.6	2.5%
16,17	(1,5)&(5,1)	256.61			244.9	4.7%
18,19	(2,5)&(5,2)	286.22			274.6	4.2%
20	(4,4)	315.83	180.9	43%	306.5	2.9%
21,22	(3,5)&(5,3)	335.57			322.8	3.8%
23,24	(1,6)&(6,1)	365.18			341.9	6.4%
-	-	-			-	-
	(20,20)	7896			3508	55%

Vibration of a plate using finite elements

If a finite element rather than a finite difference idealisation is used for the above plate vibration problem, rotation (and maybe higher order derivatives) as well as displacement variables are introduced into the analysis (Zienkiewicz 1977). Fig. 4 contrasts the 21 x 21 finite difference mesh with two finite element idealisations yielding a fairly similar number of equations. In both of these schemes all the interior nodes are assumed to have three variables each and the boundary nodes one variable each. If equations are derived using the consistent mass formulation they take the form

$$Kx = \omega^2 Mx \qquad (3.3)$$

where K and M are symmetric positive definite stiffness and mass matrices respectively. The corner node finite element formulation gives rise to K and M matrices having the same bandwidth as the finite difference matrix (assuming that the nodes are numbered consecutively by rows) but with a larger density of non-zero elements within the band. The mid-side node formulation gives an even greater density of non-zero elements in the matrix and also a greater bandwidth (see table 3). The type of idealisation of the problem therefore affects, not only the type of equations obtained but also the sparsity patterns of the matrices. It would be expected that the extra computation required to construct and solve the finite element equations will be compensated for by improved accuracy in the solution, but exactly how much extra computation is required in the

FIG. 4 Idealisations for analysis of plate vibration

(a) 21x21 finite difference mesh (400 variables)

(b) 12x12 finite element grid (407 variables)

(c) 7x7 mid-side node finite element grid (412 variables)

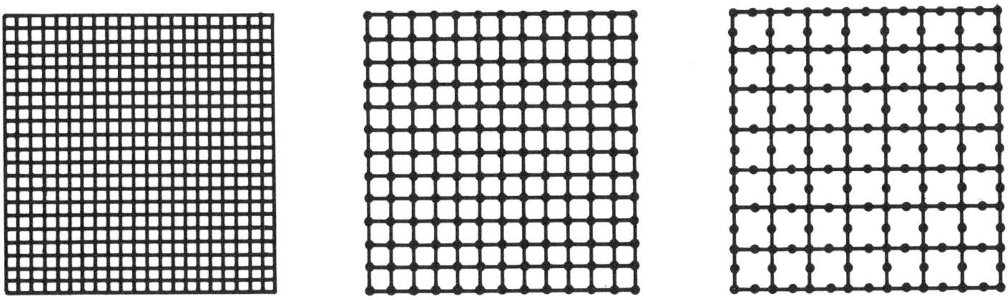

solution process depends very much on the particular eigensolution method adopted.

Other vibration problems can yield variations on the above types of equations. The following are three possible variations:

(a) When the structure or system is free to move bodily the stiffness matrix K will have as many modes of singularity as there are body freedoms (Jennings and Orr 1971).

(b) If some elements have been allocated no mass, the mass matrix will be singular.

(c) If the structure is vibrating in fluid, the inertia of the fluid may be represented by the boundary integral method giving a mass matrix which is more densely populated and has a larger bandwidth than the stiffness matrix (Murray et al 1980).

Reduction of finite element equations to standard form

Although the finite element equations may be transformed to standard form by the method shown in eq. (2.8), it is often more convenient to obtain a matrix for eigensolution which is symmetric. Thus, if $M = GG^T$, eq. (3.3) becomes

$$Kx = \omega^2 GG^T x \tag{3.4}$$

which gives the symmetric eigenvalue problem

$$\left.\begin{aligned} (G^{-1}KG^{-T})z &= \omega^2 z \\ \text{where} \quad z &= G^T x \end{aligned}\right\} \tag{3.5}$$

Table 3 Formulations of the plate bending problem

Property	Finite difference 21 x 21 mesh	Finite element 12 x 12 grid	Finite element 7 x 7 grid
n	400	407	412
No. of non-zero elements/row	12	31	42
Maximum Semi-bandwidth	41	41	68

Alternatively, if $K = LL^T$, it is possible to obtain the equations

$$\left.\begin{aligned}(L^{-1}ML^{-T})y &= \frac{1}{\omega^2}y \\ \text{where}\quad y &= L^T x\end{aligned}\right\} \tag{3.6}$$

In cases where the matrix ordinarily requiring eigensolution is singular it is possible to apply a shift to the original equations. Thus, if body freedoms are present making K singular, the choice of a small positive value for σ in the equations

$$(K + \sigma M)x = (\omega^2 + \sigma)Mx \tag{3.7}$$

permits reduction to the standard symmetric form

$$(\tilde{L}^{-1}M\tilde{L}^{-T})\tilde{y} = \left(\frac{1}{\omega^2 + \sigma}\right)\tilde{y} \tag{3.8}$$

where $\tilde{L}$ is the Choleski factor of the non-singular $(K + \sigma M)$ and $\tilde{y} = \tilde{L}^T x$.

4 TRANSFORMATION METHODS

Rutishauser's band peeling

Whereas Givens' and Householder's methods, applied to real symmetric sparse matrices, tend to produce almost complete fill-in, Rutishauser developed a variation of these methods which reduces a band matrix to tridiagonal form without causing fill-in. Considering a 14 x 14 matrix having a bandwidth of seven the elements (4,1) and (1,4) would be eliminated first by a plane rotation between variables 3 and 4. The transformation matrix N contains the submatrix

$$\begin{bmatrix}\cos\alpha & -\sin\alpha \\ \sin\alpha & \cos\alpha\end{bmatrix}$$

in rows and columns 3 and 4 with other elements corresponding to the unit matrix and the value of α chosen such that $\bar{a}_{41} = \bar{a}_{14} = 0$. Hence the effect of the transformation on the non-zero elements is to modify rows and columns 3 and 4 and to introduce new elements into positions (7,3) and (3,7) (see Fig. 5). These loose elements are then eliminated by a rotation of variables 6 and 7 which in turn introduces loose elements at positions (10,6) and (6,10). Complete chase out of loose elements outside the band involves four rotation operations with most of the non-zero elements in the matrix being modified. At this stage the same process is initiated to eliminate elements (5,2) and (2,5), the band matrix is thus gradually reduced to tridiagonal form by peeling off the outer bands one at a time. When semi-bandwidth b is such that $n \gg b \gg 2$, about $4n^2(b-2)$ flops are required for tridiagonalisation (Jennings 1977a). (A flop is a floating point operation involving an addition or subtraction and a multiplication or division). Since Householder tridiagonalisation involves approximately $2n^3/3$ flops, the band peeling method is better for cases in which $b < n/6$. Advantage can be taken of cases in which the bandwidth increases monotonically with row number but not of more general row by row variations of bandwidth. For the 21 x 21 mesh finite difference matrix the tridiagonalization alone would take about 26 Mflops of computation. The method may be applied to matrices having a sparsity structure arising from a nested dissection ordering scheme.

Crawford's method

Crawford (1973) extended the Rutishauser technique to band symmetric generalised eigenvalue problems, i.e., eq. (2.6) in which B is positive definite. The two stage process consists of transformations to reduce B to unity which will be described below followed by Rutishauser's method applied to A.

FIG. 5 First steps in Rutishauser's band peeling technique

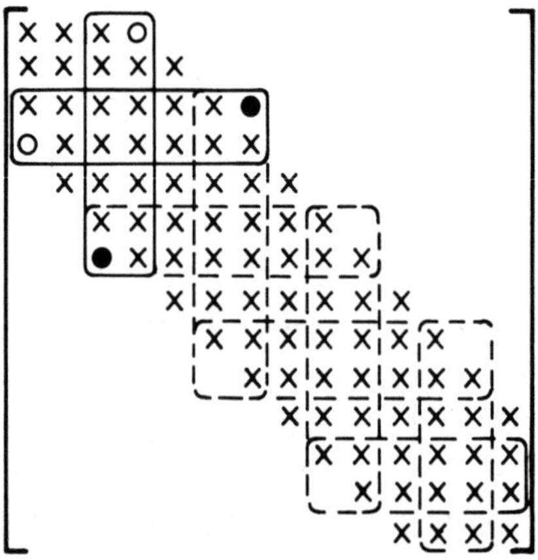

Part way through the first stage eq. (2.6) has been transformed to

$$A_k x_k = \lambda B_k x_k \tag{4.1}$$

where A_k has still its original band form and B_k has elements in the first k-1 rows replaced by the equivalent elements of the unit matrix. The next stage in the B matrix reduction is implemented through the transformation

$$B_{k+1} = \tilde{L}_k B_k \tilde{L}_k^T \tag{4.2}$$

in which $\tilde{L}_k$ is a unit matrix except for elements on the k-th column such that

$$\tilde{\ell}_{kk} = 1/b_{kk}^{\frac{1}{2}}$$

and

$$\tilde{\ell}_{ik} = -b_{ik}/b_{kk} \tag{4.3}$$

Computation of $\tilde{A}_k$ such that

$$\tilde{A}_k = \tilde{L}_k A_k \tilde{L}_k^T \tag{4.4}$$

yields transformed equations

$$\tilde{A}_k \tilde{x}_k = \lambda B_{k+1} \tilde{x}_k \tag{4.5}$$

in which $\tilde{A}_k$ and B_{k+1} have the structure shown in Fig. 6. Rotations within the first k-1 rows and columns may then be used to wash out the unwanted elements in $\tilde{A}_k$ without affecting the matrix B_{k+1} by using a backward form of Rutishauser's algorithm. Once this has been done the equations have been transformed to eq. (4.1) but with k advanced by one. When A and B both have a semi-bandwidth b and $1 \ll b \ll n$, approximately $7n^2b^2/6$ flops are required for the computation although some improvements can be made in efficiency by more elaborate sequencing of the operations (Wilkinson, 1977). The amount of computation this formula gives for the finite element idealisations of the plate bending problem are 325 Mflops (12 x 12 grid) and 916 Mflops (7 x 7 grid). These need to be compared with only about 48 Mflops to obtain a tridiagonal matrix by reduction to standard form followed by conventional Householder transformations for either grid.

Crawford's algorithm is most likely to be of use when the bandwidth of the matrices are particularly small and all or most of the eigenvalues are required.

Guyan's reduction

Guyan's reduction, otherwise known as the eigenvalue economiser or mass condensation method (Irons 1963; Guyan 1965) was the first method used to solve large order vibration problems. The displacements of the system are assumed to be a linear combination of m displacement functions represented by the columns of the nxm matrix H where m<n. If z are the variables associated with these functions

$$x = Hz \tag{4.6}$$

and equation (3.3) reduces to

$$H^T KHz = \omega^2 H^T MHz \tag{4.7}$$

Since these reduced equations are fully populated, it is necessary to find a sufficiently small number of functions to make the eigensolution process feasible without losing much accuracy in the reduction. Much of the

FIG. 6 An intermediate situation in Crawford's method. Elements modified by the $\tilde{L}$ transformation lie within the blocks.

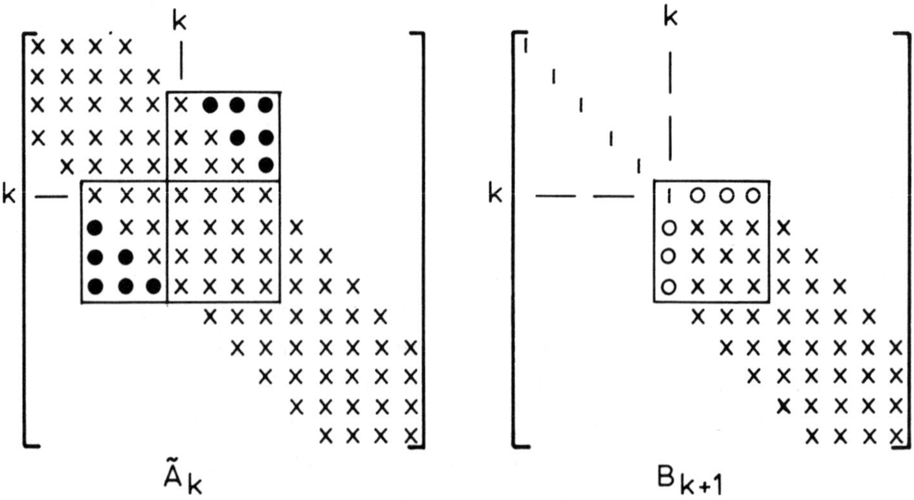

literature is therefore related to the choice of functions and the ensuing accuracy for specific examples (Shah & Raymund 1982; Thomas 1982).

5 POWERFUL VECTOR ITERATIVE METHODS

Simultaneous Iteration

Mainly in the early 1970's simultaneous iteration and subspace iteration methods were developed. They have the feature of Guyan reduction that a lower order eigenvalue problem is created by using a reduced set of functions. However, by using the results from one eigensolution to predict better functions, an iteration sequence is set up. Therefore, in contrast to Guyan reduction, results of any desired accuracy can be obtained with less trial functions and without the need for worry about the choice of trial functions and whether sufficient accuracy will be achieved (Jennings 1973).

When applied to standard unsymmetric eigenvalue problems the lop-sided simultaneous iteration method (Jennings & Stewart, W.J. 1975; Stewart, W.J. & Jennings 1981) proceeds as follows:

(a) Choose a set of m trial vectors $U = [u_1\ u_2\ \dots\ u_m]$ (m greater than the number of dominant eigenvalues required)

(b) Premultiply by the matrix to give

$$V = AU \tag{5.1}$$

(c) Set up and solve linear equations for the mxm interaction matrix H such that

$$U^T UH = U^T V \tag{5.2}$$

(d) Obtain the full eigensolution of the interaction matrix giving estimates for the dominant eigenvalues and $P = [p_1\ p_2\ \dots\ p_m]$ its matrix of right eigenvectors.

(e) Update the trial vectors by multiplying

$$W = VP \tag{5.3}$$

(f) Normalise the trial vectors W to give U*.

(g) If iteration is to continue return to (a) with U: = U*.

This method is an extension of the power method (to which it degenerates when m = 1). The matrix A is only involved in premultiplication operations and hence sparsity is completely preserved. The amount of computation required is approximately $(nmc + 2\frac{1}{2}nm^2 + f(m^3))k$ flops where c is the number of non-zero elements/row in A and k is the number of iterations. It is therefore a powerful method for cases in which both m and k can be kept small.

To understand the convergence characteristics it is necessary to represent the trial vectors as linear combinations of the m dominant right eigenvectors Q_A and the (n-m) subdominant eigenvectors Q_B according to

$$U = Q_A C_A + Q_B C_B \tag{5.4}$$

in which case

$$V = Q_A \Lambda_A C_A + Q_B \Lambda_B C_B \tag{5.5}$$

and, if $\|C_B\| << \|C_A\|$ and C_A and $U^T Q_A C_A$ are non-singular, substitution in equations (5.2) gives

$$BC_A^{-1} \simeq C_A^{-1} \Lambda_A \tag{5.6}$$

The matrix P of right eigenvectors of B is therefore an approximation for C_A^{-1} giving from eqs. (5.3) and (5.6)

$$W = Q_A \Lambda_A + Q_B \Lambda_B C_B C_A^{-1} \tag{5.7}$$

As the iteration converges $C_A \to I$ and the normalization tends to be equivalent to postmultiplication by Λ_A. Hence

$$U^* \to Q_A + Q_B \Lambda_B C_B \Lambda_A^{-1} \tag{5.8}$$

The coefficient c_{ij} of C_B denoting the component of the i-th unwanted eigenvector present in the approximation to the j-th eigenvector is multiplied by approximately λ_i/λ_j at each iteration. Thus if $|\lambda_1| \geq |\lambda_2| \geq \ldots \geq |\lambda_n|$ the error in the j-th eigenvector prediction $(j \leq m)$ reduces by approximately the factor λ_{m+1}/λ_j at each iteration. If there is good separation between the eigenvalues required and λ_{m+1} the convergence rate will be good. It is normally advantageous to carry guard vectors by making m more than the number of eigenvectors required accurately. The increased convergence rate generally offsets the greater computational effort per iteration. In the case where complex eigenvalues may be present it is possible to modify the technique so that the vectors U, V and W remain real throughout the iteration. An alternative method employing an orthogonal set of trial vectors is given by Stewart, G.W. (1976).

On the use of simultaneous iteration

If simultaneous iteration is used to determine the first six frequencies of vibration of the square flat plate, a symmetric version of the algorithm would be the most appropriate (Jennings, 1977a). It may be tempting to shift the matrix for the 21 x 21 finite difference mesh so that the required eigenvalues become the ones of largest modulus. Thus, if A is the finite difference matrix and its largest eigenvalue is computed approximately as μ, the eigenvalues of largest modulus of $A-\mu I$ must correspond to the required eigenvalues. However the shifted eigenvalues will be very close together and convergence will be very poor. For instance, if 10 trial vectors are used, three figure accuracy in all the required eigenvalues should be achieved after about 1960 iterations and requiring about 220 Mflops of computation taking account of sparsity in A. Indeed at that stage since the eigenvectors are only accurate to about 1½ figures it would be tempting to iterate further, but this will be no use, since the error in the iteration process is already much less than the error due to the finite element idealisation.

On the other hand, if simultaneous iteration is applied to the inverse matrix (which may be implemented with A in factored form), the required eigenvalues are well separated and an iteration with 10 vectors should converge to the same accuracy in about 3 iterations. This requires only about 1.2 Mflops despite the fill-in of the band through factorisation of A.

The simultaneous iteration method may be applied to sparse generalised eigenvalue problems by the transformation shown in eq. (2.8) provided that B is non-singular. The only alteration required to the simultaneous iteration procedure is that the premultiplication step, eq. (5.1) needs to be amended to

$$V = B^{-1}AU \tag{5.9}$$

In the case where A and B are sparse it is essential that V is computed as a two stage process involving the multiplication

$$W = AU \tag{5.10}$$

followed by solution of the equations

$$BV = W \tag{5.11}$$

It is possible to implement this if A is stored in sparse form and the triangular factors of B are determined. Some savings may be made if both the matrices A and B or just B are symmetric (Jennings 1977a). Allowing for these savings in the case of the finite element idealisations of the plate bending problem in which A is the consistent mass matrix and B is the stiffness matrix gives 3.3 Mflops and 4.9 Mflops as the amount of computation required for the 12 x 12 and 7 x 7 grid idealisations respectively as compared with 2.4 Mflops for the finite difference idealisation. Most of this increase is due to the higher density and bandwidth of the matrices. The storage requirement is similarly increased. The above figures for amount of computation can be reduced if iteration with individual vectors is curtailed when they reach the required accuracy. Subspace iteration (Dong et al 1972; Bathe & Wilson 1972) is effectively the same as simultaneous iteration with similar storage requirements and numerical efficiencies.

There have been differing opinions on the best way of choosing starting vectors for simultaneous iteration type methods. The author is of the opinion that random vectors should normally be used and that any capability of guessing the approximate shape of the eigenvectors should be used to check the results.

The Lanczos method

Consider the standard symmetric eigenvalue problem eq. (2.1). A sequence of vectors formed by premultiplication by A according to

$$Ap_i = p_{i+1} \tag{5.12}$$

is known as the Krylov sequence. Also from the Caley-Hamilton theorem

$$Ap_n = \Sigma\alpha_i p_i \qquad (5.13)$$

where $-\alpha_i$ are coefficients of the characteristic polynomial. Hence if $P = [p_1 p_2 \cdots p_n]$

$$AP = PH \qquad (5.14)$$

where

$$H = \begin{bmatrix} 0 & & & & \alpha_1 \\ 1 & 0 & & & \alpha_2 \\ & 1 & 0 & & \cdot \\ & & \cdot & \cdot & \cdot \\ & & & 1 & \alpha_n \end{bmatrix} \qquad (5.15)$$

Whereas P is a poor sequence of vectors for numerical computation because the higher vectors are almost linearly dependent, the equivalent mutually orthogonal vectors Y satisfying

$$P = YR \qquad (5.16)$$

where $Y^T Y = I$ and R is an upper triangular matrix, does not exhibit this disadvantage. Eliminating P between equations (5.14) and (5.16) gives

$$AY = YT \qquad (5.17)$$

where

$$T = RHR^{-1} \qquad (5.18)$$

and, hence is upper Hessenberg in form. Premultiplying equation (5.17) by Y^T gives

$$Y^T AY = T \qquad (5.19)$$

Hence T, like A, must be symmetric. Indeed it must be tridiagonal. Since also it is a similarity transformation of A, it must have the same eigenvalues. In the Lanczos method for symmetric matrices a trial vector y_1 is used to generate through (5.17) the vector sequence comprising Y and T. The completion of the analysis is effected through the eigensolution of T.

Expanding equation (5.17) gives

$$\left.\begin{aligned} Ay_1 &= \alpha_1 y_1 + \beta_1 y_2 \\ Ay_2 &= \beta_1 y_1 + \alpha_2 y_2 + \beta_2 y_2 \\ &\text{etc.} \end{aligned}\right\} \qquad (5.20)$$

where

$$T = \begin{bmatrix} \alpha_1 & \beta_1 & & \\ \beta_1 & \alpha_2 & \beta_2 & \\ & \beta_2 & \cdot & \cdot \\ & & \cdot & \cdot \end{bmatrix} \qquad (5.21)$$

If y_1 has been normalised so that $y_1^T y_1 = 1$, premultiplying the first equation by y_1^T and using the orthogonality conditions it is found that

$$\alpha_1 = y_1^T A y_1 \qquad (5.22)$$

A vector w_1 may be computed such that

$$w_1 = Ay_1 - \alpha_1 y_1 \qquad (5.23)$$

which from the first of eq. (5.20) must equal $\beta_1 y_2$. Hence using the normality condition for y,

$$w_1^T w_1 = \beta_1^2 \qquad (5.24)$$

from which β_1 can be computed. At this stage y_2 is available from

$$y_2 = (1/\beta_1) w_1 \qquad (5.25)$$

The second of eq. (5.20) can be manipulated in a similar way to yield α_2, β_2 and y_3 and subsequent equations permit the process to continue in a recursive way, with breakdown only occurring when a null y vector is encountered. The operations can be organised so that only one vector needs to be multiplied by A within each recursive cycle. Variations on this method for symmetric linear generalized eigenvalue problems arising from structural vibration are given by Newman & Pipano (1980) and Jennings (1984).

Convergence characteristics of the Lanczos method

For the solution of large sparse eigenvalue problems the most important feature of the Lanczos method is the way some of the eigenvalues often converge at early stages of the recursive process well before T has reached a size of nxn. If T_m is the mxm tridiagonal matrix obtained after m-1 recursions (m<n), the eigenvalues μ_i of T_m are the ones which minimise the function

$$h = c_1^T\left[\prod_{j=1}^{m}(\Lambda-\mu_j I)^2\right]c_1 \qquad (5.26)$$

(Jennings 1981). Here c_1 is a vector describing the magnitude of the components of the eigenvectors of A present in the first trial vector according to

$$y_1 = Qc_1 \qquad (5.27)$$

If c_{k1} is a coefficient of c_1 corresponding to the eigenvalue λ_k then

$$h = \sum_{k=1}^{n} c_{ki}^2 \prod_{j=1}^{m}(\lambda_k-\mu_j)^2 \qquad (5.28)$$

However $\prod_{j=1}^{m}(\lambda_k-\mu_j)$ is the value of the polynomial

$$f(\mu) = \prod_{j=1}^{m}(\mu-\mu_j) \qquad (5.29)$$

at $\mu=\lambda_k$ and so h is the sum of the weighted squares of the misfit of $f(\mu)$ with the n points $(\lambda_k,0)$, (see Fig. 7). The weighting factors are c_{j1}^2. Since the Lanczos method minimises h, the values μ_j are the ones which give the least squares fit to the equivalent weighted polynomial fitting problem. (Note that f = 0 is not a solution since $f(\mu)$ must have a unit value for the μ^m coefficient).

From this analogy and from Sturm sequence properties the following conclusions may be drawn:

FIG. 7 Polynomial $\prod_{j=1}^{m}(\mu-\mu_j)$ showing misfit at points $\lambda_1,\lambda_2, \ldots\lambda_n$

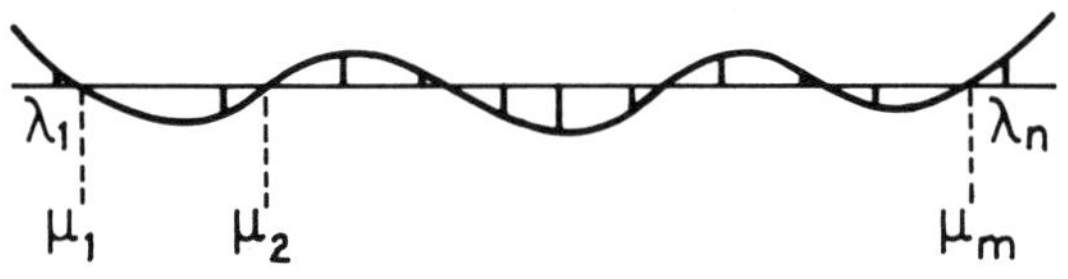

(1) The predictions for the extreme eigenvalues are likely to converge first (this is a well known feature of curve fitting when the points being fitted have roughly equal spacing).

(2) The eigenvalues will converge from inside, i.e., $\mu_1 \geq \lambda_1$, $\mu_2 \geq \lambda_2$, ..., $\mu_{m-1} \leq \mu_{n-1}$, $\mu_m \leq \lambda_n$.

(3) Only one of a set of multiple eigenvalues will be recognised.

(4) All eigenvalues should be exactly determined when m equals the number of distinct eigenvalues, provided that components of all the appropriate eigenvectors are present in y_1.

However the effect of rounding errors on the process is very significant and has led to a considerable amount of investigation. In the first instance it was found that the implicit orthogonality conditions between the Lanczos vectors (e.g. y_4 should be orthogonal to y_1 and y_2) were not being maintained. Correcting this by introducing formal orthogonalisations made the method uncompetitive when used to obtain complete eigensolutions (Wilkinson 1965) and hence, despite Lanczos's recognition of the analogy with polynomial curve fitting, interest in the method wained for some time. However Paige (1972) found that, despite loss of orthogonality in the vectors, the basic algorithm does not give incorrect results. What happens is that after an eigenvalue has been predicted accurately, rounding errors start to create another copy of the corresponding eigenvector (which, in the case of equal eigenvalues, may also contain a component of one of the hitherto undetected eigenvectors). More recently Parlett and Scott (1979) have proposed a selective orthogonalisation method which prevents unwanted multiple copies of eigenvectors arising without incurring the full penalty of performing continual explicit orthogonalisation (see also Parlett (1980) and Scott (1980)).

Furthermore the eigenvalue spectrum may be far from evenly spaced so making convergence rates either very rapid or very slow. To illustrate this the convergence rates for the eigensolution of the 21 x 21 finite difference mesh has been investigated using convergence rate bounds found elsewhere to be reasonably accurate (Jennings 1977b & 1981). The lowest six frequencies give eigenvalues of 0.00200, 0.01236 (double), 0.03158 and 0.04858 (double). Relative to the other eigenvalues which range from 0.082 to 63.28 these first six are very closely spaced. On account of this it would take about 160 Lanczos steps to obtain even the first eigenvector to about 1½ figure accuracy, the other required eigenvalues only being obtained after 400 iterations if full measures are taken to ensure adequate orthogonality between the Lanczos vectors. On the other hand, if the matrix is factorized and the Lanczos method applied implicitly to the inverse matrix, the required eigenvalues are 501.1, 80.9 (double), 31.7 and 20.6 (double) with the remaining 394 eigenvalues lying between 12.1 and 0. Since the required eigenvalues are well separated from the rest, the first four eigenvectors should be obtained to an accuracy of 1½ figures in only about 7 Lanczos steps. In order to obtain eigenvector components associated with the second eigenvalues at 80.9 and 20.6 it would be necessary to continue to 20-30 Lanczos steps depending on the word length of the computer.

As with simultaneous iteration it is much more effective to use the inverse matrix in factored form despite the fill-in incurred through factorization and the eigenvalues (apart from multiple copies) are obtained more efficiently than for simultaneous iteration. However when multiple eigenvalues are present and their presence needs to be detected it is necessary to run the recursive sequence on much longer and this advantage over simultaneous iteration will be partly or entirely lost. Note that if the Lanczos method is applied to the inverse matrix in factored form there are so few Lanczos steps that complete explicit orthogonalisation of the Lanczos vectors is not very expensive.

Some extensions of the Lanczos Method

Whereas the Lanczos method may be applied to symmetric linear generalized eigenvalue problems in which one of the matrices is positive definite, by using one of the transformation of variables shown in §3, it is also possible to proceed without transforming the variables. For instance, considering eq. (2.6) in which B is non-singular, the Lanczos tridiagonal matrix T matrix may be defined according to

$$B^{-1}AY = YT \qquad (5.30)$$

where Y comprises a set of Lanczos vectors whose orthogonality is weighted by means of the A matrix according to

$$Y^TAY = I \qquad (5.31)$$

From equations 5.30 and 5.31

$$Y^T AB^{-1} AY = T \tag{5.32}$$

Hence T is symmetric.

Equation (5.30) expands to

$$\left.\begin{aligned} B^{-1}Ay_1 &= \alpha_1 y_1 + \beta_1 y_2 \\ B^{-1}Ay_2 &= \beta_2 y_1 + \alpha_2 y_2 + \beta_2 y_2 \\ \text{etc.} & \end{aligned}\right\} \tag{5.33}$$

If y_1 is specified as the first trial vector normalised such that $y_1^T Ay_1=1$ premultiplication of the first equation by $y_1^T A$ and using the orthogonality condition gives

$$\alpha_1 = y_1^T AB^{-1} Ay_1 \tag{5.34}$$

The coefficient β_1 and the next vector y_2 may then be derived according to

$$\left.\begin{aligned} w_1 &= B^{-1}Ay_1 - \alpha_1 y_1 \\ \beta_1^2 &= w_1^T Aw_1 \\ y_2 &= (1/\beta_1) w_1 \end{aligned}\right\} \tag{5.35}$$

and the process continued in a recursive way. The operations can be organised in such a way that only one vector needs to be multiplied by A and one by B^{-1} within each recursive cycle, the latter being done by forward and back substitution operations involving the triangular factors of B.

In the block Lanczos method sets of vectors replace each single Lanczos vector of the basic method (Underwood 1975 ; Cullum 1978 ; Ruhe 1979). Consider the standard eigenvalue eq. (2.1) and let Y_1, Y_2, ... of order (nxr) be successive vector sets. The tridiagonal matrix will comprise rxr blocks, thus

$$T = \begin{bmatrix} S_1 & U_1^T & & & \\ U_1 & S_2 & U_2^T & & \\ & U_2 & S_3 & . & . \\ & & . & . & . \end{bmatrix} \tag{5.36}$$

With $Y = [Y_1Y_2...]$ eq. (5.17) gives

$$\left.\begin{aligned} AY_1 &= Y_1S_1 + Y_2U_1^T \\ AY_2 &= Y_1U_1 + Y_2S_2 + Y_3U_2^T \\ &\text{etc.} \end{aligned}\right\} \qquad (5.37)$$

In order to ensure the orthogonality of the vectors it is necessary that Y_1 be orthonormal before commencement of the process. Premultiplying the first equation by Y_1^T and using the orthogonality conditions give

$$S_1 = Y_1^TAY_1 \qquad (5.38)$$

which may be computed and used to form

$$W_1 = AY_1 - Y_1S_1 \qquad (5.39)$$

However, since

$$W_1 = Y_2U_1^T \qquad (5.40)$$

it follows that U_1 may be obtained as an upper triangular matrix by reverse Choleski factorization of $W_1^TW_1$ to satisfy

$$W_1^TW_1 = U_1U_1^T \qquad (5.41)$$

and Y_2 may be obtained by a backsubstitution process using eq. (5.40). The recursion is then continued in a similar manner. On account of the triangular form of the U matrices, T is of band form rather than block tridiagonal form.

The advantages of the block Lanczos method over the standard Lanczos method are that the convergence of multiple eigenvalues is not delayed unless the multiplicity is greater than r, and that the matrix A needs to be accessed less frequently. A disadvantage is that the eigensolution of the T matrix is no longer a relatively trivial operation.

Although the Lanczos method was originally developed for unsymmetric matrices, its application to unsymmetric eigenvalue problems in its original form was not persued because of possible numerical instability. On the other hand Arnoldi's method, which transforms a matrix to upper Hessenberg rather than tridiagonal form does not have these instability problems. Saad (1980 and 1982) and Ruhe (1983) have recently considered improvements of Arnoldi's method.

6 EIGENVALUE SEARCH TECHNIQUES

Bisection

Bisection is a method which has been used extensively for the eigensolution of symmetric tridiagonal matrices deriving from Givens or Householder transformations. The interval containing all the required eigenvalues is successively bisected with the effect that the positions of the various eigenvalues are located with increasing precision. The method uses the Sturm sequence property that, for any trial value μ, the number of eigenvalues λ_i of eq. (2.1) which are less than μ is equal to the number of changes in sign in the sequence of principal minors of $(A-\mu I)$. Peters and Wilkinson (1969) and Gupta (1970) proposed extensions of this method to find the eigenvalues of symmetric band matrices. Their method of determining the Sturm sequence derives from a factorization of $(A-\mu I)$. For instance if

$$A-\mu I = \begin{bmatrix} 3 & 1 & 5 \\ 1 & 3 & 5 \\ 5 & 5 & 1 \end{bmatrix} = \begin{bmatrix} 1 & & \\ 1/3 & 1 & \\ 5/3 & 5/4 & 1 \end{bmatrix} \begin{bmatrix} 3 & & \\ & 8/3 & \\ & & -23/2 \end{bmatrix} \begin{bmatrix} 1 & 1/3 & 5/3 \\ & 1 & 5/4 \\ & & 1 \end{bmatrix} \quad (6.1)$$

the Sturm sequence is the set of principal minors of $A-\mu I$ of increasing order starting with unity. From the factorization this appears as 1, 3, $3 \times 8/3$, $3 \times 8/3 \times (\frac{-23}{2})$ giving one change of sign and hence one eigenvalue less than μ.

One difficulty with this technique is that the matrix being factorized is normally not positive or negative definite and hence factorization without interchange may break down or yield inaccurate answers. A case where the factorization breaks down is when μ is chosen to be 3 less than in the above factorization. In such a case the leading coefficient of $A-\mu I$ is zero. Since the conventional row interchange method for factorization would not allow the principal minors to be computed, a special interchange strategy has been derived. Unfortunately more computation is required than when not interchanging rows because symmetry is destroyed. Also for band matrices, the bandwidth needs to be increased to accommodate an area of possible fill-in above the band. One factorization of the 21 x 21 mesh finite difference matrix of ∮3 will require approximately 0.34 Mflops without interchanges and about 1 Mflop using interchanges. Since it will require approximately 10 bisections to isolate one eigenvalue to 3 figure accuracy, the amount of computation required to isolate the lowest six eigenvalues of this matrix, taking into account that there are two equal pairs amongst these, is approximately 40 or 13 Mflops depending on whether or not interchanges are used. At this stage, although the multiple eigenvalues will be recognised, the corresponding eigenvectors will not have been obtained.

The bisection method may be applied almost as easily to a linear generalized eigenvalue problem as it can to a standard eigenvalue problem provided the sparsity structure of all the matrices is the same. Considering a linear generalized eigenvalue problem in the form of eq. (2.6), it is possible to show that the Sturm sequence property for the matrix $A-\mu B$ indicates the number of eigenvalues λ_i which are less than μ. If A and B have different sparsity structures, the factorization of $A-\mu B$ may be more expensive than the factorization of A or B separately, a factor which needs to be taken into account when comparing with methods which only require the factorization of A or B.

A situation in which bisection or other eigenvalue search techniques may be particularly valuable is when some interior eigenvalues within a small interval are all that is required. The Sturm sequence principle does not apply to unsymmetric matrices and search techniques for unsymmetric matrices based on the characteristic equation may be aggravated by the possible presence of complex eigenvalues.

Economisation of factorizations

The amount of computation involved in each factorization is proportional to the square of the bandwidth of the matrix and hence the above method is not efficient compared with other methods unless the bandwidth is narrow. Gupta (1973) recognised that it was only necessary to carry on the bisection until each interval contained only one eigenvalue. At that stage each eigenvalue and the corresponding eigenvector can be obtained by inverse iteration by using values of μ for inverse iteration set to the centre of each interval containing an eigenvalue. (Inverse iteration is in any case the method adopted after bisection if eigenvectors are required). Thus if μ indicates the centre of an interval containing only one eigenvalue then

$$(A-\mu I)^{-1}x = \theta x \qquad (6.2)$$

where $\theta = 1/(\lambda-\mu)$. The eigenvalue λ_i lying in the interval will correspond to the θ value of largest modulus and so power iteration with the matrix $(A-\mu I)^{-1}$ will converge to θ_i and x_i. Iteration may be carried out with $A-\mu I$ in factored form and so the same factorization technique may be adopted that is used for the bisection part of the algorithm.

Although this technique is more efficient than standard bisection, it still requires several bisections per eigenvalue on average, with some difficulties arising in special cases. For instance, when equal eigenvalues are present the bisection will continue as with the standard bisection method. Furthermore there are some situations in which convergence of the power iteration could be very slow.

An alternative technique for economising on factorizations is to base predictions on the values of the characteristic polynomial obtained from the various factorizations. Such methods have been developed by Bathe and Wilson (1973 & 1976) and Schwarz (1977).

Use of factorizations in vector iterative methods

For symmetric eigenvalue problems it is possible to combine vector iterative methods with eigenvalue search techniques to utilise the particular advantages of each. One such technique is to carry out bisections halting when no more than two or three eigenvalues lie in each interval. Then simultaneous inverse iteration using the central μ value for each interval can be used to find the eigenvalues lying in that interval. Information from the bisections can be used to determine how many simultaneous trial vectors are needed to ensure a good convergence rate for the required eigenvectors (Jennings and Agar 1977).

An alternative method of proceeding when all the required eigenvalues lie at one end of the spectrum is to start a simultaneous iteration process to determine the required eigenvalues and to monitor the convergence rate. If this becomes slow for some of the vectors, then a factorization can be adopted based on a shift value in the region of the required eigenvalues. (Jennings & Agar 1981 ; Bathe & Ramaswamy; 1980). The Lanczos method also may be accelerated by appropriate shifts which may be carried out during implementation when convergence is slow (Ericsson & Ruhe 1980; Ericsson 1983).

Such methods are of particular importance where the bandwidth (or fill-in) of the matrices is large so that most effective use should be made of each factorization, but the number of eigenvectors required is also large or they are required from an internal interval within the spectrum, so rendering vector iterative methods without shifts ineffective. Some points to note are that Sturm sequence results obtained for each factorization can be used to check that eigenvalues have not been overlooked in simultaneous iteration or the Lanczos method. This could be particularly useful to check that no multiple eigenvalues have remained undetected by a Lanczos iteration. Such methods may be applied to linear generalised eigenvalue problems.

7 RELAXATION AND RELATED TECHNIQUES

Coordinate relaxation

All of the above methods when applied to a linear generalised eigenvalue problem require the factorization of one or other matrix. Coordinate relaxation is a method which avoids any factorization (Fadeev & Fadeeva 1963). Consider the vector

$$\bar{x} = c_1 x + c_2 e_i \tag{7.1}$$

as a trial solution to a symmetric generalised eigenvalue problem in the form of eq. (2.6). Here e_i is the i-th column of the unit matrix. The condition that the Rayleigh quotient is either a maximum or a minimum is obtained by making the differentials of $\bar{x}^T(A-\lambda B)\bar{x}$ with respect to c_1 and c_2 equal to zero. This gives

$$\begin{bmatrix} x^T(A-\bar{\lambda}B)x & x^T(A-\bar{\lambda}B)e_i \\ e_i^T(A-\bar{\lambda}B)x & e_i^T(a-\bar{\lambda}B)e_i \end{bmatrix} \begin{bmatrix} c_1 \\ c_2 \end{bmatrix} = \begin{bmatrix} 0 \\ 0 \end{bmatrix} \tag{7.2}$$

If $Ax = u$ and $Bx = v$, this equation becomes

$$\begin{bmatrix} x^Tu-\bar{\lambda}x^Tv & u_i-\bar{\lambda}v_i \\ u_i-\bar{\lambda}v_i & a_{ii}-\bar{\lambda}b_{ii} \end{bmatrix} \begin{bmatrix} c_1 \\ c_2 \end{bmatrix} = \begin{bmatrix} 0 \\ 0 \end{bmatrix} \tag{7.3}$$

The technique involves generating a trial vector x, premultiplying it by A and B to form u and v, and then formulating and solving eq. (7.3) for i=1. Either the smaller or larger of the two values of $\bar{\lambda}$ are chosen depending on whether smaller or larger eigenvalues of the generalised equations are sought. If the vector is chosen such that $c_1=1$, the updated vector $\bar{x}$ and the revised u and v vectors can be easily computed to continue the relaxation with i=2 etc. Once i=n is completed the convergence test is carried out before returning, if necessary, to i=1 to restart the process.

One difficulty with this method is that convergence can be extremely slow - maybe the natural penalty for avoiding any factorizations. However a further difficulty which is more alarming is that the method frequently converges to an eigenvalue which is not the outer one. It is therefore difficult to determine which eigenvalue has been obtained.

Other relaxation type methods

The coordinate relaxation technique can be extended to include variation of more than one coordinate at each step and also to carry more than one trial vector in parallel in such a way that each vector converges onto a different eigenvector of the matrix. Both of these extensions lead to a larger eigenvalue problem to be solved in place of eq. (7.3) at each step in the process.

In parallel with methods for solution of simultaneous equations it is possible to over-relax to a limited extent. This has the effect of accelerating convergence and also makes convergence onto outer eigenvalues more certain. For such methods see Ruhe (1974) and Schwarz (1974a,b; 1977; 1981). Several related methods have been published in physics and chemistry literature (Davidson 1975; Weber et al 1980; Kalamboukis 1980a,b; Nesbet 1981). These type methods are reported to work well for problems involving molecular orbit calculations which are special in the sense that the eigenvector matrix is close to the identity (Parlett 1981). It would therefore be unwise to employ them for other problems unless similar properties are anticipated.

8 CONCLUSION

Transformation techniques can be applied to band matrices but are likely to be relatively expensive for large order sparse eigenvalue problems in which only a few eigenvalues are required. Most powerful methods in such cases are the simultaneous iteration and Lanczos type methods. However for symmetric problems Sturm sequence and related factorization techniques can be a useful alternative or adjunct to these methods. In general for rapid convergence at least one factorization will normally be required. Although methods are available which do not entail factorization they should be used with caution because of possible convergence difficulties.

REFERENCES

Bathe, K.J. & Ramaswamy, S. (1980). An accelerated subspace iteration method. Computer Methods Appl. Mech. Engng., 23, 313-331.

Bathe, K.J. & Wilson, E.L. (1972). Large eigenvalue problems in dynamic analysis. ASCE J. EM Div. 98, 1471-1485.

Bathe, K.J. & Wilson, E.L. (1973). Eigensolution of large structural systems with small bandwidth. ASCE J. EM Div. 99, 467-479.

Bathe, K.J. & Wilson, E.L. (1976). Numerical Methods in Finite Element Analysis. Englewood Cliffs, New Jersey: Prentice-Hall.

Crawford, C.R. (1973). Reduction of a band-symmetric problem. CACM 16, 41-44.

Cullum, Jane (1978). The simultaneous computation of a few algebraically largest and smallest eigenvalues of a large, sparse, symmetric matrix. BIT, 18, 265-275.

Davidson, E.R. (1975). The iterative calculation of a few of the lowest eigenvalues and eigenvectors of large real-symmetric matrices, J. Comput. Phys., 17, 87-94.

Dong, S.B., Wolf, J.A. and Peterson, F.E. (1972). A direct-iterative eigensolution technique. Int. J. Num. Meth. Engng., 4, 155-161.

Ericsson, T. (1983). Implementation and applications of the spectral transformation Lanczos algorithm. In Lecture Notes in Mathematics 973 - Matrix Pencils, ed. B. Kagstrom & A. Ruhe, pp. 177-188. Berlin:Springer-Verlag.

Ericsson, T. & Ruhe, A. (1980). The spectral transformation Lanczos method for the numerical solution of large sparse generalised symmetric eigenvalue problems. Math. Comp., 35, 1251-1268.

Fadeev, D.K. & Fadeeva, V.N. (1963). Computational methods of linear algebra. San Francisco: Freeman.

Gupta, K.K. (1970). Vibration of frames and other structures with banded stiffness matrix. Int. J. Num. Meth. Engng., 2, 221-278.

Gupta, K.K. (1973). Eigenproblem solution by a combined Sturm sequence and inverse iteration technique. Int. J. Num. Meth. Engng., 7, 17-42.

Guyan, R.J. (1965). Reduction of stiffness and mass matrices, AIAA J., 3, 380.

Irons, B.M. (1963). Eigenvalue economizers in vibration problems, J.RAeS., 67, 526-528.

Jennings, A. (1973). Mass condensation and simultaneous iteration for vibration problems. Int. J. Num. Meth. Engng., 6, 543-552.

Jennings, A. (1977a). Matrix Computation for Engineers and Scientists. London: Wiley.

Jennings, A. (1977b). Influence of the eigenvalue spectrum on the convergence rate of the conjugate gradient method. J. Inst. Maths. Applics., 20, 61-72.

Jennings, A. (1981). Eigenvalue methods and analysis of structural vibration. In Sparse Matrices and their Uses, ed. I.S. Duff, pp. 109-138, London: Academic Press.

Jennings, A. (1984). Eigenvalue methods for vibration analysis II. Shock & Vibration Digest, in press.

Jennings, A. & Agar, T.J.A. (1977). Hybrid Sturm sequence and simultaneous iteration methods, Symp. Appl. Comp. Meth. Engng., Univ. S. California, Los Angeles.

Jennings, A. & Agar, T.J.A. (1981). Progressive simultaneous inverse iteration for symmetric eigenvalue problems. Comp. Struct., 14, 51-61.

Jennings, A. & Orr, D.R.L. (1971). Application of the simultaneous iteration method to undamped vibration problems, Int. J. Num. Meth. Engng., 3, 13-24.

Jennings, A. & Stewart, W.J. (1975). Simultaneous iteration for partial eigensolution of real matrices. J. Inst. Maths. Applics., 15, 351-361.

Kalamboukis, T.Z. (1980a). Davidson's algorithm with and without perturbation corrections. J. Phys. A: Math. Gen., 13, 57-62.

Kalamboukis, T.Z. (1980b). A method for calculating the extreme eigensolution of a real symmetric matrix of high order. J. Phys. A: Math. Gen., 13, 2364-2374.

Murray, M.A., Agar, T.J.A. & Jennings, A. (1980). Evaluation of virtual mass matrices for ship hulls using fluid singularities. Int. Conf. Recent Advances in Structural Dynamics, Southampton.

Nesbet, R.K. (1981). Large matrix techniques in quantum chemistry and atomic physics. In Sparse Matrices and their Uses, ed. I.S. Duff, pp. 161-174. London: Academic Press.

Newman, M. & Pipano, A. (1980). Fast modal extraction in NASTRAN via the FEER computer program, NASA TMX-2893, 485-506.

Paige, C.C. (1972). Computational variants of the Lanczos method for the eigenproblem. J. Inst. Maths. Applics., 10, 373-381.

Parlett, B.N. (1980). The symmetric eigenvalue problem. Englewood Cliffs, New Jersey: Prentice-Hall.

Parlett, B.N. (1981). How to solve $(K-\lambda M)z = 0$ for large K and M. In Numerical Methods for Engineering. eds. Absi. et al., Proc. 2nd Int. Congress (GAMNI II). Paris: Dunod.

Parlett, B.N. & Scott, D.S. (1979). The Lanczos algorithm with selective orthogonalization. Math. Comp., 33, 217-238.

Peters, G. & Wilkinson, J.H. (1969). $Ax = \lambda Bx$ and the generalized matrix eigenvalue problem, SIAM J. Numer. Anal., 7, 479-492.

Ruhe, A. (1974). SOR methods for the eigenvalue problem with large sparse matrices. Math. Comp., 28, 695-710.

Ruhe, A. (1979). Implementation aspects of band Lanczos algorithms for computation of eigenvalues of large sparse symmetric matrices. Math. Comp., 33, 680-687.

Ruhe, A. (1983). The two-sided Arnoldi algorithm for nonsymmetric eigenvalue problems. In Lecture Notes in Mathematics 973-Matrix Pencils, ed. B. Kagstrom & A. Ruhe, pp. 104-120. Berlin: Springer-Verlag.

Saad, Y. (1980). Variations on Arnoldi's method for computing eigenelements of large unsymmetric matrices. Linear Algebra Applics., 34, 269-295.

Saad, Y. (1982). Chebyshev acceleration techniques for solving nonsymmetric eigenvalue problems. Yale Univ. Dept. of Comp. Sci. Report.

Schwarz, H.R. (1974a). The eigenvalue problem $(A-\lambda B)x = 0$ for symmetric matrices of high order. Comput. Meth. Appl. Mech. Engng., 3, 11-28.

Schwarz, H.R. (1974b). The method of coordinate over-relaxation for $(A-\lambda B)x = 0$. Numer. Math., 23, 135-151.
Schwarz, H.R. (1977). Two algorithms for treating $Ax = \lambda Bx$. Comput. Meth. Appl. Mech. Engng., 12, 181-199.
Schwarz, H.R. (1981). Simultane Interationsverfahren für Grosse Allgemaine Eigenwert probleme. Ingenieur-Archiv, 50, 329-338.
Scott, D.S. (1980). The Lanczos algorithm. In Sparse Matrices and their Uses, ed. I.S. Duff, pp. 139-159. London: Academic Press.
Shah, V.N. & Raymund, M. (1982). Analytical selection of masters for the reduced eigenvalue problem. Int. J. Num. Meth. Engng., 18, 89-98.
Stewart, G.W. (1976). Simultaneous iteration for computing invariant subspaces of non-Hermitian matrices. Numer. Math., 25, 123-136.
Stewart, W.J. & Jennings, A. (1981). A simultaneous iteration algorithm for real matrices. ACM Trans. Math. Software, 7 (2), 184-198.
Timoshenko, P.S. & Woinowsky-Krieger, S. (1959). Theory of Plates and Shells. New York: McGraw-Hill.
Timoshenko, P.S. & Young, D.H. (1955). Vibration Problems in Engineering. Princeton, New Jersey: Van Nostrand.
Thomas, D.L. (1982). Errors in natural frequency calculations using eigenvalue economization. Int. J. Num. Meth. Engng., 18, 1521-1527.
Underwood, R. (1975). An iterative block Lanczos method for the solution of large sparse symmetric eigenproblems. Report CS-75-496, Stanford Univ.
Weber, J., Lacroix, R. & Wanner, G. (1980). The eigenvalue problem in configuration interaction calculations: a computer program based on a new derivation of the algorithm of Davidson. Comp. Chem., 4, 55-60.
Wilkinson, J.H. (1965). The Algebraic Eigenvalue Problem. Oxford: Clarendon Press.
Wilkinson, J.H. (1977). Some recent advances in numerical linear algebra. In State of the Art in Numerical Analysis, ed. D.A.H. Jacobs, pp 3-83, London: Academic Press.
Zienkiewicz, O.C. (1977). The Finite Element Method. New York: McGraw-Hill.

7 SPARSE MATRIX TECHNIQUES FOR GENERAL MATRICES WITH REAL ELEMENTS: PIVOTAL STRATEGIES, DECOMPOSITIONS AND APPLICATIONS IN ODE SOFTWARE

Zahari Zlatev
Air Pollution Laboratory, Risø National Laboratory
Danish Agency of Environmental Protection
DK - 4000 Roskilde, Denmark

Abstract. If a direct method is used in the solution of linear algebraic systems with sparse matrices, then the pivotal strategy is an important tool in the efforts to preserve the sparsity of the original matrix as well as possible and to keep the round-off errors as small as possible. Normally, these two requirements cannot be satisfied simultaneously. Therefore a compromise is needed in order to obtain a balance between sparsity and accuracy. Some simple examples, where the lack of such a balance leads to catastrophic results, are demonstrated. Several pivotal strategies are described in connection with the well-known Gaussian elimination. Numerical results obtained in runs with different pivotal strategies are discussed.

Many direct methods for solving linear problems can be united in a general scheme. In this way the common properties of the methods can be studied. Moreover, some common principles, by which the efficiency of the computational process for any particular method within the general scheme is usually increased, can be formulated. This approach is described and some illustrations with well-known particular methods are given.

Applications, where sparse matrix techniques are used in the numerical treatment of mathematical models describing phenomena arising in nuclear spectroscopic resonance theory, are discussed in detail. These models contain linear systems of ordinary differential equations (ODE's). After some time discretization process a long sequence of systems of linear algebraic equations is obtained. The coefficient matrices of all systems of linear algebraic equations have many common properties (e.g. the same sparsity pattern). It is explained how this fact can be exploited in order to increase the efficiency of the numerical integration. The explanation is illustrated by many numerical results. Some of the ideas used in the case where the system of ODE's is linear can also be extended for nonlinear systems of ODE's.

1. Pivotal strategies for Gaussian elimination

1.1. Pivoting for dense matrices. Assume that the classical Gaussian elimination is used to represent a general matrix $A \in \mathbb{R}^{n\times n}$ ($n \in \mathbb{N}$) as a product of a lower triangular matrix $L \in \mathbb{R}^{n\times n}$ and an upper triangular matrix $U \in \mathbb{R}^{n\times n}$. Denote by $a_{ij} = a_{ij}^{(1)}$ ($i,j = 1(1)n$) the elements of A. The elimination process, which leads to the decomposition $A = LU$, is carried out by

$$a_{ij}^{(s+1)} = a_{ij}^{(s)} - a_{is}^{(s)} a_{sj}^{(s)} / a_{ss}^{(s)}, \tag{1.1}$$

where $s=1(1)n-1$, $i=s+1(1)n$ and $j=s+1(1)n$.

Pivotal interchanges must be applied both in an attempt to ensure that the process will not break down for regular matrix A and in an attempt to obtain an accurate decomposition LU. Let $s \in \{1,2,\ldots,n-1\}$ and consider the sets:

$$A_s = \{a_{ij}^{(s)} \; / \; i=s(1)n, \; j=s(1)n\}, \tag{1.2}$$

$$R_{is} = \{a_{ij}^{(s)} \; / \; j=s(1)n\}, \qquad i=s(1)n, \tag{1.3}$$

$$C_{js} = \{a_{ij}^{(s)} \; / \; i=s(1)n\}, \qquad j=s(1)n. \tag{1.4}$$

It is said that A_s is *the active part of matrix* A *at stage* s, R_{is} is *the active part of row* i *at stage* s, C_{js} is *the active part of column* j *at stage* s.

Pivotal interchanges at an arbitrary stage s are normally performed by using one of the following two rules.

Rule 1 - Partial pivoting. *Let* $a_{is}^{(s)}$ $(a_{sj}^{(s)})$ *be one of the largest (in absolute value) elements in* R_{is} (C_{js}). *Bring this element in position* (s,s) *by interchanging rows* i *and* s *(columns* j *and* s*).*

Rule 2 - Complete pivoting. *Let* $a_{ij}^{(s)}$ *be one of the largest (in absolute value) elements in* A_s. *Bring this element in position* (s,s) *by interchanging rows* i *and* s *as well as columns* j *and* s.

The interchanges can be described by permutation matrices (say, P and Q). If this is done, then the decomposition process with a complete pivoting can be represented in a matrix form as

$$LU = PAQ, \qquad P \in \mathbb{R}^{nxn}, \qquad Q \in \mathbb{R}^{nxn}. \tag{1.5}$$

Let I be the identity matrix in $\mathbb{R}^{nxn}$. The decomposition process with a partial pivoting may be considered as a special case of (1.5) with $Q=I$ when only row interchanges are used or with $P=I$ when only column interchanges are carried out. The decomposition process without any pivoting (which can successfully be applied for some classes of matrices; Wilkinson 1961) may also be considered as a special case of (1.5) with $P=Q=I$.

It can be proved (see e.g. Stewart 1973) that: (i) if matrix A is regular, (ii) if the computations are performed without round-off errors and <u>(iii)</u> if neither overflow nor underflow occurs, then the elimination process (1.1) will not break down and some regular decomposition LU will be calculated when either of the two rules for pivoting is in use. This means in particular that <u>all</u> pivotal elements (i.e. the elements brought in positions (s,s), s=1(1)n-1) will be <u>non-zero elements</u>. However, if the calculations needed to obtain L and U are carried out on a computer in floating-point arithmetic, then round-off errors are in general unavoidable and

$$LU = PAQ + E \tag{1.6}$$

is satisfied instead of (1.5). $E \in \mathbb{R}^{nxn}$ is a perturbation matrix, whose elements e_{ij} $(i=1(1)n, j=1(1)n)$ satisfy (see, for example, Stewart 1973):

$$|e_{ij}| \leq n\pi b_n \varepsilon \qquad (i=1(1)n, j=1(1)n), \tag{1.7}$$

where ε is the machine accuracy, π is a constant of order unity that is independent of A and (for $k=1(1)n$):

$$b_k = \max_{1\leq s\leq k, s\leq i\leq n, s\leq j\leq n} (|a_{ij}^{(s)}|). \tag{1.8}$$

It is well-known (see Stewart 1973 again) that

$$b_n \leq [n(2^1 3^{1/2} 4^{1/3} \ldots n^{1/(n-1)})]^{1/2} b_1 \tag{1.9}$$

for the complete pivoting and

$$b_n \leq 2^{n-1} b_1 \tag{1.10}$$

for the partial pivoting. Moreover, examples, where the latter bound is achieved, can be constructed (Wilkinson 1961,1963,1965). However, experience shows that this will happen very seldom. Therefore it is commonly accepted that the Gaussian elimination with partial pivoting is a stable algorithm that normally provides sufficiently accurate results. This fact, together with the fact that the complete pivoting is rather expensive, shows why the partial pivoting is used in nearly all subroutines for solving systems of linear algebraic equations with dense matrices. Of course, it is worthwhile checking the growth factor b_k/b_1 $(k=2(1)n)$ and the magnitude of the pivots. Large values of b_k/b_1 as well as small values of $|a_{ss}^{(s)}|$ $(s=1(1)n-1)$ should be considered as a signal for inaccurate results.

In the following paragraphs of this section it will be shown that all pivotal strategies for sparse matrices are generalizations of the pivotal strategies for dense matrices.

1.2. Common principles used in the construction of pivotal strategies for sparse matrices. Assume that matrix A is sparse (i.e. many of its elements are equal to zero). The Gaussian elimination can again be carried out by (1.1). It is clear that if $a_{ij}^{(s)} = 0$ but both $a_{is}^{(s)} \neq 0$ and $a_{sj}^{(s)} \neq 0$, then a new non-zero element (fill-in), $a_{ij}^{(s+1)}$ is created. One should attempt to keep the number of fill-ins small or, in other words, the sparsity of matrix A must be preserved as well as possible during the calculation of L and U. The pivotal strategy is a useful tool in the effort to avoid many fill-ins. This means that not only is the pivotal strategy for sparse matrices a device by which an attempt to calculate an accurate LU-decomposition is made, but also an attempt to calculate an LU-decomposition, which is as sparse as possible, is carried out using the pivotal interchanges. These two requirements work in opposite directions and, therefore, it is not a surprise that any pivotal strategy for sparse matrices is obtained as a compromise between sparsity and accuracy.

The pivotal strategies for sparse matrices can, roughly speaking, be divided into three groups.

Group 1 - *pivotal strategies with a priori interchanges.*

Group 2 - *pivotal strategies using the Markowitz(1957) cost-function.*

Group 3 - *pivotal strategies based on a local minimization of the the number of fill-ins.*

An arbitrary pivotal strategy, belonging to any of these three groups can formally be obtained from some pivotal strategy for dense matrices by the use of two main principles: (i) relax the accuracy requirements and (ii) introduce some additional criteria by which an attempt to preserve the sparsity better is carried out. It should be emphasized that it is necessary to keep carefully the balance between (i) and (ii).

Denote by $r(i,s)$ and $c(j,s)$ the numbers of non-zero elements in R_{is} and C_{js} respectively. Let

$$\alpha_{is} = max(|a_{ij}^{(s)}|) \quad \text{(the maximum taken for } a_{ij}^{(s)} \in R_{is}\text{)}, \qquad (1.11)$$

$$\beta_{js} = max(|a_{ij}^{(s)}|) \quad \text{(the maximum taken for } a_{ij}^{(s)} \in C_{js}\text{)}. \qquad (1.12)$$

The accuracy requirements are usually relaxed by introducing a *stability parameter* $u \geq 1$ and by choosing the pivotal element at any stage s $(s=1(1)n-1)$ among the non-zero elements of a certain *stability set* $B_s \subseteq A_s$. Set B_s depends on the pivotal strategy specified. The elements of this set satisfy at least one of the following two relations (the choice being dependent on the pivotal strategy):

$$u|a_{ij}^{(s)}| \geq \alpha_{is}, \qquad u|a_{ij}^{(s)}| \geq \beta_{js}. \tag{1.13}$$

The sparsity is preserved in a different way in the three groups of pivotal strategies. However, the following common principle can be formulated. For each pivotal strategy subsets $C_s^* \subseteq B_s$ $(s=1(1)n-1)$:

$$C_s^* = \{a_{ij}^{(s)} \,/\, a_{ij}^{(s)} \in B_s \wedge a_{ij}^{(s)} \text{ satisfies some sparsity criterion}\} \tag{1.14}$$

are determined and one of the elements of set C_s^*, *the set of candidates*, is selected as a pivot at stage s of the elimination process.

The general description of the sparse pivotal strategies given above shows why the accuracy requirements are normally relaxed; in this way the number of elements in the stability set is in general increased and, thus, the sparsity criterion is applied on a larger set.

1.3. Pivotal strategies with a priori interchanges. A typical pivotal strategy from this group can be introduced by the following definition:

Definition 1.1 Carry out the calculations in two parts as follows. *Part 1* - Order the rows (columns) of matrix A in an increasing number of non-zero elements. *Part 2* - At stage s $(s=1(1)n-1)$ of the Gaussian elimination choose as pivotal an element $a_{sj}^{(s)}$ $(a_{is}^{(s)})$ in row (column) s which has a minimal number of non-zero elements in C_{js} (R_{is}) and which satisfies the first (second) condition (1.13). Bring this element in position (s,s) by column (row) interchanges. □

If row interchanges are carried out in Part 1, then

$$B_s = \{a_{sj}^{(s)} \,/\, a_{sj}^{(s)} \in R_{ss} \wedge u|a_{sj}^{(s)}| \geq \alpha_{ss}\}, \tag{1.15}$$

$$C_s^* = \{a_{sj}^{(s)} \,/\, a_{sj}^{(s)} \in B_s \wedge (a_{sj}^{(s)} \in C_s^*,\ a_{sm}^{(s)} \notin C_s^*,\ a_{sm}^{(s)} \in B_s) \Rightarrow c(m,s) \geq c(j,m). \tag{1.16}$$

If column interchanges are carried out in Part 1, then

$$B_s = \{a_{is}^{(s)} \;/\; a_{is}^{(s)} \in C_{ss} \wedge u|a_{is}^{(s)}| \geq \beta_{ss}\} , \qquad (1.17)$$

$$C_s^* = \{a_{is}^{(s)} \;/\; a_{is}^{(s)} \in B_s \wedge (a_{is}^{(s)} \in C_s^*, a_{ms}^{(s)} \notin C_s^*, a_{sm}^{(s)} \in B_s) \Rightarrow r(m,s) \geq r(i,s)\}. \qquad (1.18)$$

It is easily seen, from (1.15) - (1.18), that any pivotal strategy among the strategies introduced by Definition 1.1 is a natural generalization of Rule 1 for dense matrices; observe that: (i) if A is dense, then Part 1 of Definition 1.1 is not needed (and one may assume that the identity matrix I is used), (ii) if A is dense, then Rule 1 will be found from Part 2 of Definition 1.1 by setting $u = 1$.

The pivotal strategies from Definition 1.1 have several advantages over the pivotal strategies from the other two groups:

(A) Only column (row) interchanges are carried out during the Gaussian elimination. The a priori interchanges can easily and efficiently be implemented by using standard algorithms; see, for example, Gustavson (1972, 1978).

(B) The development of a code based on a pivotal strategy from Definition 1.1 is more straightforward than the development of a code based on a pivotal strategy of Group 1 or Group 3.

(C) The amount of additional information, which has to be kept and updated during the Gaussian elimination,is smaller than that for the pivotal strategies from the other two groups.

(D) The code will normally be shorter than that for the strategies of the other two groups.

There is only one drawback to these strategies: *the code based on such a strategy may produce many fill-ins in the decomposition of some matrices*. This is unfortunately a very serious drawback (both the computing time and the storage are increased when many fill-ins are created; see, as an illustration only, the results given in Table 1.2). Nevertheless, subroutines based on a strategy from the first group should be included in a package for sparse matrices. The reason for this can be explained as follows. Very often many problems with quite similar matrices have to be solved. If this is so, then it may be worthwhile to test the efficiency of the available pivotal strategies for the class of matrices under consideration. If it happens that a pivotal strategy with a priori interchanges works well, then this strategy may efficiently be used in the whole computational process.

It should be mentioned that pivotal strategies (based on a priori interchanges) different from those introduced by Definition 1.1 can be applied; see Reid (1974) or Duff and Reid (1974). Such pivotal strategies will not necessarily have the advantages (A)-(D) over the pivotal strategies from Group 2 and Group 3. However, more important is the fact that always there exists a danger that many fill-ins will be produced during the decomposition by a strategy from the first group. This is so because the use of any strategy of the first group is based on the expectation that some sparsity criterion, by which the number of fill-ins produced at stage s will in general be small when this criterion is applied just before stage s $(s=1(1)n-1)$, will also be small when the sparsity criterion is applied before the beginning of the Gaussian elimination (i.e. before stage 1, taking into account the original sparsity pattern only). If the class of matrices is such that this expectation holds at each stage of the elimination process, then the strategy under consideration may successfully be used for the matrices of this class. If the class of matrices is such that this expectation does not hold, then it is better to try with some strategy from the other two groups. This shows that pivotal strategies from the first group may work well in some situations, but one should be careful when the decision whether a pivotal strategy from the first group should be used in a long simulation process or not is made.

The conclusions drawn above are valid for the case where all non-zero elements are kept in the whole elimination process. It is also possible to neglect some elements $a_{ij}^{(s+1)}$ when these become small in some sense. This possibility will be studied in Section 2. Now it should only be noted that neglecting "small" elements may change dramatically the computational process and a pivotal strategy which performs rather badly without neglecting "small" elements may sometimes become the best one when neglecting "small" elements is allowed (compare Table 1.2 and Table 2.1).

1.4. Pivotal strategies based on the use of the Markowitz cost-function.

The following definition could be used in order to introduce pivotal strategies of the second group.

<u>Definition 1.2</u> *Let* $a_{ij}^{(s)} \neq 0$ *be an arbitrary non-zero element of set* A_s. *The product*

$$M_{ijs} = [r(i,s)-1][c(j,s)-1] \qquad (1.19)$$

of the other non-zero elements in the active part of row i *at stage* s

and the other non-zero elements in the active part of column j *at stage* s *is called* ***the Markowitz cost*** *of element* $a_{ij}^{(s)}$. *The integer*

$$M_s = \min_{a_{ij}^{(s)} \in B_s} (M_{ijs}), \qquad s \in \{1,2,\ldots,n-1\}, \tag{1.20}$$

is called ***the optimal Markowitz cost at stage*** s. □

Remark 1.1 The optimal Markowitz cost at stage s depends on the stability requirements imposed. Therefere it must be emphasized that the optimal Markowitz cost at stage s may be larger than the ***minimal Markowitz cost at stage*** s (defined as $M_s^* = \min(M_{ijs})$; the minimum being taken over $\forall a_{ij}^{(s)} \in A_s \wedge \forall a_{ij}^{(s)} \neq 0$). □

Definition 1.2 Consider the set of row numbers (at any stage $s \in \{1,2,\ldots,n-1\}$*):*

$$I_s = \{i_1, i_2, \ldots, i_{p(s)}\} \quad (i_m \in \{s, s+1, \ldots, n\}, m=1(1)p(s)), \tag{1.21}$$

where

$$r(i_1,s) \leq r(i_2,s) \leq \ldots \leq r(i_{p(s)},s), \tag{1.22}$$

$$(i \notin I_s \wedge s \leq i \leq n) \Rightarrow r(i_{p(s)},s) \leq r(i,s). \tag{1.23}$$

Let the stability set B_s *and the set of candidates* C_s^* *be given by*

$$B_s = \{a_{ij}^{(s)} / a_{ij}^{(s)} \in A_s \wedge u|a_{ij}^{(s)}| \geq \alpha_{is} \wedge i \in I_s\}, \tag{1.24}$$

$$C_s^* = \{a_{ij}^{(s)} / a_{ij}^{(s)} \in B_s \wedge M_{ijs} = M_s\}. \tag{1.25}$$

The class of generalized Markowitz strategies (GMS's) contains all pivotal strategies in which an arbitrary element of set C_s^* *is chosen as pivotal at stage* s $(s=1(1)n-1)$. *The class of improved generalized Markowitz strategies (IGMS's) contains all pivotal strategies in which one of the largest (in absolute value) elements of set* C_s^* *is chosen as pivotal at stage* s $(s=1(1)n-1)$. □

Remark 1.2 Column numbers $(j_1, j_2, \ldots, j_{p(s)})$ can be considered in (1.21) instead of row numbers. If this is done, then the relations:

$$c(j_1,s) \leq c(j_2,s) \leq \ldots \leq c(j_{p(s)},s), \tag{1.26}$$

$$(j \in I_s \wedge s \leq j \leq n) \Rightarrow c(j_{p(s)},s) \leq c(j,s), \tag{1.27}$$

$$B_s = \{a_{ij}^{(s)} \,/\, a_{ij}^{(s)} \in A_s \wedge u|a_{ij}^{(s)}| \ge \beta_{js} \wedge j \in I_s\}, \tag{1.28}$$

are to be used instead of (1.22), (1.23) and (1.24) respectively. □

Remark 1.3 The class of IGMS's is a subclass of the class of GMS's. □

Remark 1.4 Consider a particular pivotal strategy P*. The expressions: "P* *is a GMS*" and "P* *is an IGMS*" will be often used as abbreviations of "P* *belongs to the class of GMS's*" and "P* *belongs to the class of IGMS's*" respectively. □

The pivotal strategies introduced by Definition 1.3 and Remark 1.2 depend on two parameters: *the stability parameter* u and *the number* p(s) *of rows (columns) among the elements of which a pivotal element is to be selected.* The original Markowitz strategy (Markowitz 1957) is a GMS with $u=\infty$ (this means that no stability restriction is imposed and, thus, $B_s=A_s$ for $\forall s$) and with $p(s)=n-s+1$ (this means that the pivotal search is carried out on the whole A_s for $\forall s$). The values of u and p(s) used in the pivotal strategies in 3 well-known codes for sparse matrices are given in Table 1.1.

The implementation of a strategy from the second group in a package of sparse matrices is much more complicated than the implementation of a strategy from the first group. The major difficulty arises from the necessity to arrange the rows or the columns (or both the rows and the columns in some implementations) in increasing numbers of non-zero elements in their active parts at the beginning of each stage s (s=1(1)n-1) of the Gaussian elimination; see (1.21) and (1.26). This process requires both extra computing time and extra storage. However, the number of fill-ins is often reduced considerably and, therefore, both the global computing time needed in the solution of $Ax=b$ and the global storage are also reduced in comparison to those for the pivotal strategies of the first group. This

Table 1.1. Basic characteristics of the pivotal strategies in 3 codes. The values of u given in the table are the values recommended in the codes, i.e. the user might specify other values. The values of p(s) in the first two codes are fixed. The value of p recommended in Y12M is 3 (there are 3 different pivotal strategies in Y12M; the pivotal strategy for general sparse matrices is considered here).*

C o d e	*Type of the strategy*	*u*	*p(s)*
MA18 (Curtis & Reid 1971)	*GMS*	$u=4$	$n-s+1$
MA28 (Duff 1977)	*GMS*	$u=10$	$n-s+1$
Y12M (Zlatev et al.1981a)	*IGMS*	$u\in[4,16]$	$\min(n-s+1,p^*)$

is demonstrated numerically in Table 1.2.

The algorithms for ordering the rows and/or the columns in increasing numbers of non-zero elements at stage s $(s=1(1)n-1)$ will not be discussed here (such algorithms are described, for example, in Duff & Reid 1979 and in Østerby & Zlatev 1983). Only the pivotal search (carried out under the assumption that some ordering has already been made in some way) will be studied for two typical cases.

Case A - $p(s)$ is small. The pivotal strategy for general matrices in package Y12M (Zlatev et al. 1981a) is an example for a pivotal strategy with a small $p(s)$. All non-zero elements in the rows (columns) of set I_s are searched. Since $p(s)$ is small and matrix A is sparse, the number of searched elements is normally small. Moreover, a pivotal strategy of the class of IGMS's can easily be specified when $p(s)$ is small. □

Case B - $p(s)$ is large. It is clear that it is not efficient to search all non-zero elements in the rows (columns) of I_s in this situation. For the case $p(s)=n-s+1$ a very elegant device, by which searching all non-zero elements in the rows (columns) of I_s is often avoided, has been developed by Curtis & Reid (1971) and improved by Duff (1977) (see also Duff & Reid 1979). Both columns and rows are searched by this device (in order of increasing numbers of non-zero elements in their active parts). If the are both rows and columns with the same number of non-zeros, then the rows ar searched before the columns. Since the non-zero elements in the code under co sideration (MA28, Duff 1977), are ordered by rows, the treatment of rows is

Table 1.2 Comparison of some characteristics obtained in runs with two versions of package Y12M with 11 matrices of order 300 produced by the matrix-generator MATRF2 (Zlatev et al. 1981b). COUNT is the largest number of non-zeros found in the LU-decomposition. The number of non-zeros in the original matrix is NZ = 300r+110. The CPU time is measured in seconds.

Density parameter	*A priori ordering of the columns*			*Markowitz cost-function*		
	COUNT	*Time*	*Accuracy*	*COUNT*	*Time*	*Accuracy*
r = 2	*1015*	*0.09*	*1.20E-2*	*710*	*0.10*	*1.15E-3*
r = 3	*15793*	*2.30*	*2.33E-3*	*1941*	*0.21*	*2.07E-3*
r = 4	*19075*	*3.28*	*2.56E-4*	*4066*	*0.49*	*2.14E-3*
r = 5	*23271*	*4.94*	*5.72E-4*	*6066*	*0.88*	*4.73E-3*
r = 6	*26838*	*8.11*	*3.32E-3*	*7409*	*1.18*	*3.49E-3*
r = 7	*27942*	*10.42*	*1.88E-3*	*10320*	*2.03*	*3.01E-2*
r = 8	*28969*	*11.41*	*1.98E-3*	*13505*	*3.23*	*3.20E-3*
r = 9	*29479*	*12.37*	*2.69E-3*	*14619*	*3.64*	*2.89E-3*
r = 10	*29823*	*12.43*	*2.40E-4*	*14569*	*3.50*	*2.34E-3*
r = 11	*30275*	*13.60*	*1.01E-4*	*17816*	*5.07*	*5.65E-3*
r = 12	*30730*	*13.29*	*8.08E-5*	*21853*	*7.49*	*3.86E-3*

much easier than that of columns and it is expected that a pivotal element will be determined while rows are searched. If this happens at stage s, then no search of columns is carried out at this stage. If a row (column) with k non-zero elements in its active part is searched at stage s and if the Markowitz cost M_{ijs} of the current element $a_{ij}^{(s)}$ is smaller than $(k-1)^2$ ($k(k-1)$), then the search is terminated. □

The efficiency of the second device (where a large $p(s)$ is used) depends crucially on the mean number of rows (columns) searched during the elimination. If this number is small, then the device performs well for the matrices under consideration. If this number is large, then the device fails to find quickly the pivotal elements and it is better to apply a pivotal strategy with a small $p(s)$. By the following examples it is shown that such classes of matrices can be constructed.

Example 1.1 Let $B \in \mathbb{R}^{nxn}$. Assume that $r(i,1) = c(j,1) = k$ for $\forall i$ and $\forall j$. Let $C \in \mathbb{R}^{nxn}$ have the same sparsity pattern as B. Consider

$$A = \begin{bmatrix} \alpha I & B \\ C & 0 \end{bmatrix}, \quad \alpha \in \mathbb{R}, \quad A \in \mathbb{R}^{2nx2n}, \quad I \in \mathbb{R}^{nxn}, \tag{1.29}$$

where I is the identity matrix. The number of non-zero elements in any of the first n rows of A is $k+1$, while the number of non-zero elements in any of the last n rows of A is k. The number of non-zero elements in any of the first n columns of A is $k+1$, while the number of non-zero elements in any of the last n columns of A is k. Assume that all non-zero elements of A satisfy the stability requirements at stage 1 of the Gaussian elimination (or, in other words, belong to B_1). Set $p(s) = n-s+1$ and use the device described in Case B during the pivotal search at stage 1. The minimal number of non-zero elements in the active parts of a row (column) at stage 1 is k. Moreover, there are n rows with k non-zero elements and n columns with k non-zero elements. The rows will be searched first. For any element $a_{ij}^{(1)} \neq 0$ with $i \in \{n+1, n+2, \ldots, 2n\}$ the Markowitz cost is

$$M_{ij1} = (k-1)k \tag{1.30}$$

and the stopping criterion of the second device is not satisfied because

$$M_{ij1} > (k-1)^2. \tag{1.31}$$

On the other hand, any $a_{ij}^{(1)} \neq 0$ with $i \in \{n+1, n+2, \ldots, 2n\}$ belongs to the set C_1^* of candidates for pivotal elements at stage 1. *This* *shows that the pivotal strategy based on the second searching device fails to find quickly the pivotal element at stage 1 not because the pivotal element is not among the searched elements but because the pivotal strategy is not able to establish the simple fact that any of the searched elements can be chosen as pivotal.* □

The above example shows that the pivotal strategy with a large $p(s)$ <u>may</u> fail to determine quickly the pivotal element at stage s of the elimination when for many elements $a_{ij}^{(s)} \in C_s^*$ the relation $c(j,s) > r(i,s)$ is satisfied. Such relations occur often when least-squares problems are solved by using augmented matrices. However, this is not the whole story. The stability restrictions may also cause difficulties for the pivotal strategies with a large $p(s)$. In these strategies the quick termination of the pivotal search at any stage s is related to the requirement that the minimal Markowitz cost belongs to the set of candidates C_s^*. If the stability requirement is such that the optimal Markowitz cost M_s is larger than the minimal Markowitz cost, then the pivotal strategy will fail to find quickly the pivotal element at the stage under consideration. A typical situation, where this may occur, is again the solution of linear least squares problems by augmentation (the augmentation is described e.g. in Björck 1967,1968 and recommended for sparse matrices in Duff & Reid 1976). An illustration is given by Example 1.2.

<u>Example 1.2</u> Consider the first stage of the decomposition of matrix A from (1.29). Let now $B \in \mathbb{R}^{m\times n}$ with, say, $m \geq 2n$. Set $C = B^T$. Assume that there are p rows with r non-zero elements among the rows of B, where p is large and r is the smallest number of non-zero elements in a row of B. Assume also that the numbers of non-zero elements in all columns of B, involved in the pivotal search of the p rows with r non-zero elements, are larger than $r+1$. Then the p diagonal elements in αI (corresponding to the p rows with r non-zero elements in B) have a minimal Markowitz cost $M_1^* = r^2$ (take into account the symmetry of A). It is well-known, however, that the choice of diagonal elements of αI as pivotal should be avoided during the first m stages of the decomposition of A (because this leads effectively to forming the normal matrix B^TB in the right-hand lower corner of A when $\alpha = 1$). Therefore it is natural to exclude the diagonal elements of αI from the stability set B_1 (and also for the next stages with $s > 1$). This can be achieved either by an appropriate choice of α and u or by incorporating a direct device for rejecting the diagonal elements of αI as pivotal during the decomposition by a

code especially designed for augmentation matrices. If this is done and if $p(s)=n-s+1$, then the code will search at least p rows and p columns (note again that A is symmetric) before the determination of the first pivot. The situation will be similar at many of the next stages. A numerical demonstration of this situation is presented in Table 1.3. It is seen from Table 1.3 that the fact that many rows and columns are searched may lead to a great increase of the computing time when the system $Ax=b$ is large.

Example 1.2 and the results presented in Table 1.3 show that: *the pivotal strategy based on the use of a large* $p(s)$ *may fail to find quickly the pivotal elements not only when* $c(j,s)>r(i,s)$ *for many non-zero elements* $a_{ij}^{(s)}$ *in the rows searched (as in Example 1.1), but also when at many stages the minimal Markowitz cost is not optimal (i.e. the elements* $a_{ij}^{(s)}$, *for which the minimal Markowitz cost is achieved, do not belong to the stability sets* B_s*).* □

Consider now the choice between a GMS and an IGMS. Since the elements of the sets of candidates for pivots, C_s^*, are also elements of the stability sets B_s, the change made to obtain an IGMS (from a GMS) does not seem to be very important for the accuracy of the results. However, it can be verified both theoretically and experimentally that there exist classes of matrices for which the use of a pivotal strategy from the class of IGMS's may be clearly superior over the use of a pivotal strategy of the class of GMS's

By Theorem 1.1 it is shown that there exist classes of matrices for which the use of any IGMS ensures stable results, while there is no guarantee that this will happen when any GMS is applied (Zlatev 1980a,pp.21-22).

Theorem 1.1 (Zlatev 1980a). Let matrix $A\in\mathbb{R}^{nxn}$ *be diagonally dominant (* $\Sigma_{j=1}^{n}|a_{ij}|<2|a_{ii}|$ *for* $\forall i$ *) and symmetric in structure (i.e.*

Table 1.3. Results obtained in the solution of a system $Ax=b$*, where A is an augmented matrix of order 2000 and B is a 1500x500 matrix produced by the matrix-generator MATRF2 (Zlatev et al. 1981b). COUNT is the largest number of non-zeros found during the LU-decomposition of A. The number of non-zeros in A is 19610. The computing time is measured in seconds.*

Compared characteristics	*Small p(s)* *Y12M; Zlatev et al. 1981a*	*Large p(s)* *MA28; Duff 1977*
Accuracy	*3.5E-5*	*1.7E-2*
Computing time	*130.15*	*1466.04*
Storage (COUNT)	*255889*	*271356*

$a_{ij} \neq 0 \Rightarrow a_{ji} \neq 0$ *for* $\forall i$ *and* $\forall j$ *). Then the process of the Gaussian elimination is stable when any pivotal strategy of the class of IGMS's is used.* □

Theorem 1.2 indicates that some of the unacceptable pivotal strategies, which may appear with a GMS, are removed when the correspondent IGMS is specified.

Theorem 1.2 (Zlatev 1980a). Assume that: (i) $Ax=b$ *is solved, (ii)* y *is the solution found, (iii)* $\|x-y\| < \delta$ *is required. Let* $G(\delta)$ *be the set of all possible and unacceptable pivotal sequences when a pivotal strategy from the class of GMS's is used (i.e. any pivotal sequence from* $G(\delta)$ *may be chosen in the decomposition of* A *and if this happens, then* $\|x-y\| \geq \delta$ *will be produced). Let* $G^*(\delta)$ *be the corresponding set when the pivotal strategy of the class of IGMS's with the same parameters* u *and* $p(s)$ *is specified. Then* $G^*(\delta) \subseteq G(\delta)$. □

The above theorems and many experiments with different classes of matrices indicate that the use of an IGMS should be preferred for matrices that satisfy the following conditions (even the use of these conditions with $s=1$ is often worthwhile):

(a). $|a^{(s)}_{i,k(i,s)}| = max(|a^{(s)}_{ij}|)$, the maximum being taken for $s \leqq j \leqq n$;

(b). $|a^{(s)}_{i,k(i,s)}| / |a^{(s)}_{ij}| \approx \bar{u} < u$, where $s \leqq i,j \leqq n$, $j \neq k(i,s)$ and only non-zero elements $a^{(s)}_{ij}$ are used;

(c). $a^{(s)}_{i,k(i,s)} \in C^*_s$;

(d). $p_{is} > 1$, where p_{is} is the number of non-zero elements in row i belonging to C^*_s.

The matrices of class $E(n,c)$ introduced in Zlatev (1980a) satisfy conditions (a)-(d) for $s=1$ with $\bar{u}=4$. Any matrix $A = E(n,c)$ is of order n and its non-zero elements are: 4's on the main diagonal, -1's on the two side-diagonals and on two bands at distance c from the main diagonal. The choice of an off-diagonal element may cause a growth by a factor $1+\bar{u}$ at the stage where this happens. Therefore if n is sufficiently large and if $u > 4$, then the results may be very inaccurate when any pivotal strategy of the class of GMS's is in use, while a good accuracy should be expected when strategies of the class of IGMS's are applied. This is demonstrated numerically in Table 1.4. Note that the fact that the solution calculated by the GMS is quite wrong may easily be detected by checking the growth factor b_n/b_1 (see §1.1). Note too that for this kind of matrices the choice of off-diagonal elements as pivotal leads to a very poor preservation of the sparsity.

The analysis and the numerical results given in this paragraph show that the use of a small p(s) in a pivotal strategy based on the use of the Markowitz cost-function has the following advantages (in comparison with the use of a large value of p(s)):

(1). It is not necessary to store and update at each stage of the Gaussian elimination the information about columns (rows) in an increasing number of non-zero elements when set I_s contains row (column) numbers.

(2). The additional storage needed in the computational process is reduced.

(3). No pivotal search among the non-zero elements of columns (rows) is needed when I_s contains row (column) numbers.

(4). The number of computational operations connected with the pivotal strategy may be reduced considerably.

(5). A pivotal strategy from the class of IGMS's can easily be implemented. The use of such a strategy may sometimes result in much more accurate solutions than those obtained by the use of a pivotal strategy from the class of GMS's.

Since only a few rows are searched, the pivotal strategy with a small p(s) may produce more fill-ins than the corresponding pivotal strategy with a large p(s). The results obtained in many experiments indicate that this will happen extremely seldom. Nevertheless, it is clear that for classes of matrices, where this happens, it is better to use a pivotal strategy with a large p(s). Therefore an option where a large p(s) could be specified may be included in the package for sparse matrices. In the new

Table 1.4. Comparison of a pivotal strategy from the class of GMS's and the corresponding pivotal strategy from the class of IGMS's in the solution of systems of linear algebraic equations whose matrices are of class E(n,c), A=E(3600,c), c=20(20)140. *The number of non-zero elements in matrix* A *is* NZ = 5n-2c-2.

	The pivotal strategy is a GMS				The pivotal strategy is an IGMS			
c	COUNT	Time	b_n/b_1	Accuracy	COUNT	Time	b_n/b_1	Accuracy
20	162788	163.98	1.85E5	2.28E0	105074	15.90	1.0	8.74E-3
40	256747	790.74	2.36E5	6.91E1	123720	24.25	1.0	2.96E-3
60	301889	839.51	1.51E5	1.80E0	148439	40.59	1.0	1.44E-3
80	329007	1347.52	3.85E6	6.97E1	131682	31.42	1.0	8.41E-4
100	308611	1383.99	2.54E5	7.80E0	138913	35.15	1.0	4.98E-4
120	271453	909.35	5.53E5	6.17E0	119119	23.62	1.0	3.59E-4
140	238675	704.89	1.90E5	1.79E0	111747	19.68	1.0	1.69E-4

version of MA28 both small and large values of p(s) may be chosen (see the Harwell Subroutine Library Bulletin, No. 16, p.2, 1983).

1.5. Pivotal strategies based on a local minimization of fill-ins. The following definitions can be used to introduce pivotal strategies of the third group.

Definition 1.4 *Let* $a_{ij}^{(s)}$ *be an arbitrary non-zero element in set* A_s *and let* F_{ijs} *be the number of fill-ins created at stage* s $(s=1(1)n-1)$ *when element* $a_{ij}^{(s)}$ *is chosen as a pivot. The integer*

$$F_s = \min(F_{ijs}) \qquad (s\in\{1,2,\dots,n-1\}, s\le i,j\le n,\ a_{ij}^{(s)}\in B_s) \quad (1.32)$$

is called the optimal fill-in cost at stage s. □

Definition 1.5 *Let* I_s *and* B_s *be the same as in Definition 1.3 or Remark 1.2. Let*

$$C_s^* = \{a_{ij}^{(s)} \,/\, a_{ij}^{(s)} \in B_s \wedge F_{ijs} = F_s\}. \qquad (1.33)$$

The class of generalized minimum fill-in strategies (GMFS's) contains all pivotal strategies in which an arbitrary element of the set C_s^* *of candidates is chosen as pivotal at stage* s $(s=1(1)n-1)$. *The class of improved generalized minimum fill-in strategies (IGMFS's) contains all pivotal strategies in which one of the maximal (in absolute value) elements of* C_s^* *is chosen as pivotal at stage* s. □

It is rather difficult to implement a pivotal strategy of this group in a code for sparse matrices. Duff & Reid (1974) have carried out some experiments using a pivotal strategy of this group. Their conclusion is that "although this algorithm is the most computationally expensive of those under consideration its cost is not totally prohibitive". Therefore, a pivotal strategy of this group should not be the only available strategy in a package for sparse matrices. However, such a strategy may be useful for some special classes of matrices. As an example only, consider the matrix whose sparsity pattern is given in Fig. 1.1. The application of any strategy based on the Markowitz cost-function will give *3 times more fill-ins* than the application of a strategy based on a local minimization of the number of fill-ins (the numbers of fill-ins being 27 and 9 respectively). Moreover, one can attempt to apply the same idea as in §1.4 and develop pivotal strategies with a small p(s). Such a strategy may perform well for matrices that give a considerably small number of fill-ins in comparison with the number of fill-ins produced by the pivotal strategies of the other groups

1.6. Choice of a pivotal strategy in a package for sparse matrices. The survey of the main principles used in the development of a pivotal strategy for the Gaussian elimination of sparse matrices indicates that no pivotal strategy will be the best one in all situations. This is so because in all of these strategies one attempts to preserve the sparsity by some criterion that preserves or attempts to preserve the sparsity during a local stage s (the attempt to preserve the sparsity globally leads to a very big combinatorial problem and seems to be prohibitive). It is well-known that each such strategy may fail to preserve the sparsity globally. Therefore *several different pivotal strategies have to be implemented in a general-purpose package for sparse matrices*. Both pivotal strategies based on the Markowitz cost-function with a small $p(s)$ and with a large $p(s)$ are available in the new version of MA28. Three different pivotal strategies (two of them for special matrices; see below) are available in Y12M. Moreover, a version with a pivotal strategy based on a priori ordering has been developed and will be an available option in the near future. There are plans to develop a pivotal strategy based on a local minimization of fill-ins by the use of a small value of $p(s)$.

Special pivotal strategies may be very useful for some classes

Figure 1.1 The sparsity pattern of a 22x22 matrix such that 9 fill-ins are produced when a pivotal strategy based on a local minimization of fill-ins is used, while the number of fill-ins produced by a pivotal strategy based on the Markowitz cost-function is 27 .

x	0	0	0	0	0	0	0	0	x	x	x	0	0	0	0	0	0	0	0	0	0
0	x	0	0	0	0	0	0	0	x	x	x	0	0	0	0	0	0	0	0	0	0
0	0	x	0	0	0	0	0	0	x	x	x	0	0	0	0	0	0	0	0	0	0
0	0	0	x	0	0	0	0	0	0	0	0	x	x	x	0	0	0	0	0	0	0
0	0	0	0	x	0	0	0	0	0	0	0	x	x	x	0	0	0	0	0	0	0
0	0	0	0	0	x	0	0	0	0	0	0	x	x	x	0	0	0	0	0	0	0
0	0	0	0	0	0	x	0	0	0	0	0	0	0	0	x	x	x	0	0	0	0
0	0	0	0	0	0	0	x	0	0	0	0	0	0	0	x	x	x	0	0	0	0
0	0	0	0	0	0	0	0	x	0	0	0	0	0	0	x	x	x	0	0	0	0
x	0	0	0	0	0	0	0	0	0	0	0	0	0	0	0	0	0	x	x	x	x
0	x	0	0	0	0	0	0	0	0	0	0	0	0	0	0	0	0	x	x	x	x
0	0	x	0	0	0	0	0	0	0	0	0	0	0	0	0	0	0	x	x	x	x
0	0	0	x	0	0	0	0	0	0	0	0	0	0	0	0	0	0	x	x	x	x
0	0	0	0	x	0	0	0	0	0	0	0	0	0	0	0	0	0	x	x	x	x
0	0	0	0	0	x	0	0	0	0	0	0	0	0	0	0	0	0	x	x	x	x
0	0	0	0	0	0	x	0	0	0	0	0	0	0	0	0	0	0	x	x	x	x
0	0	0	0	0	0	0	x	0	0	0	0	0	0	0	0	0	0	x	x	x	x
0	0	0	0	0	0	0	0	x	0	0	0	0	0	0	0	0	0	x	x	x	x
0	0	0	0	0	0	0	0	0	0	0	0	0	0	0	0	0	0	x	x	x	x
0	0	0	0	0	0	0	0	0	0	0	0	0	0	0	0	0	0	x	x	x	x
0	0	0	0	0	0	0	0	0	0	0	0	0	0	0	0	0	0	x	x	x	x
0	0	0	0	0	0	0	0	0	0	0	0	0	0	0	0	0	0	x	x	x	x

of matrices. For example, the usé of pivotal elements from the main diagonal is the best choice for some matrices. There is an option in Y12M where only diagonal elements are allowed as pivots. For some matrices it is worthwhile to suppress entirely the pivotal interchanges (there is such a possibility in Y12M). Other examples could be given. It should be stressed, however, that the user should be sure that his matrices allow the application of the special pivotal strategy which he intends to specify. The use of any special strategy is efficient only if the matrix under consideration is suitable for such a strategy.

All conclusions in this section are only valid when all non-zero elements are kept during the whole decomposition stage. If neglecting "small" non-zero elements is allowed, then the comparison between the different pivotal strategies may lead to other conclusions. This situation, in a more general context, will be discussed in the next section.

2. General scheme for solving linear algebraic problem

2.1. Definition of the general scheme. Let $m \in \mathbb{N}$, $n \in \mathbb{N}$, $b \in \mathbb{R}^{m\times 1}$ and $A \in \mathbb{R}^{m\times n}$ be given. Assume that $m \geq n$ and $rank(A) = n$. Denote $A^{\dagger} = (A^T A)^{-1} A^T$. Consider the problem: *find the vector* $x = A^{\dagger} b$. Obviously, $x \in \mathbb{R}^{n\times 1}$ can be considered: (i) as a least squares solution (i.e. as the vector that minimizes $\| r \|_2$, $r = b - Ax$) or (ii) as the solution of the system $Ax = b - r$ when $A^T r = 0$ is also satisfied. The solution of this problem is trivial when matrix A is a product of some special matrices (e.g. diagonal, tridiagonal, orthogonal, triangular). Such matrices are called *easily invertible*. A method for calculating x, in which matrix A (or a matrix B obtained in some way from A) is expressed as a product of easily invertible matrices, is called *direct*. There exist many direct methods for solving $x = A^{\dagger} b$. This is especially true for the case $m=n$. If $m=n$ and $rank(A)=n$, then $A^{\dagger} = A^{-1}$ and $r = 0$. Thus, the system $Ax = b$ is to be solved when $m=n$. Most of the direct methods can be found as special cases in the general scheme described below. It must immediately be emphasized however, that this general scheme is not a new method for solving the problem $x = A^{\dagger} b$ (though it might be used in the future for the construction of new particular methods) but rather an attempt to develop a tool for studying the common properties of the direct methods. In the case where matrix A is *sparse* this common treatment of the direct methods can be used to construct an algorithm which often leads to an improvement of the efficiency of the solution process. The usefulness of this algorithm (which may be applied in connection with any particular direct method within the general scheme) is demonstrated by numerical experiments.

Definition 2.1 *The computational process consisting of the following 3 steps is called* **a general k-stage scheme** *for solving the linear problem* $x = A^{\dagger}b$.

Step 1 - Transformation. The problem $x = A^{\dagger}b$ *is replaced by* $y = B_1^{\dagger}c$ $(B_1 \in \mathbb{R}^{p\times q},\ c \in \mathbb{R}^{p\times 1},\ p \in \mathbb{N},\ q \in \mathbb{N},\ p \geq q,\ rank(B_1) = q)$ *where: (i)* B_1 *and* c *are calculated by the use of* A *and* b, *(ii) there exists a simple relationship between* x *and* y.

Step 2 - Generalized decomposition. Calculate

$$\bar{B}_i = P_i B_i Q_i + E_i \quad (P_i \in \mathbb{R}^{p\times p},\ Q_i \in \mathbb{R}^{q\times q},\ E_i \in \mathbb{R}^{p\times q},\ i = 1(1)k,\ k \in \mathbb{N}), \tag{2.1}$$

where: (a) P_i *and* Q_i *are permutation matrices, (b)* E_i *are perturbation matrices, (c)* $\bar{B}_k$ *is a product of* **easily invertible** *matrices and (d) if* $k > 1$, *then*

$$\bar{B}_i = C_i \bar{C}_i D_i, \quad B_{i+1} = C_i^T C_i \bar{C}_i, \quad i = 1(1)k-1; \tag{2.2}$$

the matrices D_i *being* **easily invertible.**

Step 3 - Generalized substitution. Compute

$$y_1 = Hc, \quad H \overset{\text{def}}{=} \left(\prod_{i=1}^{k-1} Q_i D_i^{\dagger}\right) Q_k B_k^{\dagger} P_k \left(\prod_{i=1}^{k-1} P_i^T C_i\right) \tag{2.3}$$

(it is always assumed that a product, in which the upper index is smaller than the lower one, is equal to the identity matrix of an appropriate order). □

The above definition is meaningful only if it allows us to calculate the exact solution x of the problem $x = A^{\dagger}b$ (using the relationship between x and y assumed in Step 1) when all computations are carried out without rounding errors. This will be the case if $y_1 = y$ in absence of rounding errors. Therefore it is necessary to show that this condition is satisfied.

Lemma 2.1 (Zlatev 1980b). Consider

$$F = I - HB_1 \quad (I \in \mathbb{R}^{q\times q},\ F \in \mathbb{R}^{q\times q},\ I \text{ being the identity matrix}). \tag{2.4}$$

Assume that $\bar{B}_k$ *and* D_i $(i=1(1)k-1)$ *have a full column rank. Then*

$$F = \sum_{j=1}^{k} H_j, \tag{2.5}$$

where

$$H_j \overset{\text{def}}{=} \left(\prod_{i=1}^{k-1} Q_i D_i^{\dagger}\right) Q_k \bar{B}_k^{\dagger} P_k \left(\prod_{i=j}^{k-1} P_i^T C_i\right)^T P_j^T E_j Q_j^T \left(\prod_{i=1}^{j-1} Q_i D_i\right)^T. \tag{2.6}$$

Proof Use (2.4), (2.3), (2.1) and (2.2) to get

$$F=I-\left(\prod_{i=1}^{k-1} Q_i D_i^{\dagger}\right) Q_k \bar{B}_k^{\dagger} P_k \left(\prod_{i=1}^{k-1} P_i^T C_i\right)^T B_2 \left(\prod_{i=1}^{1} Q_i D_i^T\right)^T + H_1. \tag{2.7}$$

Use successfully (k-2 times) (2.7), (2.1) and (2.2) to transform (2.7) into

$$F=I-\left(\prod_{i=1}^{k-1} Q_i D_i^{\dagger}\right) Q_k \bar{B}_k^{\dagger} P_k B_k \left(\prod_{i=1}^{k-1} Q_i D_i^T\right)^T + \sum_{j=1}^{k-1} H_j. \tag{2.8}$$

Apply (2.1) to rewrite the last equality as

$$F=I-\left(\prod_{i=1}^{k-1} Q_i D_i^{\dagger}\right) Q_k \bar{B}_k^{\dagger} \bar{B}_k Q_k^T \left(\prod_{i=1}^{k-1} Q_i D_i^T\right) + \sum_{i=1}^{k} H_j. \tag{2.9}$$

The second term in the right-hand side of (2.9) is equal to the identity matrix (the assumption that the matrices involved in this term have a full column rank should be used in the proof of this statement) and, thus, the lemma is proved. □

Theorem 2.1 (Zlatev 1980b). If $E_i = 0$ $(i=1(1)k)$, *then* $H = B_1$. *If, moreover, the computations during Step 3 of the general k-stage scheme are carried out without rounding errors, then* $y_1 = y$. □

2.2. Application to some particular direct methods.

Four examples are given below to illustrate how the general treatment of the direct methods can be applied to some particular algorithms for solving the linear problem $x = A^{\dagger}b$. Example 3 shows that Step 1 of the general k-stage scheme is not always needed. The first two examples are 1-stage direct methods, while the last two are 2-stage methods. To the author's knowledge no k-stage direct method with $k \geq 3$ is used in practice at present. The theory given in §2.1 indicates that it is possible, in principle at least to construct such a method.

Example 2.1 Consider the method based on forming the normal equations (see, for example, Stewart 1973). This algorithm can be found from the general k-stage scheme with $k = 1$ and

$$B_1 = A^TA, \quad c = A^Tb, \quad p = q = n, \quad y = x; \tag{2.10}$$

$$\bar{B}_1 = L_1DL_1^T, \quad \bar{B}_1^{\dagger} = (L_1^T)^{-1}D^{-1}L_1^{-1}; \tag{2.11}$$

where $L_1 \in \mathbb{R}^{n\times n}$ is a unit lower triangular matrix and $D \in \mathbb{R}^{n\times n}$ is a diagonal matrix. □

Example 2.2. Consider the method based on the use of augmented matrices and Gaussian elimination (see, for example, Björck 1976). This algorithm can be found from the general k-stage scheme with k=1 and

$$B_1 = \begin{bmatrix} \alpha I & A \\ A^T & 0 \end{bmatrix}, \quad c = \begin{bmatrix} b \\ 0 \end{bmatrix}, \quad p = q = m+n, \quad y = \begin{bmatrix} \alpha^{-1}r \\ x \end{bmatrix}; \tag{2.12}$$

$$\bar{B}_1 = LU, \quad \bar{B}_1^{\dagger} = U^{-1}L^{-1}; \tag{2.13}$$

where $L \in \mathbb{R}^{q\times q}$ is a unit lower triangular matrix and $U \in \mathbb{R}^{q\times q}$ is an upper triangular matrix. □

Example 2.3. Consider the Peters-Wilkinson (1970) method. This algorithm can be found from the general k-stage scheme with k=2 and

$$B_1 = A, \quad c = b, \quad p = m, \quad q = n, \quad y = x; \tag{2.14}$$

$$\bar{B}_1 = LU, \quad C_1 = L, \quad \bar{C}_1 = I, \quad D_1 = U; \tag{2.15}$$

$$B_2 = L^TL, \quad \bar{B}_2 = L_1DL_1^T, \quad \bar{B}_2^{\dagger} = (L_1^T)^{-1}D^{-1}L_1^{-1}; \tag{2.16}$$

where $L \in \mathbb{R}^{m\times n}$ is a unit lower trapezoidal matrix, $U \in \mathbb{R}^{n\times n}$ is an upper triangular matrix, $L_1 \in \mathbb{R}^{n\times n}$ is a unit lower triangular matrix and $D \in \mathbb{R}^{n\times n}$ is a diagonal matrix. □

Example 2.4. Consider the Golub (1965) method. This algorithm can be found from the general k-stage scheme with k=2 as follows. Set $P_i = Q_i = I$ for i=1,2. Use the fact that if $rank(A) = n$, then there exist matrices $Q \in \mathbb{R}^{m\times n}$, $D \in \mathbb{R}^{n\times n}$ and $R \in \mathbb{R}^{n\times n}$ (where $Q^TQ = I \in \mathbb{R}^{n\times n}$, D is diagonal and R is upper triangular) such that $YAZ = QDR$; Y and Z being permutation matrices of appropriate orders. Then the Golub algorithm can be represented as a 2-stage direct method by

$$B_1 = A^TA, \quad c = A^Tb, \quad p = q = n, \quad y = x; \tag{2.17}$$

$$\bar{B}_1 = A^TYQDRZ^T, \quad C_1 = I, \quad \bar{C}_1 = A^TY^TQ, \quad D_1 = DRZ; \tag{2.18}$$

$$B_2 = A^TY^TQ, \quad \bar{B}_2 = ZR^TD, \quad \bar{B}_2^{\dagger} = D^{-1}(R^T)^{-1}Z^T. \quad \square \tag{2.19}$$

Remark 2.1. Simple calculations show that for the method from Example 2.4 matrix $H = ZR^{-1}D^{-2}(R^T)^{-1}Z^T$ does not depend on matrix Q. This means that in the actual computations there is no need to store Q. This fact can be exploited when matrix A is large and sparse; an example is the code LLSS01 described in Zlatev (1982b). □

Remark 2.2. All examples given in this section are applicable for $m \geq n$. Special method that are valid only for $m = n$ can also be considered. Both methods based on the Gaussian elimination process and methods based on different kinds of orthogonalization can be represented as 1-stage direct methods (see some more details in Zlatev 1980b). □

2.3. Efficient decomposition of sparse matrices. Let A be large and sparse. Then some sparse matrix technique (SMT) can be used in the process of solving $x = A^{\dagger}b$. It is well-known that during the generalized decomposition some new non-zero elements (fill-ins) are created in positions where the original elements are equal to zero. Sometimes these new non-zero elements are small. Therefore one can attempt to increase the efficiency of the process by introducing a drop-tolerance T so that if the absolute value of any element becomes smaller than T, then this element is set equal to zero (and, in fact, neglected in the further calculations; moreover, the place in the computer memory occupied by this element is considered as a free location). Of course, there is a danger that the vector y_1 computed with a large drop-tolerance (LDT) is not very accurate. Therefore one should try to improve the accuracy by the use of some iterative method (applying y_1 as a starting approximation and using the calculated decomposition to accelerate the speed of convergence). The application of the above ideas leads to the following algorithm which can be used with any particular method within the general k-stage scheme when A is sparse.

Common algorithm for sparse matrices (CASM)

1. *Use some SMT.*
2. *Specify an LDT.*
3. *Carry out some iterative process in an attempt to regain the accuracy lost because the computations are performed by using a large drop-tolerance (LDT).*

Remark 2.3 The application of the above algorithm is justified only if the sparsity of matrix A is not lost during the transformation (Step 1 in the general k-stage scheme). From the examples given in §2.2 it becomes clear that this may happen; see Example 2.1 where A^TA is formed. For the 2-stage methods the sparsity may be lost during the transition from Stage 1 to Stage 2. This may, for example, happen when matrix $L^T L$ is calculated in the Peters-Wilkinson method. Therefore the CASM should be applied in practice only when the matrices involved in the calculations are sparse and remain sparse during the computational process. □

If matrix A is sparse and if the sparsity is not destroyed in the transition from stage i to stage $i+1$ when $k > 1$, then the CASM may give savings both in storage and in computing time. Therefore the implementation of such an algorithm as an option in a code for solving $x = A^{\dagger} b$ is worthwhile. The case, where the simple iterative refinement (IR) is applied in the third step of the CASM, will be considered in the next paragraphs. Such algorithms are applied as options in two codes: LLSS01 (Zlatev 1982b, Zlatev and Nielsen 1980) and Y12M (Zlatev et al. 1981a).

It must be emphasized here that the ideas discussed in this paragraph have been proposed and discussed first by Evans (1973).

2.4. Application of IR in connection with the general k-stage scheme. Iterative refinement (IR) of the solution y_1 obtained by (2.3) can be carried out as follows:

$$r_i = c - B_1 y_i, \quad d_i = H r_i, \quad y_{i+1} = y_i + d_i \quad (i=1(1)\rho-1). \tag{2.20}$$

Different stopping criteria (see, for example, Björck 1978a, Stewart 1973) must be applied in order to terminate the IR process (2.20) after some $i = \rho - 1$. If y_ρ satisfies certain acceptability criteria, then y_ρ is used (instead of y) to obtain an approximation to x.

Consider matrix F from (2.4) and set $F^0 = I$. The following results (concerning the IR process as applied in the general k-stage scheme) can easily be derived from well-known results (see, for example, Fadeev and Fadeeva 1963).

Theorem 2.2. Let $\{y_i\}$ *be the sequence of vectors calculated by (2.3) and (2.20). Then*

$$y_i = y + F^{i-j}(y_j - y) + \left(\sum_{\nu=0}^{i-j-1} F^{\nu}\right) Hs \quad (s \overset{\text{def}}{=} c - B_1 y), \tag{2.21}$$

where $j < i$ *is a fixed positive integer.* □

Theorem 2.3. Let $\{d_i\}$ *be a sequence of vectors found by (2.20). Then*

$$d_i = F^{i-j} d_j, \tag{2.22}$$

where $j \leq i$ *is a fixed positive integer.* □

Theorem 2.4. Let λ_i (i=1(1)q) *be the eigenvalues of matrix* F. *Assume that*

$$|\lambda_j| \le |\lambda_1| < 1 \qquad (j=2(1)q). \tag{2.23}$$

Then

$$y = y_j + \sum_{i=j}^{\infty} d_i - (HB_1)^{-1} Hs \qquad (\ s \ \text{being defined as above}), \tag{2.24}$$

where j *is a positive integer.* □

Remark 2.4 Condition (2.23) can be replaced by

$$\| F \| < 1, \tag{2.25}$$

where $\| \cdot \|$ is any matrix norm induced by the vector norm used. □

The above theorems indicate that the corrections d_i can be used in the stopping criteria for the IR process applied in connection with the general k-stage scheme. This fact is well-known for the IR process applied in connection with some particular direct methods; see Björck (1978a) or Stewart (1973). For example, since (2.24) implies

$$\| d_i \| \to 0 \qquad \text{as} \qquad i \to \infty, \tag{2.26}$$

it is clear that $\| d_{i+1} \| > \| d_i \|$ indicates that the IR process should be terminated and this condition is used as a stopping criterion in the subroutins discussed in Zlatev (1982a,b) together with the requirement $i > 2$ (concerning the latter condition see Björck 1978a, p. 165).

The corrections d_i could also be used in an attempt to measure the accuracy achieved (or, in other words, in the acceptability criteria); this is also implemented in the subroutines discussed in Zlatev (1982a,b).

2.5. Relationship between the convergence of the IR process and the drop-tolerance. Assume that B_i (i=1(1)k) are well-scaled and that the drop-tolerance T_i (to be used at stage i of the general k-stage scheme) is chosen so that:

$$T_i = w_i b_i \qquad (\ 0 \le w_i < 1, \quad i=1(1)k\ ), \tag{2.27}$$

where b_i is the magnitude of the non-zero elements in B_i. Assume that at stage i the decomposition (2.1) is obtained either by some version of the Gaussian elimination or by some orthogonal transformation. Then

$$\| E_i \| \le f_i(m,n)\, \bar{\varepsilon}_i\, g_i(A) \qquad (\ \bar{\varepsilon}_i = max(\varepsilon, T_i/b_i)\ ), \tag{2.28}$$

where $f_i(m,n)$ is a function of m and n which depends on the method applied at stage i, ε is the machine accuracy and $g_i(A)$ is a function depending on the norm of matrix A corresponding to the norm used in the left-hand side of (2.28). For $T_i=0$ explicit expressions for f_i and g_i for some particular direct methods can be found in Björck (1978a), Stewart(1973), Voevodin (1977) or Wilkinson (1961,1963,1965). If

$$f(m,n) = max(f_1(m,n), f_2(m,n), \ldots, f_k(m,n)), \tag{2.29}$$

$$g(A) = max(g_1(A), g_2(A), \ldots, g_k(A)), \tag{2.30}$$

$$\bar{g}(A) = \max_{1\le i\le k}\left(\| (\prod_{j=1}^{k-1} Q_j D_j^{\dagger}) Q_k \bar{B}_k^{\dagger} P_k (\prod_{j=i}^{k-1} P_j^T C_j)^T P_i^T \| \; \| Q^T (\prod_{j=1}^{i-1} Q_j D_j^T)^T \| \right), \tag{2.31}$$

then

$$\| F \| \le k\, f(m,n)\, \bar{\varepsilon}\, g(A)\, \bar{g}(A), \qquad \bar{\varepsilon} = max(\bar{\varepsilon}_1, \bar{\varepsilon}_2, \ldots, \bar{\varepsilon}_k). \tag{2.32}$$

In the case where $k=1$ and $T_1=0$ the product $g(A)\bar{g}(A)$ can be expressed, for many particular methods, by the condition number $\kappa(A)$ (corresponding to the matrix norm used) of matrix A; see the above references.

Of course, (2.32) cannot be used for a direct estimation of the norm of matrix F. This is true also for the simplest case with $k=1$ and $T_1=0$. In all references given before formula (2.29) it is emphasized that the values of $f(m,n)$ theoretically found are rather pessimistic and give normally a very great overestimation of the norm of F. This is especially true when m and n are large. However, at least in the cases where the product $g(A)\bar{g}(A)$ can be expressed as a function of $\kappa(A)$, the bound (2.32) together with the requirement $\| F \| < 1$ (needed to ensure a convergent IR process) indicate an important relationship between $\kappa(A)$ and $T = max(T_i)$ (the maximum being taken over $\forall i$): *the drop-tolerance* T *can be chosen larger when the condition number* $\kappa(A)$ *of matrix* A *is smaller.*

The choice of T depends on the particular direct method (within the general k-stage scheme) specified. Consider the method based on the use of the normal equations (Example 2.1 in §2.2) and the method based on the use of augmented matrices (Example 2.2 in §2.2) with an optimal α (Björck 1967,1968,1978b). In the former case $g(A)\bar{g}(A)=\kappa^2(A)$, while in the latter one $g(A)\bar{g}(A)=\sqrt{2}\kappa(A)$. Therefore it is clear that the use of the normal equations imposes a very severe restriction on the choice of T compared with the use of the augmented matrices with an optimal value of α. It should be mentioned here that it is not very easy to find out an optimal value of α.

However, in many experiments good results were obtained with α varied in a quite large interval; even the simplest choice, $\alpha=1$, gives very often excellent results with quite large values of T. Many numerical results are given in Zlatev (1982a,b), Zlatev & Thomsen (1980), Zlatev et al. (1980, 1981a,b, 1982) and Østerby and Zlatev (1983).

2.6. Implementation of the CASM for some particular direct methods. The common algorithm for sparse matrices (CASM, see §2.3) has been implemented in several algorithms for solving large linear problems In package Y12M (Zlatev et al. 1981a) the simple Gaussian elimination (which is a special case with k=1 of the k-stage general scheme for treatment the case where m=n) is used to decompose a square matrix A. This package can also be used in the solution of least squares problems by the method of augmentation (Example 2.2 in §2.2) and has in fact been used to solve such prob lems; see Zlatev et al.(1982) and Østerby & Zlatev (1983).

The method described in Example 2.4 (§2.2) is implemented (with some modifications) in package LLSS01 (Zlatev 1982b, Zlatev & Nielsen 1980).

The use of a large drop-tolerance T and IR is only an option both in Y12M and LLSS01. The efficiency of this option for some problems has been verified in Zlatev et al. (1980,1981a,b,1982) by comparisons with the option where T=0 and y_1 is accepted as a solution without any attempt to use IR.

Some new subroutines for handling of symmetric matrices will be added to package Y12M. In this way package Y12M will be able to solve problems by the use of the normal equations and by the use of the Peters-Wilkinson method (see Example 2.1 and Example 2.3 in §2.2).

2.7. Testing the subroutines in which CASM is implemented. Many rules and algorithms in the sparse matrix techniques are based on heuristics. This is also true for the rules and algorithms proposed here. Therefore it is necessary to check carefully how the rules and the algorithm work in practice. In this check the following problem has to be solved: *how to choose suitable test-matrices?* One can collect matrices which often appear in practice. This leads to huge files and tapes and may cause difficulties with data integrity in the transfer between different computers. It should be mentioned also that the number of matrices in such files and tapes is not large. Nevertheless, it is necessary to test the codes on such matric because the codes must work well on practical problems. Therefore during the experiments with packages Y12M and LLSS01 test-matrices from the Harwell set (Duff & Reid 1978) together with test-matrices appearing in the solution of some partial differential equations (that govern the two-phase flow systems, Dongarra et al. 1981), in nuclear spectroscopic theory (Schaumburg et al. 1980) and in thermodynamics (Zlatev & Thomsen 1979) have been used. Moreover, several generators for test-matrices have been developed; Zlatev et al. 1981b. One of these matrix-generators, MATRF2, creates matrices dependent of 5 parameters: m, n, r, c and α so that:

(a) the number of rows in matrix A can be varied by m,

(b) the number of columns in matrix A can be varied by n,

(c) the number of non-zeros in matrix A can be varied by r,

(d) the sparsity pattern of matrix A can be varied by c,

(e) the magnitude of the non-zeros in matrix A can be varied by α ($max(|a_{ij}|)/min(|a_{ij}|) = max(10\alpha^2, (rn-n)\alpha)$ for $a_{ij} \neq 0$).

A very wide range of test-matrices can be obtained by varying these parameters (or even only some of them) and, thus, a systematic investigation of the dependence of the performance of the code on any of these parameters can be carried out in an extremely efficient way. The analysis of results obtained in several thousand runs with different test-matrices (see Zlatev 1982a,b, Zlatev & Thomsen 1980, Zlatev et al. 1980, 1981a,b, 1982, Østerby & Zlatev 1983) indicates that the implementation of a CASM option in packages Y12M and LLSS01 is very useful and leads to savings in both storage and computing time for many classes of matrices (compared with the option where T = 0 and no attempt to carry out IR is made). This shows that CASM options have to be added to the general-purpose sparse packages. However, there exist examples where the options with T = 0 and without IR perform better. This is especially true in the case where the decomposition is performed with no fill-ins (or with a few fill-ins only; see Zlatev et al. 1982) and in the case where many right-hand sides are present (see Zlatev & Thomsen 1979). Therefore the CASM's should be implemented only as options in general-purpose sparse matrix software.

Some numerical results are given in Table 2.1 - Table 2.3. These results should be compared with the results given in Table 1.2 - Table 1.4, where the same problems were solved with T = 0 and without IR. The results given in Table 2.1 are quite unexpected (at least for $r > 3$). It was expected that for the pivotal strategy with a priori ordering the values of COUNT would be larger than the corresponding numbers for the pivotal strategies based on the Markowitz cost-function but the difference would be smaller than that for the case with T = 0 and without IR (and because of this

Table 2.1. Comparison of some characteristics obtained in the runs of two versions of package Y12M (the first based on a priori ordering of the columns, the second on the use of the Markowitz cost-function with a small p(s); see §1.3 and §1.4). T = 0.1 *and IR are used in both versions (the results obtained with* T = 0 *and without IR are given in Table 1.2). The numbers of iterations are given in brackets. All matrices are stored in single precision. High accuracy is achieved because some inner product are accumulated and kept in double precision; see Zlatev (1982a), p. 395.*

Density	A priori ordering of the columns			Markowitz cost-function		
parameter	COUNT	Time	Accuracy	COUNT	Time	Accuracy
r = 2	784	0.09(6)	2.91E-12	710	0.12(8)	1.81E-12
r = 3	1331	0.15(7)	6.43E-13	1203	0.19(8)	3.38E-13
r = 4	1778	0.18(7)	5.61E-13	2027	0.26(7)	6.71E-13
r = 5	1982	0.23(9)	9.56E-13	2928	0.40(9)	8.92E-13
r = 6	2331	0.24(7)	4.97E-13	3531	0.51(9)	4.65E-13
r = 7	2707	0.30(8)	1.01E-12	4249	0.57(7)	1.02E-12
r = 8	3098	0.31(6)	7.74E-13	4923	0.75(10)	7.85E-13
r = 9	3497	0.42(10)	9.02E-13	5639	0.84(8)	8.78E-13
r = 10	3827	0.41(7)	7.39E-13	6057	0.95(8)	7.71E-13
r = 11	4191	0.45(7)	8.21E-13	6859	1.14(10)	7.96E-13
r = 12	4549	0.47(6)	5.26E-13	7247	1.18(9)	7.53E-13

Table 2.2. Results obtained in the solution of the test-problem from Table 1.3 with T = 0.1 *and IR* (T = 0 *and no IR are applied in the run described in Table 1.3). The number of iterations is given in brackets. The classical mode of computation (accumulating the results in double precision and then rounding them to single precision) is used in in the calculation of the residuals in this version of Y12M.*

Compared characteristics	Package Y12M (small p(s))
Accuracy	1.03E-6
Computing time	8.11(9)
Storage (COUNT)	55412

Table 2.3. Numerical results obtained in the solution of systems of linear algebraic equations whose matrices are A=E(3600,c). *An IGMS with a small p(s),* T = 0.001 *and IR is used in this experiment (the same problems are solved with* T = 0 *and without IR in Table 1.4). The code stops the computations when the accuracy estimation becomes smaller than a parameter PREC* (PREC = 0.0001 *has been used in these runs).*

c	COUNT	Time	Iterations	Accuracy
20	74462	30.23	114	7.92E-5
40	76572	18.57	53	7.53E-5
60	74390	12.87	27	2.67E-5
80	71521	10.23	17	5.34E-5
100	69789	8.94	12	3.05E-5
120	67260	7.87	9	6.87E-5
140	65743	7.44	8	2.00E-5

the former strategy, which is cheaper, was expected to result in slightly smaller computing times in some cases). It is seen from Table 2.1 that this is so only for $r \leq 3$. For $r > 3$ the values of COUNT (which are used to measure the storage needed by the algorithm under consideration) are considrably smaller than these for the second strategy (this fact leads also to using considerably smaller computing times). Of course, this will not be true in general. However, these results indicate that the use of a large drop- tolerance may influence in a different way the performance of the pivotal strategies and that the influence of the application of a positive drop-tolerance on the pivotal strategies deserves further investigation.

The results given in Table 2.2 should be compared with the results in Table 1.3. It is seen that the computing time is reduced by a factor approximately equal to 16 (compared with the computing time taken by Y12M with $T = 0$ and without IR; see Table 1.3). The storage is reduced by a factor approximately equal to 5.

The results given in Table 2.3 correspond to the results given in Table 1.4 (where $T = 0$ and IR is not used). It is seen that the storage is reduced by a factor of 2 (compared with the storage for the IGMS in Table 1.4). The computing time for all examples except the first one is also reduced considerably (when $c=100$ the computing time is reduced by a factor approximately equal to 4). The computing time for the first example is increased by a factor of 2. The reason for this increase is the large number of iterations needed to obtain the required accuracy. This example as well as some examples with Newmann boundary conditions from Dongarra et al. (1981) indicate that some algorithms for acceleration of the convergence speed are necessary sometimes. Some work in this direction is in progress. It should be mentioned here that the new version of MA28 (see the Harwell Subroutine Library Bulletin, No. 16, p.2, 1983) has been used to obtain the results presented in Table 2.3 (as well as these in Table 1.4).

2.8. A numerical illustration of the relationship between the drop-tolerance and the condition number. Increasing the value of parameter α in the matrix generator MATRF2 (Zlatev et al. 1981b), one should expect the condition number to increase. By calculating directly the condition numbers (using Harwell subroutines in double precision) and by calculating estimations of the condition numbers (by the algorithm proposed in Forsythe et al. 1977; such estimations are given in Table 2.4 under "COND") it has been verified that the increase of α does lead to an increase of the condition number (as expected). Small parameters m and n were used (the direct calculation of the condition number is a rather expensive process). The first 4 parameters in MATRF2 are kept fixed. At the beginning $\alpha = 2^0$ is used and after each successful run α is multiplied by 2. Under α^* the maximal value of α by which an acceptable solution has been obtained is given (for 4 different values of the drop-tolerance and two codes. It is seen from Table 2.4 that the results agree with the conclusion drawn in §2.5 (the drop-tolerance can be chosen larger when the condition number is smaller). It should be mentioned that a slightly modified version of Y12M has been used in this experiment (as well as in the experiment described in Table 1.2 and Table 2.1). In this version all inner products are accumulated in double precision (Zlatev 1982a p. 395). In this way a compatibility with the method applied in LLSS01 (Zlatev 1982b) is achieved. The calculations (as the calculations in the other experiments) have been carried out on an IBM 3033 computer, where the machine accuracy is $\varepsilon \approx 10^{-6}$. It is seen that the accuracy obtained is normally twice as great (when the number of correct digits is measured) though all matrices are stored and kept in single precision. This agrees with the analysis given by Björck (1967,1968).

A subroutine for estimating the condition number is available in Y12M (Zlatev et al. 1981a,1983c). The user may choose among 3 options. In the first option the algorithm from Forsythe et al. (1977) is implemented. The second option is based on the algorithm of Cline et al. (1979) as implemented in LINPACK (Dongarra et al. 1979). The third algorithm is based on the modification of the LINPACK estimator advocated by Grimes & Lewis (198 Note that if the condition number is known, then $T \approx (\kappa(A))^{-1}$ is in genera a good choice for T. Therefore if many problems with the same (or very silar matrices) are to be solved, then it may be useful to determine first

Table 2.4. An experiment with matrices produced by the matrix generator MATRF2 with m=n=22, r=2, c=11 and different values of α. If the drop-tolerance is larger, then the maximal value α of parameter α for which the IR is still convergent becomes smaller (for both codes). The exact solution is denoted by x, the calculated one is denoted by y.*

	Y12M			LLSS01		
T	α^*	COND	$\|\|x-y\|\|_\infty$	α^*	COND	$\|\|x-y\|\|_\infty$
0.0	16777216	4.43E+14	8.64E-14	32768	1.69E+12	7.09E- 7
0.0001	65536	6.75E+12	2.79E-11	512	4.02E+ 8	7.54E-10
0.001	8192	1.01E+11	6.51E-12	128	2.35E+ 7	4.66E-10
0.01	1024	1.69E+ 9	3.38E-12	16	2.43E+ 5	5.17E-12

the condition number and then to carry out the computations with an appropriate value of the drop-tolerance. Other methods for determining the drop-tolerance are discussed in Zlatev (1982a). A special method, which can efficiently be used in the numerical integration of stiff systems of linear ordinary differential equations, will be considered in the next section.

3. Application of sparse matrix technique in the numerical solution of stiff systems of linear ordinary differential equations

3.1. Discretization process.

An LIVP 1 (linear initial value problem for systems of first order ordinary differential equations; see Schaumburg et al. 1979) is written as

$$y' = A(x)y+b(x), \quad x \in [a,b], \quad y \in \mathbb{R}^n, \quad n \in \mathbb{N}, \quad y(a) = \eta, \qquad (3.1)$$

where $A(x) \in \mathbb{R}^{n\times n}$, $b(x) \in \mathbb{R}^n$ and $\eta \in \mathbb{R}$ are given. It is assumed that the LIVP 1 is stiff (i.e. the eigenvalues of $A(x)$ have negative real parts for $\forall x \in [a,b]$ and the modulus of at least one of them is very large). It is also assumed that some eigenvalues of $A(x)$ are located close to the imaginary axis (this is the case if, for example, nuclear spectroscopic phenomena are modelled by the above LIVP 1). Methods of Runge-Kutta (RK) type are most suitable when these assumptions are satisfied. In a code developed at RECKU (Regional Computing Centre at the University of Copenhagen) a 2-stage method from the class of the modified diagonally implicit RK methods (MDIRKM's; Zlatev 1981) is applied. This method can be described as follows. Consider the grid:

$$G_N = \{x_m \,/\, x_0=a,\ x_{m+1}=x_m+h_m,\ h_m>0,\ m=0(1)N-1,\ x_N=b,\ N \in \mathbb{N}\} \quad (3.2)$$

and let y_0 be some approximation to η. Assume that approximations y_i to the exact solution vectors $y(x_i)$ at the points $x_i \in G_N$ are already computed for $i=1(1)m$. Then an approximation y_{m+1} to $y(x_{m+1})$ ($x_{m+1} \in G_N$) is calculated by

$$\tilde{A}_m k_1(h_m) = \gamma\Big(A(x_m+\tfrac{1}{2}h_m)\,y_m + b(x_m+\tfrac{1}{2}h_m)\Big), \qquad (3.3)$$

$$\tilde{A}_m k_2(h_m) = \gamma\Big[A(x_m+\tfrac{1}{2}h_m)\Big(y_m+h_m\beta k_1(h_m)\Big) + b(x_m+\tfrac{1}{2}h_m)\Big], \qquad (3.4)$$

$$y_{m+1} = y_m + \tfrac{1}{2}h_m\Big(k_1(h_m)+k_2(h_m)\Big), \qquad (3.5)$$

where $\beta = \sqrt{2} - 1$, $\gamma = 1 - \sqrt{2}/2$, $\gamma_m = -(h_m\gamma)^{-1}$ $(m=0(1)N-1)$ and

$$\tilde{A}_m = \gamma_m I + A(x_m + \tfrac{1}{2}h_m) \quad (I \in \mathbb{R}^{nxn} \text{ being the identity matrix}) \tag{3.6}$$

Remark 3.1. Some extra computations are needed (in order to decide whether y_{m+1}, calculated by (3.5), is acceptable or not and in order to determine some optimal value for the stepsize h_{m+1} which will be used at the next step). These computations (see Schaumburg et al. 1979 or Zlatev 1981) will not be discussed in this paper because they do not change in the transition from some *dense matrix technique* (DMT) to some *sparse matrix technique* (SMT). The implementation of an SMT in the process described by (3.3) to (3.6) is the main topic of this section. □

Remark 3.2. It should be emphasized that two systems of linear algebraic equations with the same matrix $\tilde{A}_m$ have to be solved at each step m when a 2-stage MDIRKM is used. The application of any other 2-stage semi-implicit RK method leads to the solution of two systems of linear algebraic equations with different matrices at each step m. This is also true for the 2-stage diagonally implicit RK methods (the DIRK methods; Alexander 1977). □

3.2. Computational operations at an arbitrary step m.

calculations at step m can be carried out as follows: (i) factorize matrix $\tilde{A}_m$ using the Gaussian elimination ($L_mU_m = P_m\tilde{A}_mQ_m + E_m$, where L_m is a unit lower triangular matrix, U_m is an upper triangular matrix, P_m and Q_m are permutation matrices and E_m is a perturbation matrix), (ii) solve system (3.3), (iii) use the calculated solution $k_1(h_m)$ to compute the right-hand side of (3.4), (iv) solve system (3.4) and (v) use (3.5) to obtain an approximation y_{m+1} to $y(x_{m+1})$. This algorithm can successfully be applied when the number of equations, n, is small (Zlatev et al. 1983a) If the number of equations is large (and this will be the case when sparse matrix techniques are in use), then it is more profitable to attempt to use some old factorization L_jU_j (calculated at some step $j < m$) at step m. Let $k_1^{[0]}(h_m)$ be some starting approximation ($k_1^{[0]}(h_m)$ may be calculated by solving $P_j^TL_jU_jQ_j^Tk_1^{[0]}(h_m) = b_m^*$, b_m^* being the right-hand side of (3.3), but it is also possible to set $k_1^{[0]}(h_m) = k_2(h_{m-1})$). Consider the iterative process:

$$r_m^{[i]} = b_m^* - \tilde{A}_m k_1^{[i]}(h_m), \qquad d_m^{[i]} = Q_j(L_jU_j)^{-1}P_jr_m^{[i]},$$

$$k_1^{[i+1]} = k_1^{[i]}(h_m) + d_m^{[i]} \tag{3.7}$$

If $P_j^T L_j U_j Q_j^T$ is considered as an approximate factorization of $\tilde{A}_m$, then the iterative process (3.7) is iterative refinement (IR). The quasi-Newton iterative process (QNIP) is often used in the solution of non-linear systems of ODE's (ordinary differential equations) $y' = f(x,y)$. It is easily seen that the QNIP is equivalent to IR when f is a linear (in y) function.

The IR process is terminated if for some $i=p$ any of the following stopping criteria is satisfied

$$\| d_m^{[p]} \| \leq c\varepsilon \| k_1^{[p]}(h_m) \| \quad (c > 1,\ \varepsilon \text{ being the machine accuracy}), \quad (3.8)$$

$$\| d_m^{[p]} \| \geq \| d_m^{[p-1]} \| \wedge p > 1, \quad (3.9)$$

$$p - \text{MAXIT} \quad (\text{MAXIT being the maximal number of iterations allowed}). \quad (3.10)$$

If the acceptability criterion:

$$\| d_m^{[p-1]} \| \leq \text{RELERR} \, \| k_1^{[p]}(h_m) \| \quad (\text{RELERR being given}) \quad (3.11)$$

is satisfied, then $k_1(h_m)$ is set equal to $k_1^{[p]}(h_m)$ and a similar procedure is applied to calculate $k_2(h_m)$. If the acceptability check (3.11) fails, then a new factorization $P_m^T L_m U_m Q_m^T$ is computed. If this is done, then $k_1(h_m)$ and $k_2(h_m)$ are set equal to $Q_m(L_m U_m)^{-1} P_m b_m^*$ and $Q_m(L_m U_m)^{-1} P_m \tilde{b}_m$ respectively; $\tilde{b}_m$ being the right-hand side of (3.4). Assume that the approximation $k_1^{[p]}(h_m)$ has been accepted and that the IR process (3.7), applied to calculate $k_2(h_m)$, is terminated for $i=q$. If the acceptability criterion (3.11), applied with $k_1^{[p]}(h_m)$ replaced by $k_2^{[q]}(h_m)$, is satisfied, then $k_2(h_m)$ is set equal to the last approximation obtained in the iterative process $k_2^{[q]}(h_m)$. If the modified acceptability check (3.11) fails, then a new factorization $P_m^T L_m U_m Q_m^T$ is calculated and $k_2(h_m)$ is set equal to $Q_m(LU)^{-1} P_m \tilde{b}_m$.

The following fact is very important in connection with the IR process. *It is not necessary to keep a copy of matrix* $\tilde{A}_m$ because this matrix can easily be calculated using matrix $A(x_m + \frac{1}{2}h_m)$; see (3.6). This is discussed in a more general context in Zlatev (1981), but it has been used in the codes described by Schaumburg et al. (1979,1980). It should only be noted here that: (i) only $2n$ extra arithmetic operations are needed to obtain $\tilde{A}_m$ from $A(x_m + \frac{1}{2}h_m)$ and (ii) matrix $A(x_m + \frac{1}{2}h_m)$ must in any case be stored because it is needed after the calculation of $k_1(h_m)$; see (3.4).

3.3. Implementation of sparse matrix technique. Package Y12M (Zlatev et al. 1981a) has been attached to the solver of LIVP's 1 based on the MDIRKM described in §3.1. The principles used in the transition from DMT to SMT are rather simple: (i) the two-dimensional arrays, in which the elements of matrices $A(x)$ and $\tilde{A}_m$ are kept must be replaced by one-dimensional arrays A and AJAC, (ii) some additional INTEGER arrays must be declared (the row and the column numbers of the non-zero elements of the two matrices $A(x)$ and $\tilde{A}_m$ as well as some pointers are stored in these arrays). At the beginning of each step m the non-zero elements of matrix $A(x_m+\frac{1}{2}h_m)$ (in an arbitrary order) must be stored in the first NZ locations of array A (NZ being the number of the non-zero elements). If an element $a_{ij}(x_m+\frac{1}{2}h_m) \in A(x_m+\frac{1}{2}h_m)$ is stored in A(K), then RNR(K) = i and SNR(K) = j must also be initialized ($1 \le K \le NZ$; RNR and SNR being INTEGER arrays). This is not the most efficient structure (this structure cannot be used efficiently in the factorization and, therefore, the non-zero elements have to be reordered before the beginning of the Gaussian elimination). However, this is the easiest way of performing the input operations at step m (only the non-zero elements together with their column and row numbers are required; the order in which the non-zero elements are stored being arbitrary). The subroutines of package Y12M reorder the elements by rows and carry out the factorization, the substitutions and the IR process. Several changes in the DMT driver subroutine (DENS1; Zlatev et al. 1983b) were necessary. The most important of these changes is the calculation of expressions containing matrix-vector products. Assume that the expression $z = Ax+y$ has to be calculated. If vectors z, x and y are stored in arrays Z, X and Y respectively and if the non-zero elements of matrix A are stored in the first NZ locations of array A (and their column and row numbers in the arrays SNR and RNR as described above), then the computations are carried out by the following code (N being the order of matrix A):

```
      DO 10 I = 1,N
   10 Z(I) = Y(I)
      DO 20 I = 1,NZ
   20 Z(RNR(I)) = Z(RNR(I)) + A(I)*X(SNR(I))
```

The SMT driver subroutine in the package for solving LIVP's 1, obtained by using the principles described above is called SPAR1 and will be fully documented in the near future.

3.4. Storage requirements. The arrays, used in SPAR1 instead of the two two-dimensional arrays for the elements of $A(x_m+\frac{1}{2}h_m)$ and $\tilde{A}_m$ in DENS1, are given in Table 3.1. It is seen that $S = 3NZ+2NN+NN1+14N$ locations are used in SPAR1 instead of the $D = 2N^2$ locations used in DENS1 (assuming that the integers and the real numbers occupy the same storage in the computer memory). Both NN and NN1 must be larger than NZ (in order to ensure free locations for fill-ins). A requirement $NN \geq 2NZ$ is imposed in package Y12M. Experience shows that $NN1 = 0.6NN$ is often a good choice. Set $NN = \nu NZ$ and $NN1=0.6\nu NZ$ in the expression for S. Then

$$S \leq D \iff NZ/N^2 \leq g(\nu,N) \tag{3.12}$$

with

$$g(\nu,N) \overset{def}{=} (2-14/N)/(3+2.6\nu). \tag{3.13}$$

Set the minimal value of ν $(\nu=2)$ in (3.12) and let $N \to \infty$. Use the fact that $g(\nu,N)$ is a decreasing function in ν when N is fixed. The result is

$$g(\nu,N) < g(2,N) < \lim_{N\to\infty} g(2,N) = \frac{2}{8.2} < 0.244. \tag{3.14}$$

From (3.14) it follows immediately that the use of a DMT is more efficient (with regard to storage requirements) than the use of the SMT when more than 1/4 of the elements of matrix $A(x)$ are non-zero elements. The converse is not true. The efficiency of the SMT depends both on N (which is entirely determined by the particular problem solved) and on ν (which depends also on our expectation for fill-ins and, thus, on the way in which storage is reserved for the run under consideration). Values of function $g(\nu,N)$ for some values of ν and N are given in Table 3.2.

Table 3.1. The arrays used in the SMT driver of the solver for LIVP's 1 instead of the two two-dimensional arrays that are used in the DMT driver

Name of the array	*Type*	*Length of the array*
A	*REAL*	*NZ*
SNR	*INTEGER*	*NZ*
RNR	*INTEGER*	*NZ*
AJAC	*REAL*	*NN*
SNRJAC	*INTEGER*	*NN*
RNRJAC	*INTEGER*	*NN1*
HA	*INTEGER*	*13N*
PIVOT	*REAL*	*N*

Remark 3.3. Function $g(\nu,N)$ is machine dependent (an assumption that the integer numbers occupy the same storage in the computer memory as the real numbers was made). If INTEGER*2 statements are available on the computer under consideration, then $S = 2NZ+1.5NN+0.5NN1+7.5N$, $g(\nu,N) = (2-7.5/N)/(2+1.8\nu)$ and $\lim g(2,N) = 0.357$ as $N \to \infty$. If double precision is in use and if INTEGER*2 statements are available, then $S = 1.5NZ+1.25NN+0.25NN1+4.25N$, $g(\nu,N) = (2-4.25/N)/(1.5+1.4\nu)$ and $\lim g(\nu,N) = 0.465$ as $N \to \infty$. □

3.5. Removing "small" non-zero elements. Since the Gaussian elimination is a special case (with k=1) of the general k-stage scheme considered in Section 2, it is possible to apply the CASM (the common algorithm for sparse matrices; see §2.3) in connection with the computational process carried out at step m (m=1(1)K). A drop-tolerance has to be introduced in order to achieve this. Using the notation of this sectic the drop-tolerance can be applied as follows. Let a non-negative parameter T (the drop-tolerance) be fixed. In the process of the Gaussian eliminatior all elements which become (at some stage s) smaller in absolute value than T are removed from the array AJAC (and their row and column numbers from the arrays RNRJAC and SNRJAC). In this way both storage and computing time (in the factorization process) can often be saved. The computing time needed for the IR process may be increased (more iterations are needed in order to regain the accuracy lost because inaccurate factors L and U are in use). However, since the factorization process is rather expensive, the global computing time is normally also reduced. Two numerical examples are given in order to demonstrate the usefulness of the drop-tolerance in the solution of problems arising in nuclear spectroscopic theory (see Table 3.3 and Table 3.4). A system of 63 linear ODE's is solved in the first exampl (Table 3.3) with T=0 (no non-zero element is removed) and with T=2. It

Table 3.2. Values of function $g(\nu,N)$ *for different* ν *and* N.

N \ ν	2	3	4	5	6
100	0.227	0.172	0.139	0.116	0.100
250	0.237	0.180	0.145	0.122	0.105
1000	0.242	0.184	0.148	0.124	0.107
∞	0.244	0.185	0.149	0.125	0.108

is seen from the tables that the numbers of function evaluations (the evaluations of the right-hand side of the system of ODE's), the numbers of integration steps and the numbers of factorizations are practically the same (slightly smaller when T=0). The number of substitutions for T=0 is smaller than that for T=2 (as should be expected because the IR process converges in general faster in the former case). Both the storage and the computing time are reduced considerably. The same is seen in the second table where results obtained in the solution of a medium spectroscopic problem (255 equations) with T=0 and with T=1 are given. It is also seen that the reductions of storage and computing time are much larger than those for the small spectroscopic problems. It should be noted that reductions both in storage and in computing time are much more desirable when the problems are large.

3.6. Removing "small" elements when the shifted Jacobian matrix is formed. In the examples presented in §3.5 "small" elements are removed only in the course of the Gaussian elimination. However, in many cases some of the elements of the shifted Jacobian matrix

Table 3.3. Comparison of some characteristics obtained in runs of a small spectroscopic problem (63 equations) with two values of the drop-tolerance.

Compared characteristics	*T = 0*	*T = 2*
Function evaluations	*1597*	*1547*
Integration steps	*774*	*749*
Factorizations	*82*	*85*
Substitutions	*5038*	*5364*
Non-zeros in the beginning (NZ)	*839*	*839*
Maximal number of non-zeros kept (COUNT)	*1972*	*896*
Computing time	*63.09*	*35.21*

Table 3.4. Comparison of some characteristics obtained in runs of a medium spectroscopic problem (255 equations) with two values of the drop-tolerance.

Compared characteristics	*T = 0*	*T = 1*
Function evaluations	*979*	*985*
Integration steps	*473*	*476*
Factorizations	*67*	*68*
Substitutions	*3015*	*3499*
Non-zeros in the beginning (NZ)	*3323*	*3323*
Maximal number of non-zeros kept (COUNT)	*20170*	*7706*
Computing time	*606.52*	*161.37*

$\tilde{A}_m$ from (3.6) are also small. Therefore it is worthwhile to try to scan the elements of $\tilde{A}$ during the process of forming this matrix and to remove all non-zero elements of $\tilde{A}_m$, which are smaller in absolute value than the drop-tolerance before the beginning of the Gaussian elimination process. Such a scan is carried out in subroutine SPAR1. The examples from §3.5 have also been run removing the small elements of $\tilde{A}_m$ before the beginning of the factorization. The results are given in Table 3.5 and in Table 3.6 (for comparison the results obtained with the same value of the drop-tolerance but without removing non-zero elements before the beginning of the Gaussian elimination are also given in these tables). It is seen that the first three of the characteristics compared do not change very much in the two corresponding runs (for the medium spectroscopic problem, Table 3.6, they do not change at all). The number of substitutions increases when a scan for removing small elements is carried out. Both the storage and the computing time are reduced when a scan of the non-zero elements of $\tilde{A}_m$ is carried out and when some of the non-zero elements of $\tilde{A}_m$ are removed before the beginning of the Gaussian elimination

It should be emphasized here that $\tilde{A}_m$ is recomputed at each iteration at step m (see the end of §3.2) because it is needed in the cal-

Table 3.5. Comparison of some characteristics obtained in runs without a sca and with a scan of the non-elements of $\tilde{A}_m$ before the Gaussian elimination

Compared characteristics (63 equations)	*Without a scan*	*With a scan*
Function evaluations	1542	1583
Integration steps	749	768
Factorizations	85	84
Substitutions	5364	5407
Non-zeros in the beginning (NZ)	839	336
Maximal number of non-zeros kept (COUNT)	896	518
Computing time	35.21	29.24

Table 3.6. Comparison of some characteristics obtained in runs without a sca and with a scan of the non-zeros of $\tilde{A}_m$ before the Gaussian elimination.

Compared characteristics (255 equations)	*Without a scan*	*With a scan*
Function evaluations	985	985
Integration steps	476	476
Factorizations	68	68
Substitutions	3499	3588
Non-zeros in the beginning (NZ)	3323	2171
Maximal number of non-zeros kept (COUNT)	7706	6523
Computing time	161.37	144.21

culation of the residual vectors; see (3.7). No non-zero element is removed from the shifted Jacobian matrix when it is used in the calculation of the residual vectors (i.e. the scan described above is carried out only before the Gaussian elimination).

3.7. Automatic determination of the drop-tolerance.

In the numerical experiments described in §3.5 and §3.6 an optimal value of the drop-tolerance has been found by several trials. Of course, this is not very efficient. While it is not very difficult to determine a good drop-tolerance (i.e. a drop-tolerance which will give better results than a run with T=0), this is not the case if one is interested in finding the optimal value of the drop-tolerance. However, when systems of ODE's are solved it is possible to apply a device by which the code will attempt to determine values of the drop-tolerance which are "nearly" optimal. Moreover, the attempt will be carried out automatically in the course of the numerical integration. This device can be described as follows. Let $a = max(|\tilde{a}_{ij}|)$, where $\tilde{a}_{ij} \in \tilde{A}_1$. Set $T_{initial} = 0.1a$. Begin the integration process with $T = T_{initial}$ and set $T = T/2$ each time when the iteration process fails to converge at steps where the shifted Jacobian matrix has been recomputed (i.e. at steps m

Table 3.7. Comparison of some characteristics obtained in runs of the small spectroscopic problem with different application of the drop-tolerance option.

Compared characteristics	*Constant T*	*Variable T*
Function evaluations	1583	1595
Integration steps	768	774
Factorizations	84	92
Substitutions	5407	5984
Non-zeros in the beginning (NZ)	336	315
Maximal number of non-zeros kept (COUNT)	518	475
Initial drop-tolerance	2.0	97.40
Final drop-tolerance	2.0	2.17
Computing time	29.24	30.12

Table 3.8. Comparison of some characteristics obtained in runs of the medium spectroscopic proble with different applications of the drop-tolerance option.

Compared characteristics	*Constant T*	*Variable T*
Function evaluations	985	1003
Integration steps	476	484
Factorizations	68	81
Substitutions	3588	3852
Non-zeros in the beginning (NZ)	2171	2283
Maximal number of non-zeros kept (COUNT)	6523	6634
Initial drop-tolerance	1.0	55.54
Final drop-tolerance	1.0	0.87
Computing time	144.21	148.60

with $m \in \{1,2,\ldots,K\}$, where $\tilde{A}_m$ has been factorised and where L_m and U_m have been used in the iterative process). If the drop-tolerance has successively been changed J times (J=10 in SPAR1), then T=0 is used in the further computations. This algorithm is described in Zlatev (1982a), but it has been used in a code (Y12NB) discussed by Schaumburg et al. (1980).

The examples used in the previous two paragraphs have also been run with the device described above. The results are given in Table 3.7 and Table 3.8. It is seen that the code is able to determine a nearly optimal drop-tolerance. The computing times are slightly larger than those for the case where an optimal value of the drop-tolerance was used (but remember that this value was determined after several trials). The difference is very small even when the final drop-tolerance is smaller than the optimal one because on a part of the integration interval the drop-tolerance used is larger than the final drop-tolerance. Taking into account that the user will in general not be able to find easily an optimal value of the drop-tolerance (especially when the problem is very large), the device for an automatic determination of an optimal drop-tolerance must be considered as very efficient. It should be emphasized that this conclusion is drawn after many different tests (not only with spectroscopic problems). A work for further refinement of the device for an automatic determination of the drop-tolerance is in progress.

<u>3.8. Exploitation of the common properties of the matrices</u> $\tilde{A}_m$. The matrices $\tilde{A}_m$ have the same structure. This property can be exploited in the calculations by using the same pivotal sequence during many integration steps. There is a danger that the results may become inaccurate after several steps due to the variation of the magnitude of the elements of matrix $A(x)$. The fact that an IR process is carried out can be used to decide when a new pivotal sequence is needed. If a factorization of matrix $\tilde{A}_m$ ($m \in \{2,3,\ldots,K\}$) is carried out using a pivotal sequence determined in a previous step $j < m$ and if the convergence is slow (or if the IR does not converge at all), then a new pivotal sequence has to be calculated.

The exploitation of another common property of the matrices $\tilde{A}_m$ has been discussed at the end of §3.2.

<u>3.9. Extension of the results for systems of non-linear ODE's</u>. The technique used in this section can easily be extended for non-linear ODE's. If the system of ODE's is non-linear, then a QNIP should be used instead of the IR process applied in this section.

The shifted Jacobian matrix $\gamma_m I + \partial f/\partial y$ has to be factorized instead of (3.6) in this case. The computation of $\partial f/\partial y$ may be very expensive in this case, but when this matrix is calculated one can (in principle at least) use the same ideas as in the previous paragraphs of this section.

If the storage needed is a crucial factor, then the storage requirements can sometimes be reduced by exploiting some machine-dependent facilities such as paging, multibanking etc. A version of package Y12M (Zlatev et al. 1981a) that exploits the multibanking facility on UNIVAC has been developed by Wasniewski et al. (1982). This version may be applied to reduce the storage needed in the solution of linear ODE's when UNIVAC computers of the series 1100 are in use. These ideas can also be extended for non-linear ODE's.

References

Alexander, R. (1977). *Diagonally implicit Runge-Kutta methods for stiff ODE's.* SIAM J. Numer. Anal., 14, 1006-1021.

Björck, Å. (1967). *Iterative refinement for linear least squares solutions, I.* BIT, 7, 257-278.

Björck, Å. (1968). *Iterative refinement for linear least squares solutions, II.* BIT, 8, 8-30.

Björck, Å. (1976). *Methods for sparse linear least squares problems.* In Sparse Matrix Computations, eds. J.Bunch and D.Rose, pp. 177-199. New York-London: Academic Press.

Björck, Å. (1978a). *Comment on the iterative refinement of least squares problems.* J. Amer. Stat. Assoc., 73, 161-166.

Björck, Å. (1978b). *Numerical algorithms for linear least squares problems.* Report No. 2/78. Department of Mathematics, University of Trondheim, Trondheim, Norway.

Curtis, A.R. and Reid, J.K. (1971). *The solution of large sparse unsymmetric systems of linear equations.* J. Inst. Maths Applics, 8, 344-353.

Cline, A.K., Moler, C.B., Stewart, G.W. and Wilkinson, J.H. (1979). *An estimate of the condition number of a matrix.* SIAM J. Numer. Anal., 16, 368-375.

Dongarra, J.J., Bunch, J.R., Moler, C.B. and Stewart, G.W. (1979). *LINPACK - user's guide.* Philadelphia: SIAM.

Dongarra, J.J., Leaf, G.K. and Minkoff, M. (1981). *A preconditioned method for solving a class of non-symmetric matrices.* Report No. ANL-81-71. Argonne National Laboratory, Argonne, Illinois, USA.

Duff, I.S. (1977). *MA28: A set of Fortran subroutines for sparse unsymmetric linear equations.* Report No. R8730. A.E.R.E., Harwell, England.

Duff, I.S. and Reid, J.K. (1974). *A comparison of sparsity orderings for obtaining a pivotal sequence in Gaussian elimination.* J. Inst. Maths Applics, 14, 281-291.

Duff, I.S. and Reid, J.K. (1976). *A comparison of some methods for the solution of sparse overdetermined systems of linear equations.* J. Inst. Maths Applics, 17, 217-224.

Duff, I.S. and Reid, J.K. (1978). *Performance evaluation of codes for sparse matrices.* Report No. CSS66. A.E.R.E., Harwell, England.

Duff, I.S. and Reid, J.K. (1979). *Some design features of a sparse matrix code.* ACM Trans. Math. Software, 5, 18-35.

Evans, D.J. (1973). *The analysis and application of sparse matrix algorithms in the finite element method.* In The Mathematics of Finite Elements and Applications, ed. J.R.Whiteman, pp. 427-447. London-New York: Academic Press.

Fadeev, D.K. and Fadeeva, V.N. (1963). *Computational methods of linear algebra.* London-San Francisco: Freeman.

Forsythe, G.E., Malcolm, M.A. and Moler, C.B. (1977). *Computer Methods for Mathematical Computations.* Englewood Cliffs, N.J.: Prentice-Hall.

Golub, G.H. (1965). *Numerical methods for solving linear least squares problems.* Numer. Math., 7, 206-216.

Grimes, R. and Lewis, J.C. (1981). *Condition number estimation for sparse matrices.* SIAM J. Sci. Statist. Comput., 2, 384-388.

Gustavson, F.G. (1972). *Some basic techniques for solving sparse systems of linear equations.* In Sparse Matrices and Their Applications, eds. D.J.Rose and R.A.Willoughby, pp. 41-52. New York: Plenum Press.

Gustavson, F.G. (1978). *Two fast algorithms for sparse matrices: multiplication and permuted transposition.* ACM Trans. Math. Software, 4, 250-269.

Harwell Subroutine Library Bulletin (1983), ed. I.S.Duff, No. 16, p. 2. A.E.R.E. Harwell, England.

Markowitz, H.M. (1957). *The elimination form of inverse and its application to linear programming.* Management Sci., 3, 255-269.

Peters, G. and Wilkinson J.H. (1970). *The least squares problem and pseudo-inverses.* Computer J., 13, 309-316.

Reid, J.K. (1974). *Direct methods for sparse matrices.* In Software for Numerical Mathematics, ed. D.J.Evans, pp. 29-48. London-New York: Academic Press.

Schaumburg, K., Wasniwski, J. and Zlatev, Z. (1979). *Solution of ordinary differential equations with time-dependent coefficients. Development of a semiexplicit Runge-Kutta algorithm and application to a spectroscopic problem.* Comput. Chem., 3, 57-63.

Schaumburg, K., Wasniewski, J. and Zlatev, Z. (1980). *The use of sparse matrix technique in the numerical integration of stiff systems of linear ordinary differential equations.* Comput. Chem., 4, 1-12.

Stewart, G.W. (1973). *Introduction to matrix computations.* New York-San Francisco-London: Academic Press.

Voevodin, V.V. (1977). *Computational bases of the linear algebra.* Moscow: Nauka; *in Russian.*

Wasniewski, J., Zlatev, Z. and Schaumburg, K. (1982). *A method for reduction of the storage requirements by the use of some special computer facilities; application to linear systems of algebraic equations.* Comput. Chem., 6, 181-192.

Wilkinson, J.H. (1961). *Error analysis of direct methods of matrix inversion.* J. Assoc. Comput. Mach., 8, 281-330.

Wilkinson, J.H. (1963). *Rounding Errors in Algebraic Processes.* Notes in Applied Mathematics No. 32. London: H.M.S.O.

Wilkinson, J.H. (1965). *The Algebraic Eigenvalue Problem.* London: Oxford University Press.

Zlatev, Z. (1980a). *On some pivotal strategies in Gaussian elimination by sparse matrices.* SIAM J. Numer. Anal., 17, 18-30.

Zlatev, Z. (1980b). *On solving some large linear problems by direct methods.* Report No. PB-111. Computer Science Department, Århus University, Århus, Denmark.

Zlatev, Z. (1981). *Modified diagonally implicit Runge-Kutta methods.* SIAM J. Sci. Statist. Comput., 2, 321-334.

Zlatev, Z. (1982a). *Use of iterative refinement in the solution of sparse linear systems.* SIAM J. Numer. Anal., 19, 381-399.

Zlatev, Z. (1982b). *Comparison of two pivotal strategies in sparse plane rotations.* Comput. Math. Applics, 8, 119-135.

Zlatev, Z. and Nielsen, H.B. (1980). *Least squares solution of large linear problems.* In Symposium in Applied Statistics 1980, eds. A. Höskuldsson, K. Conradsen, B. Sloth Jensen and K.Esbensen, pp. 17-52. NEUCC, Technical University of Denmark, Lyngby Denmark.

Zlatev, Z., Schaumburg, K. and Wasniewski, J. (1980). *Implementation of an iterative refinement option in a code for large and sparse systems.* Comput. Chem., 4, 87-99.

Zlatev, Z. and Thomsen, P.G. (1979). *Application of backward differentiation methods to the finite element solution of time dependent problems.* Internat. J. Numer. Meth. Engng, 14, 1051-1061.

Zlatev, Z. and Thomsen, P.G. (1980). *Sparse matrices - efficient decomposition and applications.* In Sparse Matrices and Their Uses, ed. I.S.Duff, pp. 367-375. London-New York: Academic Press.

Zlatev, Z., Wasniewski, J. and Schaumburg, K. (1981a). *Y12M - solution of large and sparse systems of linear algebraic equations.* Lecture Notes in Computer Science, Vol. 121. Berlin: Springer.

Zlatev, Z., Wasniewski, J. and Schaumburg, K. (1981b). *A testing scheme for subroutines for solving large linear problems.* Comput. Chem., 5, 91-100.

Zlatev, Z., Wasniewski, J. and Schaumburg, K. (1982). *Comparison of two algorithms for solving large linear systems.* SIAM J. Sci. Statist. Comput., 3, 486-501.

Zlatev, Z., Wasniewski, J. and Schaumburg, K. (1983a). *On some useful options in a code for solving stiff systems of linear ordinary differential equations.* Report No. 83/5. Regional Computing Centre at the University of Copenhagen, Copenhagen, Denmark.

Zlatev, Z., Wasniewski, J. and Schaumburg, K. (1983b). *Subroutine DENS1 for solving stiff systems of linear ordinary differential equations (basic algorithms, documentation, demonstration programs).* Report No. 83/7. Regional Computing Centre at the University of Copenhagen, Copenhagen, Denmark.

Zlatev, Z., Wasniewski, J. and Schaumburg, K. (1983c). *Condition number estimators in a sparse matrix software.* Report No. 83/13. Regional Centre at the University of Copenhagen, Copenhagen, Denmark.

Østerby, O. and Zlatev, Z. (1983). *Direct Methods for Sparse Matrices.* Lecture Notes in Computer Science, Vol. 157. Berlin: Springer.

8 SPARSITY IN TRANSPORTATION PROBLEMS

A. Brameller
Power System Laboratory, University of Manchester Institute of Science & Technology, PO Box 88, Manchester M60 1QD, England

Introduction

In general a transportation problem is a problem where a product is to be shipped in amounts S_1 S_2, ... S_m from each of m shipping origins and received in amounts D_1 D_2, ... D_n by each of n destinations. The cost of shipping a unit amount from the i-th origin to the j-th destination is c_{ij} and is known for all (i,j). The problem is to determine the amounts x_{ij} to be shipped over all routes (i,j) so as to minimise the total cost of transportation.

The method is applicable to that sub-class of linear programming problem in which the requirements and resources are expressed in terms of only one kind of unit. When such problems are expressed in the standard LP form, [A][X]=[b], the matrix [A] has a simple and special structure. It consists of +1 and 0 elements and a given variable appears in two and only two of the constraints; therefore in any column of [A] only two elements differ from zero.

Because of the special structure of [A] it is possible to develop algorithms for solution of transportation problems which are much more efficient computational procedure than the simplex method. In general the transportation problem can be divided into four categories as follows:

(a) Standard transportation problem

This problem is characterised by the following:

1. From each origin there is only direct route to any destination
2. There is no direct route between any two origins or between any two destinations
3. Total amount of product at all origins is equal to the total amount required at all destinations
4. There is no restriction on the amount shipped by any route.

A standard transportation problem may be represented by a radial network as shown below.

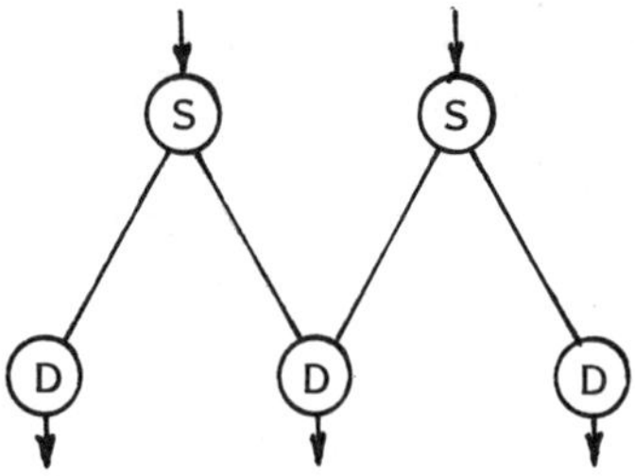

(b) Trans-shipment problem

When from any origin i there is more than one direct route to any destination j or if there are direct routes between origins or destinations or a direct route from destination to origin then the standard transportation problem is defined as a trans-shipment problem.

(c) Capacitated problem

When some or all of the routes have limits on the amount of product that can be shipped then the problem is defined as a capacitated problem.

(d) Generalised problem

When the total amount of product at all destinations is less than the total amount of product at all origins then the problem is defined as a generalised transportation problem.

A trans-shipment problem may be represented by a network as shown below:

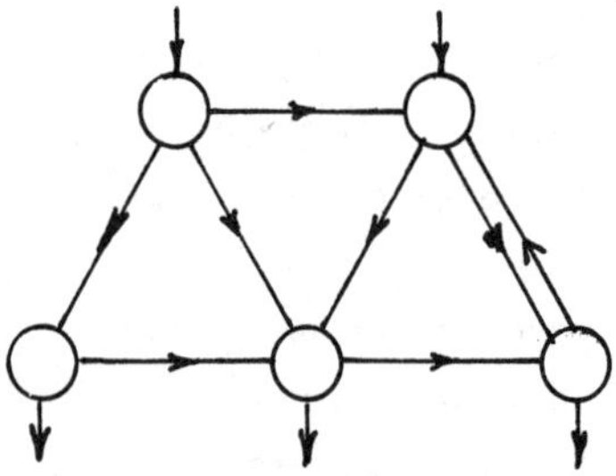

Summary of linear programming equations

$$z = [C][X] \tag{1}$$

$$[A][X] = [b] \tag{2}$$

$$[X] > [0] \tag{3}$$

$$\begin{bmatrix} [A_B] & \vdots & [A_N] \end{bmatrix} \begin{bmatrix} [X_B] \\ \hdashline [X_N] \end{bmatrix} = [b] \tag{4}$$

$$[A_B][X_B] + [A_N][X_N] = [b] \tag{5}$$

$$z = [C_B][X_B] + [C_N][X_N] \tag{6}$$

$$[X_B] = [A_B]^{-1}[b] - [A_N][X_N] \tag{7}$$

for

$$[X_N] = [0], \quad [X_B] = [A_B]^{-1}[b] \tag{8}$$

$$z = [C_B][A_B]^{-1}[b] + [C'_N][X_N] \tag{9}$$

$$[C'_N] = [C_N] - [C_B][A_B]^{-1}[A_N] \tag{10}$$

Incidence matrices

A - Branch nodal

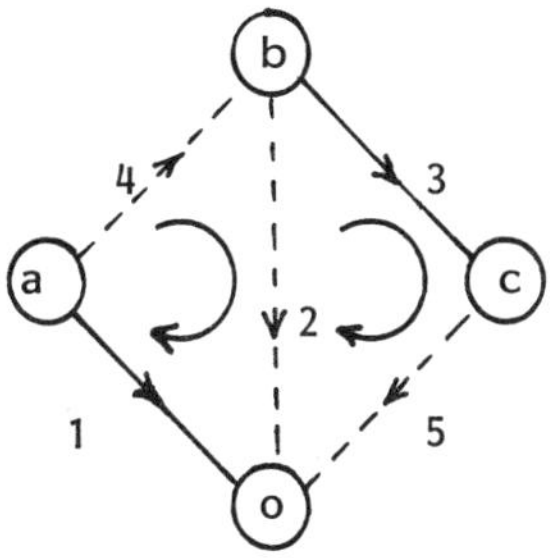

$$[A] = \begin{array}{c|ccc|cc|} & 1 & 2 & 3 & 4 & 5 \\ \hline a & 1 & & & 1 & \\ b & & 1 & 1 & -1 & \\ c & & & -1 & & 1 \\ \hline \end{array} = [A_t]\,[A_c]$$

t = tree branches

c = co-tree branches

B – Branch path

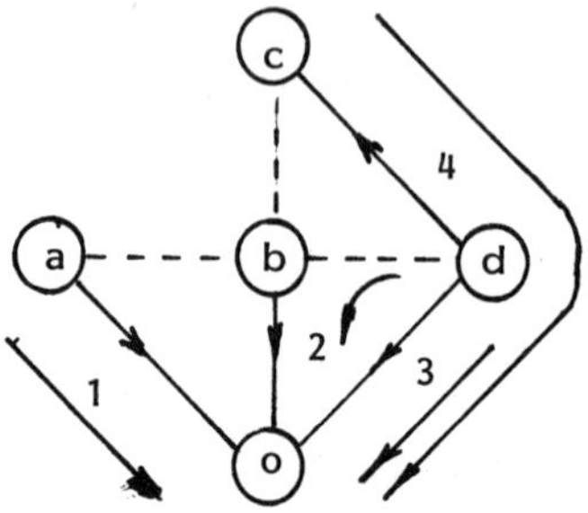

[B] =

	1	2	3	4
a	1			
b		1		
c			1	−1
d			1	

$[A_t]$ =

	1	2	3	4
a	1			
b		1		
c				−1
d			1	1

$$[A_t][B^T] = [U] \qquad (11)$$

$$\therefore \quad [B^T] = [A_t]^{-1} \qquad (12)$$

D – Branch loop

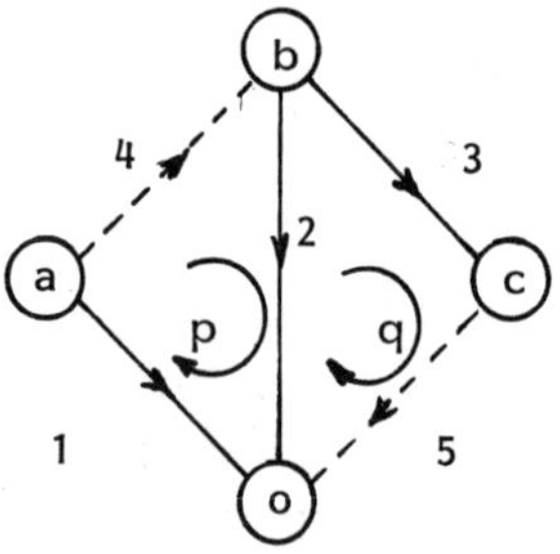

[D] =

	1	2	3	4	5
p	−1	1		1	
q		−1	1		1

= $[D_t] \mid [U]$

$$[D_t]^T = -[A_t]^{-1}[A_c]$$

Since $[A_t]^{-1} = [B]^T$ from equ. (12)

$$[D_t]^T = -[B^T][A_c] \qquad (13)$$

Network solution of a trans-shipment problem

A transportation problem with unit costs as defined in a matrix form shown below

destinations

sources	0	1	2	3	availability
0		3	5		15
1			1		0
2					0
3		6	4		10
demands	0	0	25	0	

may be represented by a network as shown below:

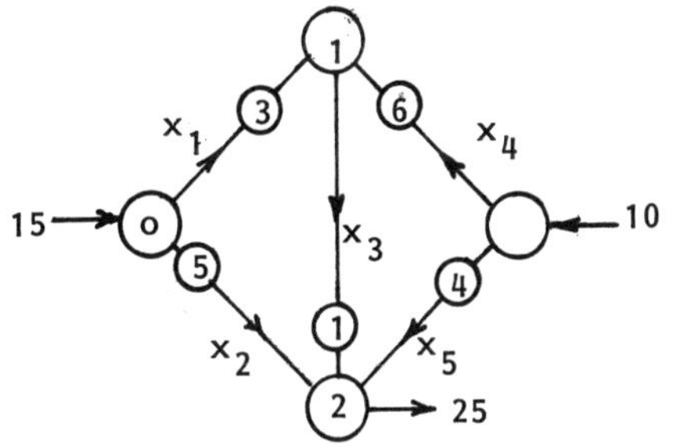

x_i = amount of commodity transported along route - i

c_i = cost of transporting unit commodity along route - i

Using Kirchhoff's laws

$$x_1 + x_2 = 15$$

$$-x_1 + x_3 - x_4 = 0$$

$$-x_2 - x_3 - x_5 = -25$$

$$x_4 + x_5 = 10$$

Choosing node 0 as reference the above equations can be stated in matrix form as

$$\begin{bmatrix} -1 & & 1 & -1 & \\ & -1 & -1 & & -1 \\ & & & 1 & 1 \end{bmatrix} \begin{bmatrix} x_1 \\ x_2 \\ x_3 \\ x_4 \\ x_5 \end{bmatrix} = \begin{bmatrix} 0 \\ -25 \\ 10 \end{bmatrix}$$

or as

$$\min z = [C][X]$$

subject to

$$[A][X] = [b]$$

where

$$[C] = \begin{bmatrix} 3 & 5 & 1 & 6 & 4 \end{bmatrix}$$

$[A]$ = branch-nodal incidence matrix

Basic feasible solution

From equation (7)

$$[X_B] = [A_B]^{-1} ([b] - [A_N][X_N])$$

where

$[A_B] = [A_t]$ = non-singular basis matrix corresponding to a tree

$[X_B] = [X_t]$ = flows in tree branches

$[X_N] = [X_c]$ = flows in co-tree branches

from equation (12).

$[A_t]^{-1} = [B^T]$ = branch-path incidence matrix

therefore $[A_B]^{-1} = [A_t]^{-1} = [B^T]$ can be obtained by tracing a tree, e.g.

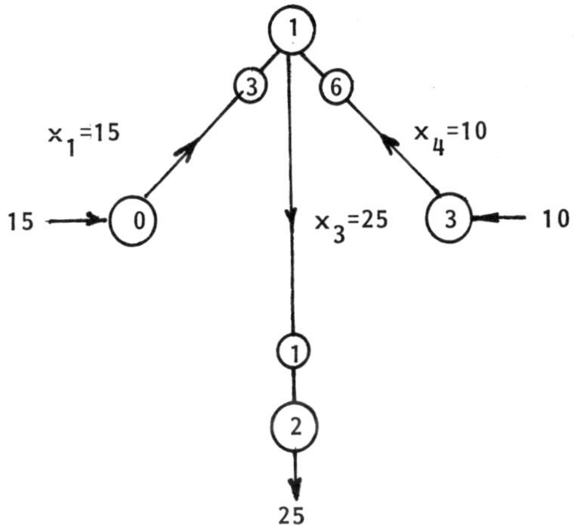

<u>Improving a basic feasible solution (selection of a co-tree to enter the basis)</u>

From equation (10)

$$[C'_N] = [C_N] - [C_B][A_B]^{-1}[A_N]$$

or

$$[C'_N] = \begin{bmatrix} [C_N] & \vdots & [C_B] \end{bmatrix} \begin{bmatrix} [U] \\ \hdashline -[A_B]^{-1} \; [A_N] \end{bmatrix}$$

from equation (12) and (13)

$$[D^T] = [B^T][A_c] = -[A_t]^{-1}[A_c]$$

therefore

$$[D^T] = -\,[A_B]^{-1}[A_N]$$

and

$$[C'_N] = [C][D] \qquad \text{since } [D] = [D_t] \;\; [U]$$

C'_N algebraic sum of the cost of the branches around the basic loops

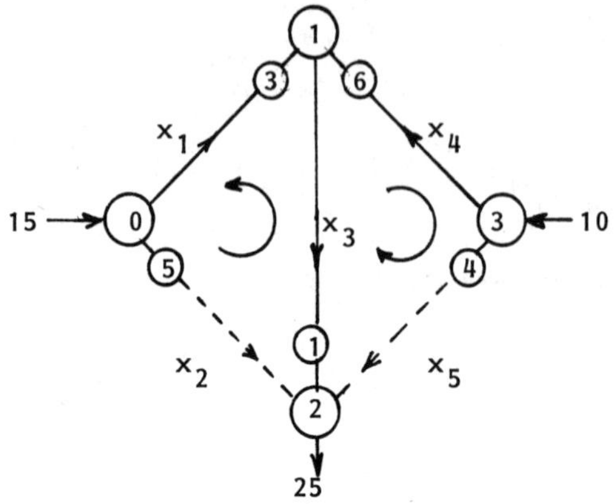

$c'_2 = (c'_p) = 5 - 1 - 3 = 1$

$c'_5 = (c'_q) = 4 - 1 - 6 = -3$

By analogy to the simplex method, the best improvement is obtained when the co-tree branch corresponding to the most negative cost coefficient enters the tree. Therefore branch 5 must change to a tree branch.

Taking variables out of the basis

The criterion used in the simplex method is that for matrix A_B to remain non-singular as required; the vector in the basis must remain linearly independent after bringing in a variable which was previously outside the basis. It follows, therefore, that the branch to go out of the basis must be one which forms a loop with the new co-tree

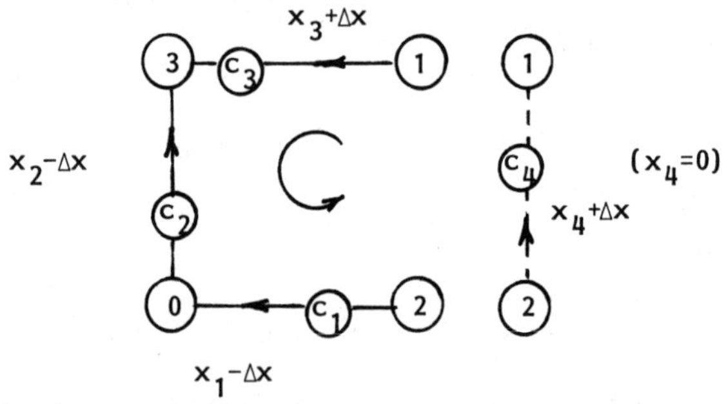

If a tree-branch "j" goes out of the tree

$$x_j = \Delta x$$

Δx must be positive, since the flow in branch 4 was zero ($x_4=0$). To achieve this, only those branches with direction opposite to the loop direction need be considered. For feasibility (so that remaining flows do not change directions) choose the branch with minimum flow.

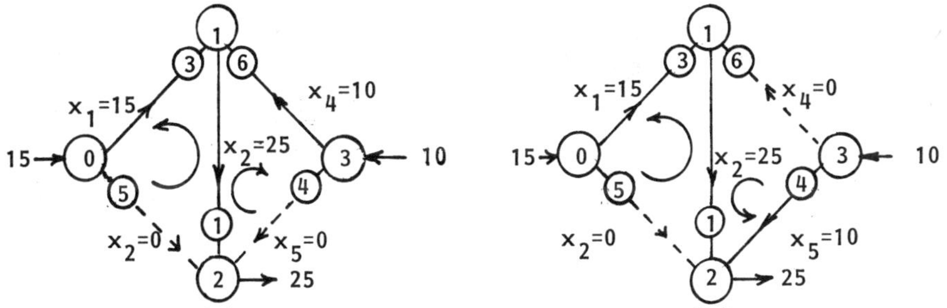

(a) (b)

Branch 5 (x_5) enters basis, branch 4 with minimum flow (x_4=10) leaves the basis resulting in the new solution shown in (b) above.

Since both cost coefficients are positive

$$c'_5 = 5 - 1 - 3 = 1$$

$$c'_4 = 6 + 1 - 4 = 3$$

the new solution is optimal

$$x_1 = 15, \; x_2 = 0, \; x_3 = 15, \; x_4 = 0, \; x_5 = 10$$

<u>Capacitated trans-shipment problems</u>

In general, a capacitated trans-shipment problem is a problem where it is required to minimise the cost of transporting a certain product from a set of supplies to a set of destinations along specified unidirectional routes. The cost of shipping a unit amount along the i-th destination is c_i and is known for all routes. The problem is to determine the amount x_i to be shipped over all routes to minimise the total cost of transportation where some or all the variables x_i have an upper limit. The problem can be stated in matrix form as:

$$\min z = [C][X]$$

subject to

$$[A][X] = [b]$$

$$[0] \leq [X] \leq [X]^{max}$$

where [A] = branch-nodal incidence matrix.

It is possible to solve a linear programming problem with upper limits on all or some of the variables without explicitly adding the upper limit constraints to the tableau. Likewise, a capacitated trans-shipment can also be solved using a network model, which does not explicitly include the upper limit constraints. By analogy to the simplex method and the network solution of transportation problems the solution process can be established as follows:

The non-basic variables can be legitimate corresponding to co-tree branches at zero level or slack corresponding to co-tree branches at upper limits. Therefore, the effect on the objective function of introducing a non-basic legitimate variable into the set of basic variables can be calculated in the same way as for the uncapacitated case. Introducing a non-basic slack variable into the set of basic variables is equivalent to reducing the value of the associated legitimate variable, i.e. reducing the flow in the corresponding co-tree branch which is at its upper limit. Therefore, the effect on the objective function can be found by tracing the loop in a direction opposite to that of the co-tree branch.

Consequently, to determine the variable that would have to go into the basis to effect the maximum improvement on the objective function, the costs around the basic loops defined by the co-tree branches are found, tracing each loop in the appropriate direction, and the co-tree branch associated with the most negative cost is identified.

If the variable chosen for entering the set of basic variables is legitimate (at zero level) one of the three cases can arise:

(a) It will replace its associated slack variable, i.e. will reach its upper limit in which case the co-tree branch remains in the co-tree but at its upper limit;

(b) It will replace another legitimate variable in which case the co-tree branch becomes a tree branch at partial level and the replaced tree branch becomes a co-tree branch at zero level;

(c) It will replace a slack variable associated with another legitimate variable in which case the co-tree branch becomes a tree branch at partial level and the tree branch becomes a co-tree branch at its upper limit.

Consequently, the flow in the co-tree branch corresponding to the legitimate variable chosen for entering the set of basic variables is increased until it reaches its upper limit, or until the flow in a tree branch becomes zero or reaches its maximum.

Similarly, if the variable chosen for entering the set of basic variables is a slack variable (co-tree branch at its upper limit), one of three cases can again arise:

(a) It will replace its associated legitimate variable;

(b) It will replace another legitimate variable;

(c) It will replace another slack variable.

Consequently, the flow in the co-tree branch is decreased until it becomes zero, or until the flow in a tree branch becomes zero or reaches its upper limit. In the first case the co-tree branch remains co-tree; in the other two cases it becomes a tree branch and the tree branch becomes a co-tree branch.

The steps of solution can now be summarised as follows:

Initial feasible solution

Establish an initial feasible tree for the system. This may be obtained by adding artificial branches with a high cost and a high upper limit from a fictitious reference to each node with the following convention for flow direction:

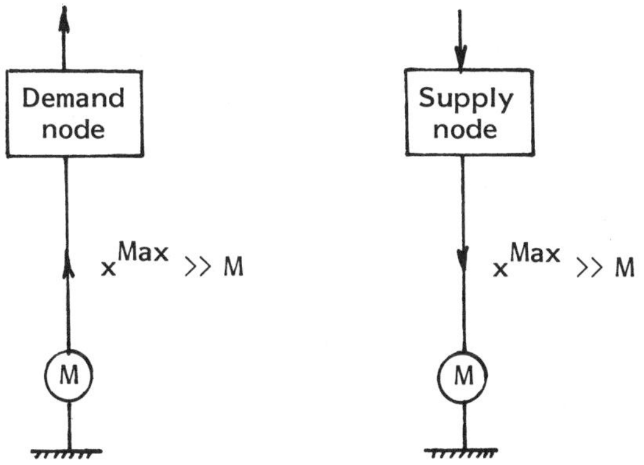

Improving a basic feasible solution

1. Find the cost around the loop defined by each co-tree branch, and determine the co-tree branch that gives the maximum improvement (most negative cost for a minimisation problem). If the flow in the co-tree branch is zero, the loop is traced in the direction of the co-tree branch. If, however,

it is at maximum level, the loop is traced in the opposite direction. If all costs are positive without artificial branches, the solution is optimal.

2. For the loop "i" corresponding to the minimum cost:

(a) consider first only branches with direction opposite to the loop direction and find

$$\theta_1 = \min x_j$$

(b) consider next only branches with the same direction to the loop direction and find

$$\theta_2 = \min(u_k - x_k)$$

(c) find

$$\theta = \min(\theta_1, \theta_2, u_i)$$

where u_i = upper limit of co-tree branch i
u_k = upper limit of tree branch k

3. Increase flows (in the sense of loop direction) in all branches by an amount = θ

(a) If $\theta = \theta_1$ (i) branch x_i enters the basis
(ii) branch x_j leaves the basis

(b) If $\theta = \theta_2$ (i) branch x_i enters the basis
(ii) branch x_j leaves the basis (at its upper limit)

(c) If $\theta = u$ (i) branch x_i remains a co-tree branch (at its upper limit)

Repeat steps 1, 2, 3 until optimal solution is obtained.

Optimality condition 2(a) ensures that all variables $x \geq 0$, and corresponds to the trans-shipment problem.

Optimality condition 2(b) ensures that all variables $x \leq u$ (except for the co-tree branch i which is included in 2(c).

Optimality condition 2(c) ensures that

$$0 \leq x \leq u$$

The tracing of a loop in the opposite direction to the direction of the co-tree branch, when the co-tree branch is at its upper bound, ensures that the co-tree branch level can only be decreased.

Flow diagram of steps of solution

The steps of solution can be summarised as a flow diagram.

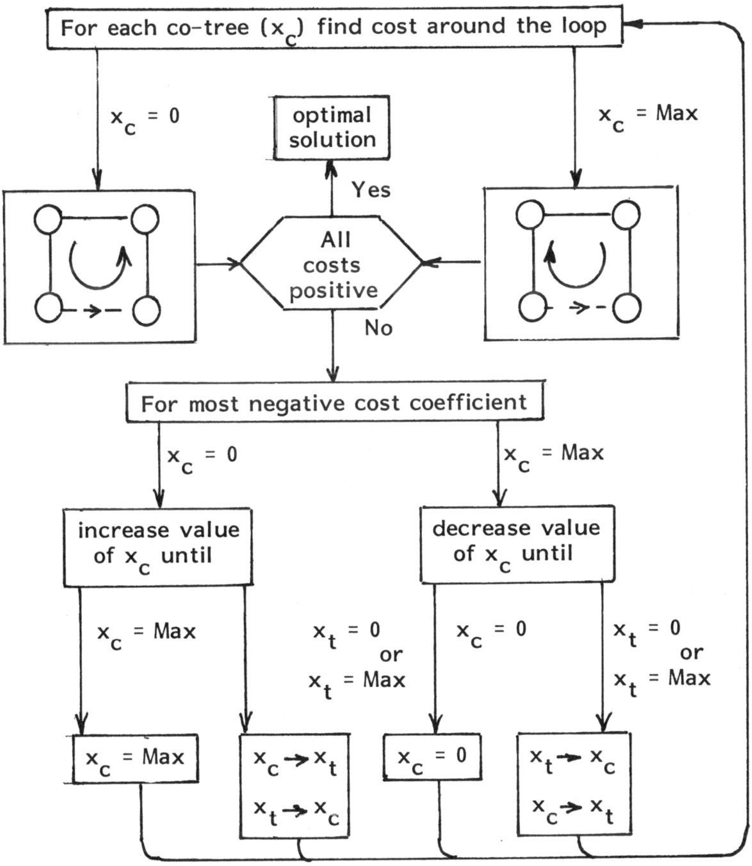

x_c = variable in co-tree list

x_t = variable in tree list

x = 0, variable at zero level

x = Max, variable at maximum level

Post-optimisation

In many engineering applications, cases may arise where slight alterations are introduced on an already solved trans-shipment problem, and the solution of the resulting problem is required. An example arises when it is required to establish the sensitivity of the optimal solution to minor variations in the problem parameters or when a series of closely related problems are generated, as may be the case in a branch and bound method. In such cases it is much more efficient to post-optimise the available solution rather than solve the whole problem from the beginning.

The network approach to optimisation of transportation problems enables post-optimisation to be carried out easily and efficiently.

Changing upper limits

When the upper limits are changed, one of two cases of interest may arise. The first is when the flow in a tree or co-tree branch exceeds the new upper limit. In this case the flow in the branch under consideration is decreased to correspond to the upper limit, and the surplus is carried by an artificial branch, added to the co-tree list, in parallel with the original branch and in the same direction. The second case is when the flow in a co-tree branch is less than the new upper limit. Here the flow in the branch is increased to correspond to the new limit, and the slack is taken by an artificial branch, added to the co-tree list, in parallel with the original branch in the opposite direction.

Changing the requirement vector

The change in the requirement vector at each node is assigned to the artificial branch leading from the reference to the node concerned in such a direction as to balance the change. If the artificial branch was in the tree at zero level it remains in the tree but at positive level. If it was in the co-tree it remains so but its upper limit is changed to correspond to the flow assigned to it.

9 METHODS FOR NETWORK CHANGES AND COMPENSATED NETWORK SOLUTIONS

A. Brameller
Power System Laboratory, University of Manchester Institute of Science & Technology, PO Box 88, Manchester M60 1QD, England

Basic Solution Method

Sometimes it is desirable to obtain a solution to a particular network and then to change the value of a branch admittance and find a new solution. If the inverse of the admittance matrix is known, the new solution may be easily obtained by modifying it using the Woodbury formula. Such a formula can be simply obtained from diakoptical consideration.

When sparsity techniques are used, the inverse of the admittance matrix is not explicitly available, although any one column can be easily obtained from the solution of the equations with the right-hand vector set to zero except for unity in the position corresponding to the desired column of the inverse matrix. From this consideration, a new solution may be obtained from the knowledge of previous solution without the availability of the full inverse matrix as follows:

Assuming that V_0 is known from the solution of the original equations:

$$Y_0 V_0 = I_0$$

where Y_0 represents the original admittance matrix, and that it is required to find new value V_1 after one or more branches in the network have been modified.

Consider the network shown in Figure 1, where only one branch is modified by δy_1 from $\frac{1}{y_0} = z_0$ to $\frac{1}{y_1} = 0$

Figure 1 - Network changes

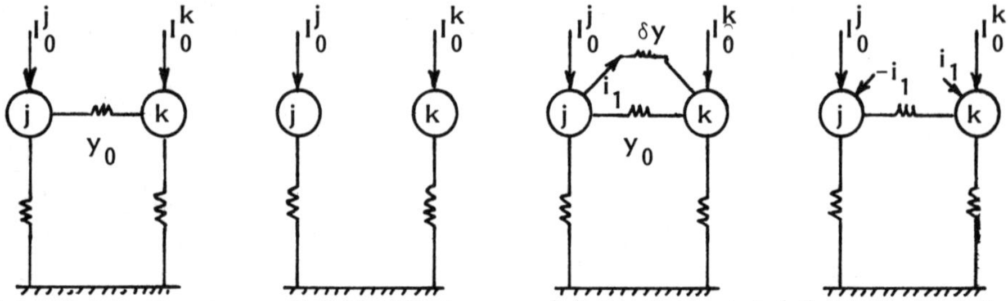

(a) Original $Y_0V_0=I_0$ (b) Modified $Y_1V_1=I_0$ (c) Equivalent 1 $(Y_0+Y_1)V_1=I_0$ (d) Equivalent 2 $Y_0V_1=I_0+I_1$

From Figure 1(c)

$$\delta y_1 = -y_0$$

$$i_1 = \delta y_1(v_1^i - v_1^k)$$

or in matrix form $i_1 = \delta y_1 \; C_1^t \; V_1$ (1)

where $C_1^t = \begin{bmatrix} & & & & 1 & & -1 \end{bmatrix}$ (entries 1 in column j, −1 in column k)

From Figure 1(d)

$$I_1 = -C_1 \; i_1 \tag{2}$$

(1) → (2) $I_1 = -C_1 \; \delta y_1 \; C_1^t \; V_1$ (3)

Also from Figure 1(d)

$$Y_0V_1 = I_0 + I_1$$

$$V_1 = Z_0 \; I_0 + Z_0 \; I_1 \tag{4}$$

where $Z_0 = Y_0^{-1}$

From Figure 1(a)

$$V_0 = Z_0 I_0 \tag{5}$$

(5) → (4) and (3) → (4)

$$V_1 = V_0 - Z_0 C_1 \delta y_1 C_1^t V_1 \tag{6}$$

Multiplying by C_1^t

$$C_1^t V_1 = C_1^t V_0 - C_1^t Z_0 C_1 \delta y_1 C_1^t V_1$$

$$(1 + C_1^t Z_0 C_1 \delta y_1) C_1^t V_1 = C_1^t V_0 \tag{7}$$

$$(\delta z_1 + C_1^t Z_0 C_1) \delta y_1 C_1^t V_1 = C_1^t V_0 \quad \text{where } \delta z_1 = (\delta y_1)^{-1}$$

let $$d_1 = (\delta z_1 + C_1^t Z_0 C_1)^{-1} \tag{8}$$

∴ $$\delta y_1 C_1^t V_1 = d_1 C_1^t V_0 \tag{9}$$

(7) → (6) $$\boxed{V_1 = V_0 - Z_0 C_1 d_1 C_1^t V_0} \tag{10}$$

<u>Solution steps for a single change</u>

Let $W = Z_0 C_1$ (dif. of column i & j of Z_0)

$d_1 = (\delta z_1 + C_1^t W)^{-1}$ ($C_1^t W$ = diff. of elements i & j of W)

$b_1 = d_1 C_1^t V_0$ ($C_1^t V_0$ = dif. of elements i & j of V_0)

$$\boxed{W = Z_0 C_1}$$

↓

$$\boxed{d_1 = (\delta z_1 + C_1^t W)^{-1}} \quad \text{(equation 8)}$$

↓

$$\boxed{b_1 = d_1 C_1^t V_0}$$

↓

$$\boxed{V_1 = V_0 - W b_1} \quad \text{(equation 10)}$$

Note 1

$$d_1 = (\delta z_1 + C_1^t Z_0 C_1)^{-1} = (\delta z_1 + z_{ii} + z_{kk} - z_{ik} - z_{ki})^{-1}$$

z_{ik} = i, k element of Z_0

Note 2

$$V_1 = Z_1 I_0 = Z_0 I_0 - Z_0 C_1 d_1 C_1^t Z_0 I_0$$

$$\therefore \quad \boxed{Z_1 = Z_0 - Z_0 C_1 d_1 C_1^t Z_0}$$

for a single change $\quad Z_0 C_1 = Z_0^i - Z_0^k$ = dif. of column i and i of Z_0

$$\therefore \quad \boxed{Z_1 = Z_0 - (Z_0^i - Z_0^k)\, d_1\, (Z_0^i - Z_0^k)^t}$$

Note 3

Equation (10) can be expressed in the classical diakoptical form as:

$$\boxed{V_1 = (Y_0^{-1} - Y_0^{-1} C_1 d_1 C_1^t Y_0^{-1})\, I}$$

Note 4

Z_1 may be obtained in the following steps:

$$\left.\begin{aligned} W &= Z_0 C_1 \\ d_1 &= (\delta z_1 + C_1^t W)^{-1} \end{aligned}\right\} \quad \text{as for solution}$$

$$b_1 = d_1 C_1^t Z_0 \qquad \text{(for symmetrical } Z_0,\ C_1^t Z_0 = W^t)$$

$$Z_1 = Z_0 - W b_1$$

Example on branch removal

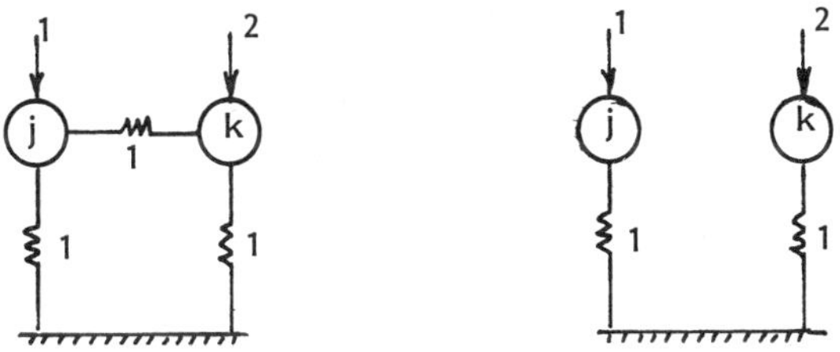

Original Modified

$$Y_0 = \begin{bmatrix} 2 & -1 \\ -1 & 2 \end{bmatrix} \qquad Z_0 = \frac{1}{3}\begin{bmatrix} 2 & 1 \\ 1 & 2 \end{bmatrix} \qquad V_0 = Z_0 I_0 = \frac{1}{3}\begin{bmatrix} 4 \\ 5 \end{bmatrix}$$

$$\delta y_1 = -1 \qquad C_1^t = \begin{array}{cc} i & k \\ \end{array} \begin{bmatrix} 1 & -1 \end{bmatrix}$$

$$d_1 = (-1 + \frac{2}{3} + \frac{2}{3} - \frac{1}{3} - \frac{1}{3})^{-1} = -3$$

$$C_1^t V_0 = \frac{1}{3}(4 - 5) = -\frac{1}{3}$$

$$Z_0 C_1 = \frac{1}{3}\begin{bmatrix} 1 \\ -1 \end{bmatrix}$$

$$V_1 = V_0 - (Z_0 C_1)\, d_1\, (C_1^t V_0)$$

$$= \frac{1}{3}\begin{bmatrix} 4 \\ 5 \end{bmatrix} - \frac{1}{3}\begin{bmatrix} +1 \\ -1 \end{bmatrix} (-3)(-\frac{1}{3}) = \frac{1}{3}\begin{bmatrix} 4 \\ 5 \end{bmatrix} - \frac{1}{3}\begin{bmatrix} +1 \\ -1 \end{bmatrix} = \begin{bmatrix} 1 \\ 2 \end{bmatrix}$$

****

Simultaneous changes

Equation (10) is a general equation for any number of modifications or outages if the scalars are considered as matrices. When m branches are modified simultaneously, δy becomes an (mxm) diagonal matrix of the admittance changes, and matrix C_1 has m columns, each with entries

+1 and −1 in the relevant position (a shunt only has the +1 entry). For a double change

$$\delta Y = \begin{bmatrix} \delta y_1 & \\ & \delta y_2 \end{bmatrix} \qquad C^t = \begin{array}{ccccccccc} & i & & k & & p & & q & \\ \hline & 1 & & -1 & & & & & \\ & & & & & 1 & & -1 & \end{array}$$

<u>Singularity or islanding</u>

Singularity of equation (8)

$$d_1 = (\delta z_1 + C_1^t \, Z_0 \, C_1)^{-1}$$

indicates that the modification disconnects at least one node from the remaining nodes, including ground.

<u>Example on double fault and islanding</u>

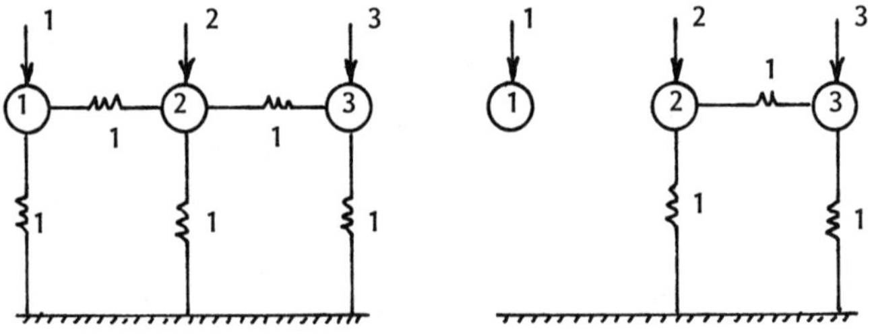

Original Modified

$$Y_0 = \begin{bmatrix} 2 & -1 & \\ -1 & 3 & -1 \\ & -1 & 2 \end{bmatrix} \qquad Z_0 = \frac{1}{8}\begin{bmatrix} 5 & 2 & 1 \\ 2 & 4 & 2 \\ 1 & 2 & 5 \end{bmatrix} \qquad V_0 = \frac{1}{2}\begin{bmatrix} 3 \\ 4 \\ 5 \end{bmatrix}$$

$$\delta y = \begin{bmatrix} -1 & \\ & -1 \end{bmatrix} \qquad C^t = \begin{array}{ccc} 1 & 2 & 3 \\ \hline 1 & -1 & \\ 1 & & \end{array} \qquad \delta z = \delta y$$

$$d = (\delta z + C^t \, Z_0 \, C)^{-1}$$

$$Z_0\,C = \frac{1}{8}\begin{bmatrix} 3 & 5 \\ -2 & 2 \\ -1 & 1 \end{bmatrix} \qquad C^t\,Z_0\,C = \frac{1}{8}\begin{bmatrix} 5 & 3 \\ 3 & 5 \end{bmatrix}$$

$$d = \begin{bmatrix} -1 & \\ & -1 \end{bmatrix} + \frac{1}{8}\begin{bmatrix} 5 & 3 \\ 3 & 5 \end{bmatrix} = \frac{3}{8}\begin{bmatrix} -1 & 1 \\ 1 & -1 \end{bmatrix}$$

$$\det(d) = 0$$

****

Solution of the non-isolated part of the network may be obtained by placing a shunt branch of arbitrary size at one of the isolated groups of nodes.

Node merger

Modifications are easily applied when two or more nodes are short - circuited (merged). Although the order of the network matrix is reduced, by retaining the original order, a new solution can be obtained from standard equation 9, be it that some of the voltages will have the same value.

Node addition

The network modification process can be used to modify existing network solution as well as to network extension. This can be best illustrated from the following example.

Example 5 on node addition

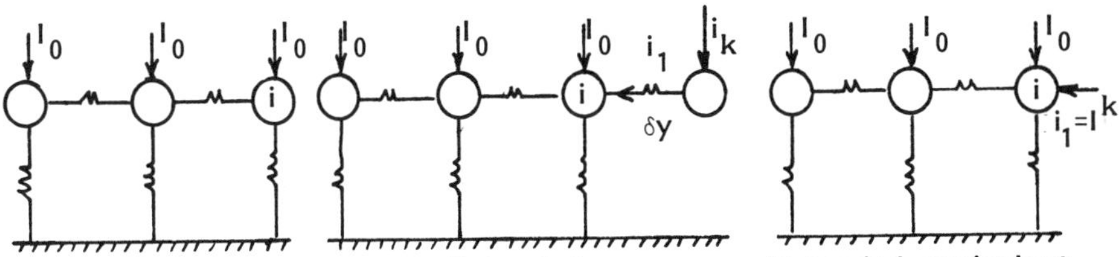

Original $Y_0 V_0 = I_0$ Extended $Y_1 V_1' = I_1$ Extended equivalent $Y_0 V_1 = I_0 + I_1$

$$i_1 = I_1^k$$

$$V_1 = Z_0\, I_0 + Z_0\, I_1$$

$$V_1 = V_0 + Z_0\, I_1$$

$$V_1 = V_0 + Z_0\, C_1\, i_1 \qquad \text{but } i_1 = I_k$$

$$V_1 = V_0 + Z_0\, C_1\, I^k \qquad \text{and } V_1^k = V_1^i + \delta y I_1^k$$

****

Asymmetry

Most electrical problems are symmetrical; therefore only the equivalent of the upper or lower triangular matrix needs to be calculated and stored. It is sometimes necessary to analyse problems which have asymmetrical coefficient matrix. Such problems arise, for example, when transformers with quadrature taps are represented. In general, problems such as asymmetry occur in only a few elements but necessitate the calculations and storage of the full coefficient matrix. If symmetry is to be preserved, the asymmetry can be transformed as an injected current and the problem solved iteratively.

A better alternative is to calculate the solution of a symmetrical problem first and represent the asymmetry as network modification.

Recursive (sequential) changes

Simultaneous multiple network changes can be evaluated as an extension of a single change if the scalars are considered as matrices. To avoid the inversion of a matrix of an order corresponding to the number of changes, the changes can be carried out one at a time in a sequential manner. The process is carried out in a similar manner as for a single fault.

Solution steps for recursive changes

The solution steps for a single and double change can be extended to a general multiple simultaneous change in a recursive manner as follows:

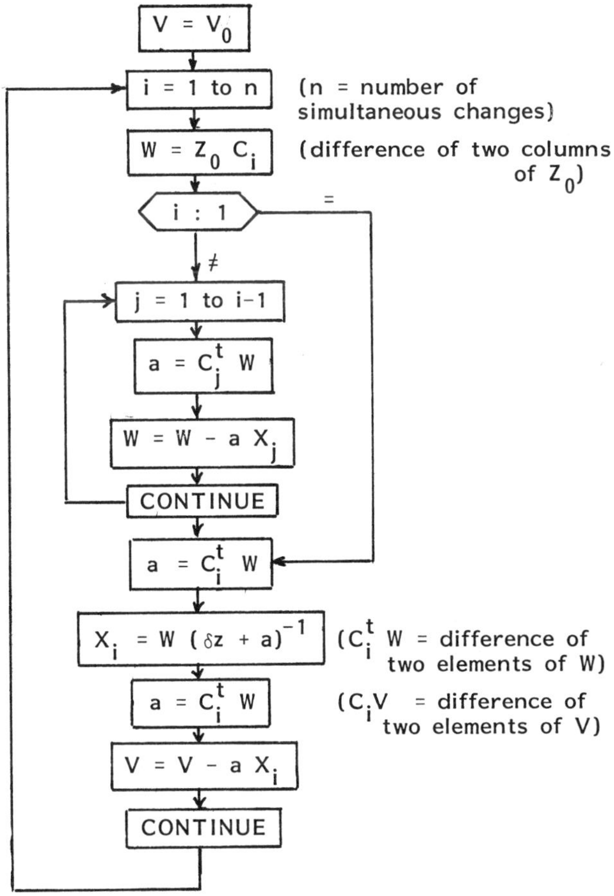

Note 1

$$\delta z_i = \begin{cases} -\dfrac{1}{y_{ij}}, & \text{Open circuit between nodes i and j} \\ +\dfrac{1}{y_{ij}}, & \text{New circuit between nodes i and j} \\ 0, & \text{Short circuit} \end{cases}$$

Note 2 X_i must be stored, W and a are working stores.

Application to non-linear elements

If the removed element between nodes i and j is linear, the current flowing in it is defined by:

$$i^1 = \delta y^1 \, C_1^t \, V^1 \qquad \text{(equation 1)}$$

where 1 indicates the modified solution or new values
and 0 " " original " " " "

If the removed element is non-linear, then the current flowing in it can be defined by the general relation:

$$i^1 = f(\delta y^1, V^1) \qquad (11)$$

From equation (4)

$$V^1 = Z^0 \, I^0 + Z^0 \, I^1$$

(5) → (4) and (2) → (4)

$$V^1 = V^0 - Z^0 \, C \, i^1 \qquad (12)$$

Writing explicitly the p-th and q-th equation:

$$\left.\begin{aligned} v_p^1 &= v_p^0 - (z_{pp}^0 - z_{pq}^0)\, i^1 \\ v_q^1 &= v_q^0 - (z_{qp}^0 - z_{qq}^0)\, i^1 \end{aligned}\right\}$$

The above equations in conjunction with equation (11) can now be solved as a set of two simultaneous equations in two variables (v_p^1, v_q^1) and hence for i^1 from equation (11). Finally, solution for V^1 can be obtained from equation (12)

$$V^1 = V^0 - Z^0 \, C_1 \, i^1$$

Application to non-linear optimisation

The solution to non-linear problems can be obtained from the unconstrained Lagrangian function:

$$L = f + \lambda^t G$$

The necessary conditions for a stationary point are:

$$\frac{\partial L}{\partial \lambda} = G = 0$$

$$\frac{\partial L}{\partial X} = \frac{\partial f}{\partial X} + \frac{\partial G^t}{\partial X} \lambda$$

$$\frac{\partial L}{\partial U} = \frac{\partial f}{\partial U} + \frac{\partial G^t}{\partial U} \lambda$$

The original solution can be obtained from the following iterative steps:

1. $U_r = U^0$

2. $G_r^0 = 0$ — load flow solution

3. $f_r = \begin{cases} f_r & \text{no violation on X} \\ f_r + p_r & \text{violation on X} \end{cases}$

4. $\lambda_r = -\left\{\frac{\partial G^0}{\partial X}\right\}^{-1} \frac{\partial f}{\partial X}$

5. $\nabla f_r = \frac{\partial f}{\partial U} + \frac{\partial G^t}{\partial U} \lambda$

6. $\nabla U_r = -c\nabla f$

7. $U_{r+1} = \begin{cases} U_r + \Delta U & \text{no violation on U} \\ U^{limit} & \text{violation on U} \end{cases}$

8. $\nabla f = \varepsilon$ → Convergence (otherwise return to step 2)

where p_r = penalty function

r = iterative step

L = Lagrangian function

f = objective function

G = constraints

X = dependent variables

U = independent variables

λ = Lagrangian multipliers

Outages or modification

A new solution for the non-linear problem with an outage or line change can be obtained from the original one as follows:

1. $U_r = U^0$

2. $G_r^1 = 0$ load flow solution with outage

No overloads on $|I|$ etc. (used for security)

3. $$f_r^1 = \begin{cases} f_r^1 & \text{no violation on } X \\ f_r^1 + p_r^1 & \text{violation on } X \end{cases}$$

4. $$\lambda_r^0 = -\left(\frac{\partial G^0}{\partial X}\right)^{-1} \frac{\partial f^1}{\partial X}$$

$$\lambda_r^1 = \lambda_r^0 - \left\{\left(\frac{\partial G^0}{\partial X}\right)_i^{-1} - \left(\frac{\partial G^0}{\partial X}\right)_j^{-1}\right\} b$$

5. $$\nabla f_r^1 = \frac{\partial f}{\partial U} + \left(\frac{\partial G^0}{\partial U}\right)^t \lambda^1$$

etc. as per original solution

Convergence

Note:

$$\lambda = -\left(\frac{\partial G}{\partial X}\right)^{-1} \frac{\partial f}{\partial X}$$

$$= ZI \qquad Z = -\frac{\partial G}{\partial X}^{-1} \qquad I = \frac{\partial f}{\partial X}$$

$$\lambda^m = \lambda^0 - (Z_i - Z_j)\, b$$

$G^m = 0$ can be solved without forming G^m

see application to load flows.

10 ALGEBRAIC MULTIGRID (AMG) FOR SPARSE MATRIX EQUATIONS

A. Brandt
Weizmann Institute of Science, Israel

S. McCormick
J. Ruge
Institute for Computational Studies, Colorado State University

This work was supported in part by the Air Force Office of Scientific research (Grant No. AFOSR82-0240) and the National Bureau of Standards.

ABSTRACT

Many matrix equations are either inherently discrete (e.g., in geodesy) or for certain practical purposes remote from their origin (e.g., a finite element discretization on a preselected irregular grid). AMG is an algorithm designed to solve such problems by using information contained only in the matrix while at the same time basing itself on multigrid principles. This paper introduces the basic AMG concepts, develops its foundations, and describes current AMG strategies.

1. INTRODUCTION

The class of problems to which multigrid methods have been applied successfully is constantly growing. So far, most of them have been derived from continuous problems, where knowledge of the underlying geometry is used as a guide to provide several uniform diescretizations of the domain. Each of these so-called grids or levels can be used as a uniform coarsening of the next finer one. The solution process, which involves relaxation sweeps on each grid, fine-grid-to-coarse-grid transfers of residuals, and coarse-to-fine interpolations of corrections, constitutes a very fast solver for the finest-grid equations. In fact, whether the equations are linear or not, a solution with algebraic error smaller than truncation error is typically obtained in four to ten *work units*, where a work unit is the amount of computer operations required to *express* the equations (e.g., the work of a matrix multiply for linear equations).

Although this process seems to rely heavily on the geometry and continuous nature of the problem, the principles involved in solving the fine grid matrix system can be abstracted and applied to various classes of matrix problems. The aim of this paper is to develop such an algorithm, algebraic multigrid (AMG), that results from basing multigrid concepts solely on information contained in the matrix. The potential benefits of AMG include:

1. 'Black box' multigrid software. The virtues here are obvious, especially in light of the present state of multigrid software and the substantial human design effort needed for many multigrid applications.
2. Applications to unorganized grids. When the given grid is not topologically piecewise uniform, as in the case of finite element discretizations with arbitrary, irregular triangulations, conventional multigrid design may be faced with difficulties. Determination of coarsening (i.e., the coarse grids and their associated operators) may not be practical. A further difficulty in such problems is to choose both relaxation and coarsening when the problem exhibits certain directional properties. This includes anisotropic operators, which require either some form of block relaxation or semi-coarsening. Especially severe cases arise when the computational mesh is highly stretched as a result of coordinate transformations or Lagrangian

discretizations. These difficulties disappear in the AMG approach.

3. Pathological coefficient applications. Even for uniform discretizations, conventional coarsening may not be able to account for certain pathologies in the equation coefficients. Such is the case, for example, with five-point discretizations of two- dimensional diffusion problems where the diffusion coefficients are distributed in some particular patterns (see [1; Sec. 8])]. Again, AMG would have no difficulty in such cases since it would select much better coarse grids than the 'natural' ones.

4. Algebraic problems. AMG may be applied to large sparse linear and nonlinear systems which are not derived from continuous problems, including the geodetic application treated in [3; Sect. 6] and [7]. Many of these problems are such that each unknown is associated with a point in a low dimensional space (e.g., dimension 2 or 3) and that most of the points exhibit pairwise couplings (i.e., matrix entries) that either diminish or become smooth as the distances between the points increase. It is this feature of an algebraic problem that should allow for efficient solution by AMG.

Multilevel processing for the solution of problems not derived from continuous, nor even geometrically-based, systems has for a long time been used in economics (cf., [4] and the survey in [6]). In a sense, nobody 'invented' it; multilevel organization is simply the way society itself found efficient. This concept led to iterative aggregation algorithms (cf., [6] and [10]) in which the coarsening (aggregation) procedures are supposedly given by the nature of the problem just as they are in usual multigrid algorithms and other geometrically-based multilevel processing. The main point of AMG, by contrast, is the complete automation of the coarsening procedures, including the selection of the coarser levels and the inter-level transfers, basing them on the given matrix only. The efficiency obtained by AMG is similar to the typical multigrid efficiency, which is not approached by the less sophisticated aggregation methods.

The AMG structure provides more than a fast linear solver. It can, for example, be used to solve with similar speed nonlinear problems, to update for negligible work the entire solution as the problem changes locally, to solve large problems in small storage and to give inexpensive representation and computation of the inverse of large sparse matrices. (See [3; Secs. 6.7 and 7].)

This paper is intended first to provide a *brief* description of AMG including its basic concepts and foundation. We further treat *in brief* several important issues and AMG strategies that have arisen since our first report [3]. However, even as we write this account, new attempts are being made at fundamental improvements in these strategies.

Brief has special significance here. During the course of our work on AMG an overwhelming collection of observations, concepts, and algorithmic variations suggested themselves. By necessity, we used experience, analysis, and some numerical experiments in our pursuit of efficient and robust AMG techniques. Many alternatives to the AMG processes described below are easy to conceive, but their abundance prevents us from discussing them here.

This paper is organized as follows: Section 2 describes the algebraic properties of the iteration error that characterize slow convergence of relaxation methods; Section 3 motivates the basic principles of AMG; Section 4 describes the basic AMG processes; Section 5 includes a note on numerical performance; Section 6 describes a few major areas of further research; and the Appendix contains a somewhat detailed description of the current AMG coarsening strategy.

2. PROPERTIES OF SLOWLY CONVERGING ERRORS

2.1 Algebraic property: small residuals

It is first important to understand, in general algebraic terms, what type of errors become dominant when basic iterative schemes are slow to converge. A common feature of those iterative schemes is that at each step some correction to the approximate solution is calculated based on the current magnitute of certain residuals. Hence, the convergence must be slow if the residuals are in some sense small compared with the errors. The converse is also true: if convergence of a *proper* relaxation scheme is slow, the residuals must in some sense be small compared with the errors.

To see this, consider for example Kaczmarz relaxation applied to a general real or complex matrix equation

(2.1) $AX=b$.

We suppose the current approximation x is converging to a solution X. (To treat this most general case, where A need not even be square, we consider Kaczmarz method which converges whenever (2.1) is solvable [9].) Denoting by $e = x-X$ the current error vector, by $r = Ae = Ax-b$ the current residual vector, and by a_i the i-th row of A, then the Kaczmarz step corresponding to that row is to replace x by $x - (r_i/a_i a_i^t)a_i^t$, thereby forcing r_i to zero. A full Kaczmarz sweep is the employment of such a step for each row of A, in the natural ordering. Let

$$\text{(2.2)} \quad \tilde{E} = \Sigma |e_i|^2 \text{ and } \tilde{R} = \sum_i \left(|r_i|^2 / \sum_j |a_{ij}|^2 \right)$$

be square norms for errors and residuals, respectively, scaled as is evident so that they are comparable. One can then prove the following result [2; Thm. 3.4]:

Theorem 2.1. Let $\tilde{R}^0$ and $\tilde{R}^1$ denote the values of $\tilde{R}$ before and after one sweep of Kaczmarz relaxation, respectively, and let

$$\tilde{\gamma}_0 = [(1+\tilde{\gamma}_-)(1+\tilde{\gamma}_+)]^{-1},$$

$$\tilde{\gamma}_1 = [\tilde{\gamma}_-\tilde{\gamma}_+]^{-1},$$

$$\tilde{\gamma}_- = \max_i \left(\Sigma \Big| \sum_{j<i\lambda} a_{i\lambda} \bar{a}_{j\lambda} \Big| \right) / \sum_\lambda |a_{i\lambda}|^2, \text{ and}$$

$$\tilde{\gamma}_+ = \max_i \left(\Sigma \Big| \sum_{j<i} a_{i\lambda}\bar{a}_{j\lambda} \Big| / \sum_\lambda | a_{i\lambda}|^2 \right.$$

Then the decrease in $\tilde{E}$ over that sweep is at least $\tilde{\gamma}_0\tilde{R}^0$; it is also at least $\tilde{\gamma}_1\tilde{R}^1$.

This theorem says in essence that *slow convergence can occur when $\tilde{R}$ is small compared with $\tilde{E}$.* (Observe than in case A arises from the discretization of differential equations, or is similarly structured, $\tilde{\gamma}_i$ are completely local quantities; they do not depend on the size of A.)

For special classes of matrices A, simpler (and more effective) relaxation schemes can achieve the same goal. For example, for symmetric positive-definite matrices, an effective method is *Gauss-Seidel relaxation*, where now the step corresponding to the i-th equation is to replace x_i by $x_i - r_i/a_{ii}$, leaving the rest of x unchanged, again forcing r_i to zero. The error and residual square norms in this case are

$$\text{(2.3)} \qquad E = e^t r \qquad \text{and } R = \Sigma\, r_i^2 / |a_{ii}|,$$

for which we have the following result [2; Thms 3.1 and 3.2]:

Theorem 2.2. Suppose A is symmetric with positive diagonal. Let R^0 and R^1 denote the values of R before and after one Gauss-Seidel sweep and let

$$\gamma_0 = [(1+\gamma_-)(1+\gamma_+)]^{-1},$$

$$\gamma_1 = [\gamma_-\gamma_+]^{-1},$$

$$\gamma_- = \max_i \sum_{j<i} |a_{ij}| / a_{ii}, \text{ and}$$

$$\gamma_+ = \max_i \sum_{j>i} |a_{ij}| / a_{ii}$$

Then the decrease in E over that sweep is at least $\gamma_0 R^0$; it is also at least $\gamma_1 R^1$.

The constants γ_i appearing in this theorem are sharp in many cases, while in others they are not far off and optimal constants can be derived by local mode analyses. Similar results hold for all other familiar relaxation schemes, such as Jacobi relaxation with a suitable under-relaxation parameter, block relaxation schemes (with a suitable change in the form of R), etc. (See [2;Sec. 3].)

2.2 Geometric and algebraic smoothness

When (2.1) represents a discretization of a differential equation LU = f, then small residuals r = Ae imply that e is locally an approximate solution to LE = 0. When L is elliptic, this implies that e is a smooth function. (A specialized sense of smoothness is implied when L is not elliptic.) This smoothness is the basis for the usual 'geometric' multigrid algorithms in which the smooth error functions produced by relaxation are subsequently approximated on coarser grids. To develop such a foundation for AMG, we must free the concept of smoothness from its geometric context.

Since E and R (or $\tilde{E}$ and $\tilde{R}$) were scaled to be comparable, it is only for special error components that the slow-convergence condition

(2.4) $$R \ll E \qquad (\text{or } \tilde{R} \ll \tilde{E})$$

can be satisfied. The deeper (2.4) is satisfied, the more special must be the error, and hence the fewer is the number of parameters needed to approximate it. Thus, broadly speaking, relaxation reduces the information content of the error and makes it approximateable by lower dimensional spaces ('coarser grids').

To see more specifically how these lower dimensional spaces should be constructed, consider first the special class of symmetric *positive-type* matrices, which form the prototype for our initial development of AMG. We call the matrix $A = (a_{ij})$ positive-type if $a_{ij} \le 0$ for all $i \ne j$, whereas, $\sum_j a_{ij} \ge 0$ for all i. (This is equivalent to the assumption that A is a diagonally dominant M - matrix.) In this case,

$$E = 1/2 \sum_{i,j} (-a_{ij})(e_i - e_j)^2 + \sum_i (\sum_j a_{ij}) \; e_i^2$$
$$1/2 \sum_{i,j} (-a_{ij})(e_i - e_j)^2.$$

By the Cauchy-Schwarz inequality, $E^2 \le R \sum a_{ii} e_i^2$,

so the slow-convergence condition (2.4) implies

(2.5) $$\sum_{i,j} (-a_{ij})(e_i - e_j)^2 \ll \sum_i a_{ii} e_i^2.$$

Since $a_{ij} \le 0$ for $i \ne j$, this means that $e_i - e_j$ is on the average small compared with $\|e\|$ whenever a_{ij} is comparable with either $\max_{k \ne i} |a_{ik}|$ or $\max_{k \ne i} |a_{jk}|$, i.e., whenever i and j are *strongly coupled*. Thus, we conclude that Gauss-Seidel relaxation (and similarly other schemes) *smoothes the error along strongest couplings* (and couplings comparable to the strongest).

Block relaxation can give stronger smoothing. By including the strongest couplings in the blocks, the error will also be smoothed along next-strongest couplings [2;Sec.3.5]. In our AMG development, however, block relaxation is not usually needed; smoothing along strongest couplings is enough because the coarser grid can be selected so that interpolation is always done along such couplings. (Similarly, in geometric multigrid, full smoothing, e.g., by line relaxation in case of anisotropic operators, is not needed if semi-coarsening is used; see [11;Sec. 4.2.1].)

The concept of smoothing along strong couplings can be generalized to *sparsely positive-definite* matrices (see [2]), and perhaps to most other matrix types of interest. In practice, the sense in which a given relaxation scheme smoothes for a given matrix A can algorithmically be detected by techniques described in Section 6.1 below. How that smoothing is exploited by AMG will be described in subsequent sections.

3. MULTIGRID PRINCIPLES

3.1 Two-level schemes

As we noted above, smooth errors can be well represented by fewer variables. To see how this can be done, we first introduce the notion of a two-level multigrid scheme. To this end, let G be a given ('fine') grid, that is, a finite set of points in R^d, where d is some positive integer. Suppose that a ('coarse') grid G^c is given that represents some 'coarsening' of G (e.g., G^c may be a subset of G) and let I_c: $G^c \to G$ ('interpolation') and I^c:$G \to G^c$ ('restriction') denote the intergrid transfers. Let A: $G \to G$ and A^c : $G^c \to G^c$ be linear operators. (We identify each grid with its associated space of grid functions.) Suppose b in G is given and let X be a solution of the fine grid equation

(3.1) $AX = b.$

With x the current approximation to X, then one multigrid cycle consists of applying, say, several Gauss-Seidel sweeps followed by a residual equation transfer to G^c via

(3.2) $A^cV^c = I^c(b - Ax).$

This problem is then solved (with present *two*-level case) and used to correct the fine grid approximation according to

(3.3) $x \leftarrow x + I_cV^c.$

The reason for the efficiency of such a multigrid scheme is that, with proper choice of the grid transfer and coarse grid operators, the error that cannot be eliminated by the coarse grid correction (3.4) is effectively reduced by relaxation. To ensure that the coarse grid correction actually does what it is intended to do, it is necessary to define A^c in the right way. Note that (3.2) and (3.3) imply that the coarse grid correction is given by

$$(3.4) \qquad x \leftarrow x + I_c {A^c}^{-1} I^c(b - Ax).$$

Letting $e = x - X$ denote the error, then (3.4) can be rewritten as

$$(3.5) \qquad e \leftarrow (I - I_c {A^c}^{-1} I^c A)e.$$

Hence, the goal of the correction is roughly to eliminate error in $R(I_c)$, the range of the interpolation operator. Thus, if $e = I_c v^c$ for some v^c, it is desirable that

$$(I - I_c {A^c}^{-1} I^c A)\, I_c v^c = 0$$

or, since I_c should be full rank, that

$$(3.6a) \qquad A^c = I^c A I_c.$$

Note that, with this choice if A^c, the coarse grid correction becomes a projection.

For the special case that A is a real symmetric positive definite matrix, we can derive this same form for A^c together with a prescription for I^c by way of a variational condition for the energy norm $\|e\|_E = e^t Ae$. Specifically, by seeking a vector v^c in G^c that minimizes $\| e - I_c v^c \|_E$, we are led to (3.6) with

$$(3.6b) \qquad I^c = I_c^{\,t}.$$

These 'variational relations' (3.6) are in fact used in AMG for such matrices to form I^c and A^c once I_c has been determined.

In many applications for geometric multigrid algorithms, the coarse grid approximates smooth fine grid vectors in a fairly strong sense. Specifically, suppose E and R represent the error measures in either (2.2) or (2.3), which we use interchangeable with $\| e \|_E^2$ and $\| e \|_R^2$, respectively, when we need to depict the error e on which they depend. Then we are often assured that, for any e in G, there exists an e^c in G^c so that

(3.7) $\| e - I_c e^c \|_E \leq \gamma \| e \|_R$.

(Here, as in Section 2, we use γ generically to denote quantities that depend only locally on A.) Let e_1, e_2, and e_3, respectively, be the errors in such a multigrid application just after the previous coarse grid correction, after relaxation has slowed, and after the subsequent coarse grid correction. Then the e_2^c for e_2 guaranteed by this approximation condition, together with the variational property of the coarse grid correction, the slow-convergence relation (2.4), and the fact that E decreases in relaxation, all imply that

$$\| e_3 \|_E \leq \| e_2 - I_c e_2^c \|_E \leq \gamma \| e_2 \|_R \ll \| e_2 \|_E \leq \| e_1 \|_E.$$

Thus, E must be substantially reduced in each full cycle.

Efficient two-level algorithms can in fact be obtained under much weaker conditions. Thus, in terms of the norm $\| e \|_W = \Sigma a_{ii} e_i^2$, it is enough that, for any e in G, there exists an e^c in G^c satisfying

(3.8) $\| e - I_c e^c \|_W \leq \gamma \| e \|_E$.

(See [2; Thm. 4.1].) Such a condition, which is weaker than (3.7), can be guaranteed by interpolation along strong connections. For example, in our prototype symmetric positive-type case, interpolation can be (and is) taken to be regular averaging, i.e.,

(3.9) $I_c = (w_{ik}), \quad w_{ik} \geq 0, \quad \sum_k w_{ik} = 1,$

in which case (3.8) is satisfied if, for every e in G, there exists an e^c in G^c so that

$$(3.10) \qquad \sum_{i,k} a_{ii} w_{ik} (e_i - e_k{}^c)^2 \leq \frac{\gamma}{2} \sum_{i,j} (-a_{ij})(e_i - e_j)^2 .$$

A simple way to satisfy (3.10) is to regard G^c as a subset of G and let the interpolation coefficient w_{ik} be proportional to the strength of connection a_{ik}. More precisely, let *C* denote the indices that correspond to G^c in G and let *F* denote the other indices. Then $v = I_c v^c$ is defined so that

$$(3.11) \qquad v_i = \begin{cases} v_i^c & i \text{ in } C \\ \dfrac{\sum_{j \in C} (-a_{ij}) v_j^c}{\sum_{j \in F} (-a_{ij})} & i \text{ in } F \end{cases}$$

Then it is easy to see [2; Sec. 4.3] that (3.10) is satisfied by

$$(3.12) \qquad \gamma = 1/2 \min_{i \in C} \sum_{j \in F} (-a_{ij}) / a_{ii} .$$

The important conclusion here is that, if there is sufficient total *direct* strength to *C* points for every *F* point, then by (2.3) we are guaranteed that a full coarse-grid correction followed by one relaxation sweep must reduce the error measure E by a substantial factor independent of the matrix size. Actually, for satisfying (3.10), we do not need to be so restrictive as to guarantee that the strength to the *C* points is *direct*, nor do we need to interpolate from all available *C* points; but we defer these issues to Section 4.

3.2 Multi-level scheme

The two-level algorithm described in the previous section in effect reduces the problem of solving an equation in A to that in A^c. To solve the latter effectively, we must continue to process to yet coarser levels. Thus, we define a multi-level (multigrid) algorithm by applying the two-level algorithm recursively.

In particular, the *set-up* process for a multi-level algorithm starts with $G^0 = G$ and $A^0 = A$ and determines G^c, A^c, I_c, and I^c from G and A according to the principles of Section 3.1. Let $G^1 = G^c, A^1 = A^c$, $I_1 = I_c$, and $I^1 = I^c$. Then this coarsening process continues by determining G^i, A^i, I_i, and I^i from G^{i-1} and A^{i-1} for each i = 2, 3, ..., M in turn according to the same procedures. Thus, A^i is the grid G^i operator and $I_i:G^i \to G^{i-1}$ and $I^i:G^{i-1} \to G^i$ are the intergrid transfers.

To describe the multi-level *solution* process, let the number ν of relaxation sweeps per grid be specified. Given f^i on grid G^i and an approximation, x^i, on G^i to the solution of

$$A^i X^i = f^i, \tag{3.13}$$

then one multi-level V-cycle on grid G^i is denoted by $x^i \leftarrow MGV(i,x^i,f^i)$ and is defined recursively by:

If $i = M$, set $x^i \leftarrow {A^i}^{-1} f^i$.

Otherwise, do the following:

perform ν Gauss-Seidel sweeps on (3.13);

$x^{i+1} \leftarrow 0$ and $f^{i+1} \leftarrow I^{i+1}(f^i - A^i x^i)$;

$x^{i+1} \leftarrow MGV(i+1, x^{i+1}, f^{i+1})$; and

$x^i \leftarrow x^i + I_{i+1} x^{i+1}$.

4. AMG FOR SYMMETRIC POSITIVE-TYPE MATRICES

The two primary tasks in developing an AMG solver for a given symmetric matrix A are the determination of *C*, the coarse grid variables, and the computation of I_c, the interpolation formula. The fine-grid-to-coarse-grid transfer I^c and the coarse grid operator A^c are then determined by the variational relations (3.6). Other AMG processes, such as the relaxation ordering (*C-F*) and cycling schemes, are usually straightforward (cf.[3]). In this section, we discuss issues that relate to these two primary tasks for symmetric positive-type matrices.

4.1 Prototype AMG and its limitations

For each i in *F*, let $S^I(i)$ denote the set of points in *C* that are used in the interpolation formula for i, that is, $S^I(i) = \{k \text{ in } C : w_{ik} \neq 0\}$. As we noted at the end of Section 3.1 for our prototype matrices (i.e., symmetric positive-type), to develop an AMG with good two-grid convergence rates, it is enough to ensure that every point i in *F* has sufficient total *direct* strength to $S^I(i)$. In our early efforts, this premise was used to develop a prototype AMG which was based on ensuring that each i in *F* had at least a certain fraction of its total direct strength to $S^I(i)$.

A major difficulty with this prototype AMG is its tendency to exhibit fairly high *total relative complexity* , where by this we mean the total number of non-zero entries of all grid matrices divided by the number of nonzeros in A. This is because this complexity depends very sensitively on the number of connections used in interpolation - and this is large when AMG is forced to achieve sufficient direct strength.

A second difficulty with our prototype is that it can in fact exhibit slow V-cycle convergence. That is, even though it achieves the approximation conditions in (3.8) that guarantee acceptable two-grid convergence rates, the stronger conditions (3.7) needed for V-cycle convergence may be violated. Simple geometric examples in [2; Sec. 4.7] show in fact that this prototype AMG can produce low-order interpolation that results in slow V-cycle rates.

4.2 Indirect strength

Condition (3.10) is easily satisfied by taking sufficiently strong direct connections into account in interpolation, but it can also be satisfied through indirect connections. For example, certainly

$$w_{ik}\,(e_i - e_k^c)^2 \leq 2\,w_{ik}\,(e_i - e_j)^2 + 2\,w_{ik}\,(e_j - e_k^c)^2.$$

Suppose point i strongly depends on point j which in turn strongly depends on point k; we say that point i is strongly but indirectly coupled to point k. Such indirect strength can thus be used to satisfy (3.10). This means that, even when large neighborhoods of strong couplings exist, suitable approximation can still be achieved provided the neighbors of the points in $S^I(i)$ largely overlap the neighbors of i. (See [2; Sec.4.4].)

To take indirect strength into account, we first modified our prototype algorithm, using what we call [3] 'p-point interpolation'.

More specifically, we adopted the new objective of requiring essentially that each point i in F have its p strongest connections to $S^I(i)$, p a small fixed integer. This limited the connections to $S^I(i)$ so that complexity was better controlled (cf.[3] and [8]), but it did not go far enough to ensure total strength (direct and indirect) to $S^I(i)$. Consider the geometric example of a fully uniform five-point discretization of the LaPlacian in two dimensions with p = 2. There is no means for AMG as such to decide which of the four neighbors of a given i to place in $S^I(i)$, yet only by interpolating from points on two opposing sides of i can linear interpolation be obtained. Such 'two-sided' interpolation is the only way here to achieve the stronger approximation conditions (3.7).

Our *current* version of AMG places no artificial bound on the number of points in $S^I(i)$, but rather attempts to make $S^I(i)$ as small as possible while ensuring that the total direct and indirect strength is sufficient. Specifically, the objectives of the point choice strategy for our current algorithm are:

1) For every point i in F, each point j on which it strongly depends most either be in $S^I(i)$ or must itself depend strongly on a point in $S^I(i)$.
2) Total relative complexity should be minimized, that is, there should be as few C points as possible and $S^I(i)$ should be as small as possible.
3) The set-up processes should be fast. Its speed is limited by the time, t, it takes to compute A^c once G^c and Ic have been defined, so the goal here is to keep the set-up time at about 2t. (t is determined by how well Objective 2 is

achieved and the speed of the solution process depends on this and the nature of the approximation by the coarse grid.)

Note that Objective 1 overcomes some of the difficulty with the previous algorithms of achieving the stronger approximation conditions (3.7). For instance, for the five-point example of the previous paragraph, our current strategy produces a red-black coarsening so that all neighbors of each point are in $S^I(i)$.

In order to achieve speed, a quick first approximation to the coarse grid is obtained with Algorithm A, a brief description of which is included in the Appendix. This distributes the C points over the grid in such a way that few C-point-to-C-point strong connections exist and the F points have connections to at least one (though usually more) C points. For geometric operators with fairly uniform strong connections, this tends to give a uniform distribution of C points suitable for interpolation. The C point density is controlled by a parameter ε_1.

Of course, this algorithm does not guarantee that Objective 1 is achieved. However, this may be tested and corrected at the same time the interpolation coefficients are being computed. This is done in Algorithm B, which is also described in the Appendix. Here, an F point is interpolated from a subset of the strong connections, as defined by another parameter, ε_2 (usually $\varepsilon_2 < \varepsilon_1$), and points not used in interpolation must be strongly connected to $S^I(i)$. There are several ways that this strength can be measured, but in any case, a third parameter, ε_3, is used to control it. If, while computing the coefficients for a point i in F, a point j also in F is found not to be strongly connected to $S^I(i)$, j is itself made a 'conditional' *C* point. If, after making j such a point, there is yet another strong connection of i which is not a *C* point and is not strongly connected to the new set $S^I(i)$ (including j), then i itself is made a *C* point and j regains its status as an *F* point. (It is not worth the added complexity to keep i an *F* point if it requires more than one additional *C* point to do it.)

4.3 Improved interpolation formulae

This revised coarse grid point choice strategy described in the previous section lends itself to an improvement in the interpolation formula. To be specific, suppose i is in *F*. Then the motivation for the operator interpolation depicted in (3.11) starts with the assumption for relaxed errors that $r_i = 0$ (approximately). Thus, for such errors,

$$(4.1) \quad e_i = \sum_{k \neq i} (-a_{ik})\, e_k \qquad \text{(approximately).}$$

The points not in $S^I(i)$ in this summation are then eliminated by using the smoothness condition which, for our prototype matrices, is written as

$$(4.2) \quad \sum_{j \in F} a_{ij}\, e_i = \sum_{j \in F} a_{ij} e_j \quad \text{(approximately).}$$

This leads directly to (3.11). However, the revised *C* point choice strategy provides more information at points not used in interpolation. Thus, for each point j not in $S^I(i) \cup \{i\}$ it can be assumed that

$$(4.3) \quad e_j = \Sigma\, a_{jk} e_k \;/\; \Sigma\, a_{jk},$$

where the summation here is take over all k in $S^I(i) \cup \{i\}$. That is, to determine w_{ik}, any a_{ij} in (4.1) can be replaced by changing the values of a_{ik} according to the stencil for point j. The values w_{ik} that result from this need not be represented explicitly here, although it is important to note that they are easily computed in practice. (See Algorithm B in the Appendix.)

5. NOTES ON NUMERICAL PERFORMANCE

The purpose of this paper is to describe the current status of AMG, not to report extensively on its performance characteristics. Numerial results for earlier AMG algorithms have been reported elsewhere (cf. [3] and [8]) so that its potential is well established. However, the current algorithm is an improvement in several important ways that bear on its performance:

Complexity. The present strategy exhibits total relative operator complexities that are typically about 2.5, well below earlier AMG values.

Convergence rates. Asymptotic convergence factors are more consistent and usually better than those reported in [3] and [8].

Robustness. The current algorithm is more robust in that its performance does not degrade as it did for certain pathological cases.

However, even the current strategy is not without room for improvement and is currently undergoing basic modification. (See the next Section.) We therefore defer reporting on numerical results to future efforts.

6. FURTHER DEVELOPMENTS

There are many aspects that are currently under study for AMG, but the main objective for the basic algorithm that will have general impact on AMG is to *ensure* that error smoothness is properly exploited. We discuss below two of the main approaches that we are now considering in this direction. They will be discussed briefly in the context of positive-type problems, although their generalization to other cases (especially systems) may provide an important vehicle for extending AMG to broader classes. Both approaches can be viewed as attempts to obtain information on coupling strength beyond direct connections. The first attempts to do this implicitly, while the second takes a more explicit approach.

6.1 Pre-relaxation.

For simplicity, imagine first that A is a symmetric M-matrix but is scaled so that some of its row sums are negative. Though a diagonal scaling exists that can convert A to positive-type, thus making AMG application straightforward, it can generally be too expensive to precisely determine. However, A need only be *essentially* positive-type [3] so that such scaling need only be approximate. The point here is that we would like to automatically determine an approximate sense of smoothness, that is, a sense of the nature of errors that produce small residuals for A.

The concept of pre-relaxation is to attempt this by performing relaxation on the homogeneous equation

(6.1) $AE = 0$

and examining the resulting approximate solutions, which of course are now also the actual errors. For the simplest case, pre-relaxation starts with *one* initial guess e that, without further information, is probably best chosen as the inverse square roots of the diagonal entries of A. Several Gauss-Seidel relaxation sweeps are then performed on e for (6.1) with the result (possibly normalized) defining entries of the diagonal matrix D_e. A is then replaced by $D_e A D_e$. The first level of coarsening is determined, and the pre-relaxation process is continued until all levels are chosen.

Initial experiments with a modified version of the above algorithm have proved very successful. In fact, just a few pre-relaxation sweeps starting with a constant vector seem to be enough to restore AMG efficiency for positive-type matrices that have been mis-scaled by random powers of 10 between 1 and 20.

More sophisticated vector pre-relaxation techniques would use a simultaneous set of several vectors that are forced to maintain local independence, are subjected to several relaxation sweeps for (6.1), and are subsequently examined to determine interpolation formulae that optimally represent the local subspaces that these vectors generate.

6.2. Total strength of connection

Simply stated, it is important to ensure for each F point that the strength to points $S^I(i)$ used to interpolate to it is a reasonable fraction of its total dependence. However, this strength to the points in $S^I(i)$ should account not only for the *direct* couplings, a_{ij}, but for indirect ones as well. Thus, if i is connected to j which is not in $S^I(i)$, then we should assess the contribution to the strength of point j via its connections to $S^I(i)$. The objective here is to measure the degree of 'overlap' (described in [2; Sec. 4.4]) to ensure a suitably large γ in (3.8). Precisely how this should be done is a subject of our current efforts.

Appendix. Current AMG Coarsening Strategy

The matrix $A = (a_{ij})_{nxn}$ is assumed to be symmetric and of positive-type. We use the following notation:

$G = \{1, \ldots, n\}$ (the 'grid' points for A)

$N(i) = \{j \neq i: \ a_{ij} \neq 0\}$ (the neighborhood of a point i)

$S(i, \varepsilon) = \{j \neq i: -a_{ij} \geq \varepsilon . \max_{k \neq i} - a_{ik}\}$ (the set of points on which i strongly depends relative to the parameter ε)

$S^T(i, \varepsilon) = \{j \neq i: i \text{ in } S(j, \varepsilon)\}$ (the set of points which strongly depend on i)

In addition, it is useful to have a measure of dependence of a point i on a set of points P. This can be defined in several different ways, the simplest of which is the following:

$$d(i, P) = \max_{j \in P} (-a_{ij})/\max_{k \neq i} (-a_{ik}).$$

(An alternative is $\sum_{j \in P} (-a_{ij})/\max_{k \neq i}(-a_{ij})$.) Using this function, the set of strong connections of a set of points, relative to ε, can be defined as follows: $S^T(P, \varepsilon) = \{j \notin P: d(j,P) \geq \varepsilon\}$. Finally, the set of points which are used to interpolate to a point i in is denoted by $S^I(i)$. In this algorithm, $S^I(i) \subset S(i, \varepsilon_2)$.

ALGORITHM A

(Initial C point choice)

Parameter ε_1 defines strong connections.

1. Set $F \leftarrow \phi, C \leftarrow \phi$, U (undecided) $\leftarrow G$. Set η_i to the number of elements of $S^T(i, \varepsilon_1)$ for all i.
2. Pick i in U with maximal η_i. If $\eta_i = 0$ or $U = \phi$, stop.
3. Put i in C.

   For each j in $S^T(i, \ \ ) \cap U$, do the following:

Put j in F.

For each k in $S(j,\varepsilon_1) \cap U$, set $\eta_k \leftarrow \eta_k + 1$

4. Go to 2.

ALGORITHM B

(Final C point choice and

interpolation definition)

Parameters: ε_2 and ε_3

Let F and C be the fine and coarse points resulting from Algorithm A. Let T be the set of F points which have been tested. Initially, $T = \phi$. Any ti a point is put in C, it is automatically removed from T and, of course, F

1. If $F \subset T$, stop. Otherwise, pick some i in $F-T$. Set $S^I(i) = S(i,\varepsilon_2) \cap C$. Set $T \leftarrow T \cup \{i\}$ and $m = 0$.
2. For each k in $S^I(i)$, set $w_{ik} = a_{ik}$. Set $w_{ii} = a_{ii}$. Set $P = N(i) \cap F$.
   a. If $P = \phi$, go to 3. Otherwise pick j in P. Set $P \leftarrow P - \{j\}$. If j is not in $S(i,\varepsilon_2)$, go to e. Otherwise, proceed with b.
   b. If $d(j, S^I(i)) \geq \varepsilon_3^*$, go to e. Otherwise, set $m \leftarrow m+1$. If $m = 1$, go to c. Otherwise, go to d.
   c. Set $j_p = j$, $S^I(i) \leftarrow S^I(i) \cup \{j\}$, and go to 2. (j becomes a conditional C point.)
   d. Put i in C. Leave j_p in F and go to 1.
   e. (Distribute the weight for point j.)

      Let $b = a_{ij} + \sum_{k \in S^I(i)} a_{jk}$.

      If $b = 0$, set $w_{ii} \leftarrow w_{ii} + a_{ij}$. Otherwise, for all λ in $S^I(i) \cup \{i\}$, set $w_{i\lambda} \leftarrow w_{i\lambda} + a_{ij} \cdot a_{i\lambda} / b$.

      Go to a.

3. For all k in $S^I(i)$, set $w_{ik} \leftarrow -a_{ik}/w_{ii}$. If m = 1, set $C \leftarrow C \cup \{j_p\}$.

Go to 1.

*An alternative criterion here is

$$d(j, S^I(i)) \geq \varepsilon_3 . (-a_{ij}) / \max_{k \neq i} (- a_{ik}).$$

This reduces complexity somewhat since it says that the dependence of j on $S^I(i)$ need not be large if dependence of i on j is not large.

The results of this algorithm are the final C/F splitting and the interpolation formula, which for any i in F and e^c in $G^c \sim C$, is given by

$$e_i = \sum_{k \varepsilon S^I(i)} w_{ik} e_k^c .$$

REFERENCES

1. R. W. Alcouffe, A. Brandt, J. E. Dendy and J. W. Painter, 'The multigrid methods for the diffusion equation with strongly discontinuous coefficients,' SIAM J. Sci. Stat. Comput. 2 (1981), 430-454.

2. A. Brandt, 'Algebraic multigrid theory: the symmetric case (a prelimina version)', Prelim. Proc.Int'l MG Conf., Copper Mountain, April, 1983

3. A. Brandt, S. McCormick and J. Ruge, 'Algebraic multigrid (AMG) for automatic multigrid solution with application to geodetic computatic report, Inst. Comp. Studies, Colo. State Univ. 1982.

4. G. Dantzig and P. Wolfe, 'Decomposition principle for linear programs,' Operations Research 8 (1960), 101-110.

5. J. E. Dendy, 'Black box multigrid,' Proc. Symp. Num. Boundary Cond. Procedures and Multigrid Meths., NASA-Ames Res. Center, 1981 (H. Lor ed.), NASA Conf. Publ. 2202.

6. L. M. Dudkin ad E. B. Yershov, 'Interindustry input-output models and th material balances of separate products,' Planned Economy 5 (1965), 54-63.

7. J. Ruge, 'Algebraic multigrid (AMG) for geodetic survey problems,' Prelim. Proc. Int'l. MG Conf., Copper Mountain, April, 1983.

8. K. Stuben, 'Algebraic multigrid (AMG): experiences and comparisons,' Prelim. Proc. Int'l MG Conf., Copper Mountain, April, 1983.

9. K. Tanabe, 'Projection methods for solving a singular system of linear equations and its applications,' Num. Math. 17 (1971), 203-214.

10. I. Y. Vakhutinsky, L. M. Dudkin and A. A. Ryvkin, 'Interative aggregati new approach to the solution of large-scale problems,' Econometrica 47 (1979), 821-841.

11. A. Brandt, 'Guide to multigrid development,' Proc. Conf. at Koln-Porz, Nov., 1981 (W. Hackbusch, U. Trottenberg, eds.), Lecture Notes Math 960, Springer-Verlag, Berlin, 1982.

11 PARALLELISM AND SPARSE LINEAR EQUATIONS

Frans J. Peters
Eindhoven University of Technology
5600 MB Eindhoven
The Netherlands.

Abstract. Recent interest in parallel computers has motivated the development of parallel algorithms to solve many types of problems.
Although efficient codes to solve sparse sets of equations are in existence nowadays, one may wonder whether processing time may be reduced by performing some tasks simultaneously, i.e. in parallel. As far as direct solution of sparse sets of linear equations is concerned, two types of parallelism may be distinguished: parallelism resulting from the sparseness of the equations and parallelism inherent in the computation of vector innerproducts. The latter is called type d parallelism, the former type s parallelism. It will be shown that no matter how many processors are available, exploitation of type s parallelism yields only a speed-up factor of about five. Further speed-up can only be obtained by exploiting type d parallelism.

0. Introduction

Recent interest in parallel computers has motivated the development of parallel algorithms to solve many types of problems.

Although efficient codes to solve sparse sets of equations are in existence nowadays, one may wonder whether processing time may be reduced by performing some tasks simultaneously, i.e. in parallel. As far as direct solution of sparse sets of linear equations is concerned, two types of parallelism may be distinguished: parallelism resulting from the sparseness of the equations and parallelism inherent in the computation of vector innerproducts.

In Section 1 we review some well-known results of sparse matrix research required for a good understanding of the subsequent sections. Section 2 outlines the parallelism resulting from the sparseness of the equations. It is shown in Section 3 that the speed-ups which may be expected are disappointing. Even in ideal circumstances (i.e. no restriction with respect to the number of processors and no delays due to interprocessor communication) they are usually less than about a factor 5.

Finally, in Section 4 we show how parallelizing the computations of vector innerproducts may speed-up the computations. The LU-decomposition of an $N \times N$ matrix associated with a regular $\sqrt{N} \times \sqrt{N}$ grid requires only $\mathcal{O}(\sqrt{N})$ time.

1. LU-decomposition

Consider the set of linear equations

$$Qx = y \tag{1}$$

to be solved by some form of Gaussian elimination or LU-decomposition, i.e. a lower triangular matrix L and an upper triangular matrix U are to be computed such that Q = LU. The solution of (1) can then be obtained by forward substitution, i.e. solving y' from Ly' = y, followed by backward substitution, i.e.e solving x from Ux = y'. Let q_{ij}, ℓ_{ij} and u_{ij} $(1 \le i,j \le n)$ denote the coefficients of Q,L and U respectively. The following relations hold (Wilkinson 1965):

$$\ell_{jj}u_{jj} = q_{jj} - \sum_{t=1}^{j-1} \ell_{jt}u_{tj} \qquad (1 \le j \le n)$$

$$\ell_{ij} = \left(q_{ij} - \sum_{t=1}^{j-1} \ell_{it}u_{tj}\right) \cdot u_{jj}^{-1} \tag{2}$$

$$(1 \le j < i \le n)$$

$$u_{ji} = \ell_{jj}^{-1} \cdot \left(q_{ji} - \sum_{t=1}^{j-1} \ell_{jt}u_{ti}\right)$$

If Q is symmetric and the coefficients u_{jj} are chosen to be 1, then we have

$$u_{ji} = \ell_{jj}^{-1}\ell_{ij} \qquad (1 \le j < i \le n) \ . \tag{3}$$

Hence it is in that case not necessary to compute U; substituting (3) in (2) we get

$$\ell_{ij} = \left(q_{ij} - \sum_{t=1}^{j-1} \ell_{it}\ell_{tt}^{-1}\ell_{jt}\right) \qquad (1 \le j \le i \le n) \ .$$

In this exposé we will concentrate on the computation of the coefficients of L, thereby assuming Q to be symmetric. If Q is nonsymmetric coefficients of U have to be computed as well; it may however be shown that for those computations the same results apply.

The coefficients of L must obviously be computed in a certain (partially prescribed) order. To compute ℓ_{ij} $(i \geq j)$ the values of ℓ_{it} $(1 \leq t < j)$ must be known.

Defining:

$$q_{ij}^{(k)} = q_{ij} - \sum_{t=1}^{\min(k,j-1)} \ell_{it} u_{tj} \qquad (1 \leq j \leq i \leq n)$$

it then follows from (2)

$$q_{ij}^{(0)} = q_{ij} \qquad (1 \leq j \leq i \leq n)$$

and for $k \geq 1$

$$q_{ij}^{(k)} = q_{ij}^{(k-1)} - \ell_{ik} u_{kj} \qquad (k < j \leq i \leq n)$$

$$q_{ij}^{(k)} = \ell_{ij} \qquad (1 \leq i \leq n,\ 1 \leq j \leq \min(i,k)) .$$

Hence

$$q_{ij}^{(n)} = \ell_{ij} \qquad (1 \leq j \leq i \leq n) .$$

The computation of L may be thought of to be composed of the processing of pivots. By "processing the kth pivot" we shall mean computing for all i, j such that $k < j \leq i \leq n$ the products $\ell_{ij} u_{kj}$ and subtracting those values from $q_{ij}^{(k-1)}$. The computation of L then consists of the processing of the n pivots of Q.

In the following we will also need the notation of processing a block of pivots. Suppose the matrix Q is partitioned into blocks as follows:

$$Q = \begin{pmatrix} Q_{II} & Q_{IJ} \\ Q_{JI} & Q_{JJ} \end{pmatrix}$$

where Q_{II} is a $k \times k$ matrix (and hence Q_{JJ} has dimensions $(n-k) \times (n-k)$).

A partial decomposition of Q with Q_{II} as a block-pivot is defined as the decomposition obtained by processing the first k pivots. Hence partial decomposition with Q_{II} as block-pivot yields the following decomposition of Q (see also (Bunch & Rose 1974))

$$Q = L_I U_I + \begin{pmatrix} O_{II} & O_{IJ} \\ O_{JI} & Q_I^r \end{pmatrix}$$

with O_{II}, O_{IJ} and O_{JI} zeromatrices,

$$L_I = \begin{pmatrix} L_{II} \\ L_{JI} \end{pmatrix}, \quad U_I = (U_{II}, U_{IJ})$$

where L_{II} is lower triangular and U_{II} upper triangular such that

$$L_{II}U_{II} = Q_{II} \qquad L_{II}U_{IJ} = Q_{IJ}$$

$$L_{JI}U_{II} = Q_{JI} \qquad Q_I^r = Q_{JJ} - L_{JI}U_{IJ} .$$

A matrix is called sparse if many of its coefficients are zero. It is well-known that a sparse matrix usually suffers fill-in, that is to say the matrix L + U contains more nonzeros than Q. (As usual we neglect numerical cancellation: ℓ_{ij} $(j \le i)$ is considered to be a zero coefficient only if $q_{ik} = 0$ and all products $\ell_{it}u_{tj}$ in (2) are zero.) It is equally well-known that the amount of fill-in depends upon the ordering of rows and columns of Q. Instead of solving (1) one usually selects a permutation matrix to rearrange the equations and variables; that permutation matrix is then chosen so as to minimize fill-in. For more details see for instance Rose & Tarjan (1975).

By its very nature it is convenient to view a sparse matrix as a graph (Parter 1961; Rose 1971). Because the order in which the pivots are processed is of importance we shall consider ordered undirected graphs. An undirected graph G = (V,E) consists of a finite set V of vertices or nodes and a finite set E of unordered vertex pairs, called edges. For a graph G = (V,E) with $|V| = n$ an ordering ∂ of V is a bijection $\partial : \{1,2,\dots,n\} \to V$. $G = (V,E,\delta)$ denotes an ordered undirected graph. If v is a node of the graph then $\partial^{-1}(v)$ is the number of v. Let Q be a symmetric $n \times n$ matrix. Define the vertex set V by $V = \{1,\dots,n\}$ and the edge set E by $(v,w) \in E$ iff $q_{vw} \ne 0$ and $(v \ne w)$. Let moreover the ordering ∂ be defined by $\partial(i) = i$ $(1 \le i \le n)$. Then the graph $G = (V,E,\partial)$ is called the connection graph of Q. If G is the connection graph of a matrix, then we make no distinction between a vertex v of the graph and its number $\partial^{-1}(v)$.

For a definition of the notion elimination graph we need the following. Let $G = (V,E)$ be a graph. If $W \subset V$, then the section graph $G(W)$ is the subgraph $(W,E(W))$, where $E(W) = \{(v,w) \in E \mid v,w \in W\}$. For a vertex v of the graph $G = (V,E,\partial)$ the elimination graph G_v is defined as $G_v = (V \setminus \{v\}, E(V \setminus \{v\}) \cup D(v))$, where $D(v) = \{(x,y) \mid (x,v) \in E, (v,y) \in E, x \neq y, (x,y) \notin E\}$. $D(v)$ is called the deficiency of v in G. Let now $G = (V,E,\partial)$ be the connection graph of a symmetric matrix Q. Processing the first pivot of Q gives:

$$Q^r = \begin{pmatrix} q_{22} - \ell_{21}u_{12} & \cdots & q_{2n} - \ell_{21}u_{1n} \\ \vdots & & \vdots \\ q_{n2} - \ell_{n1}u_{12} & \cdots & q_{nn} - \ell_{n1}u_{1n} \end{pmatrix}.$$

Since $q^r_{ij} \neq 0$ if and only if ($q_{ij} \neq 0$ or ($q_{i1} \neq 0$ and $q_{1j} \neq 0$)), it is easily seen that the elimination graph $G_{\partial(1)}$ is precisely the connection graph of Q^r. The deficiency $D(1)$ consists of those vertices (i,j) with $q_{ij} = 0$ and $q^r_{ij} \neq 0$. Let G^0 be G, then G^i is recursively defined by: G^i is the elimination graph $(G^{i-1})_{\partial(i)}$. The <u>fill-in</u> $D_\partial(G)$ is defined by $D_\partial(G) = \bigcup_{i=1}^{n-1} D^{i-1}(\partial(i))$, where $D^{i-1}(\partial(i))$ is the deficiency of $\partial(i)$ in G^{i-1}.

In order to characterize the fill-in, we introduce the notion rt-path. A path $v_0, v_1, \ldots, v_k$ in the graph $G = (V,E,\partial)$ is called an rt-path if $\partial^{-1}(v_i) < \min(\partial^{-1}(v_0), \partial^{-1}(v_k))$ for $0 < i < k$. In other words: a path from v_0 to v_k is an rt-path if it passes only through nodes with a number less than v_0 and v_k.

See for instance Figure 1 which shows a symmetric 5×5 matrix and its connection graph. In that graph 2, 1, 3 is an rt-path, whereas 4, 5, 3 and 2, 5, 3 are not. In fact the graph contains only one rt-path.

From Rose & Tarjan (1975) we know the following lemma:

<u>Lemma 1</u>: Let $G = (V,E,\partial)$ be the connection graph of the matrix Q with LU-decomposition $Q = LU$. Then:

$\ell_{ij} \neq 0$ iff $i \leq j$ and there is an rt-path between i and j.

As mentioned above the order in which the pivots are processed is of importance. Not only the fill-in, but in general also the numerical stability depends upon the ordering chosen. However, if the matrix is sufficiently sparse then the prescribed pivot order may be lessened without affecting the fill-in or in the numerical stability. See for instance the connection graph in Figure 1. That graph has one fill-in edge. The nodes are numbered according to the prescribed pivot order.

By inspection of the matrix it can be seen that in order to process pivot 4 none of the values obtained by processing pivot 1 are required. Hence the order of processing pivots 1 and 4 is irrelevant; they may be processed simultaneously. In order to process pivot 3, results of processing pivot 1 are required, but because of fill-in results of processing pivot 2 are needed as well. Hence processing pivot 3 must wait until the processing of pivots 1 and 2 is finished. Only if the matrix is sparse some pivots may be processed in parallel. If all coefficients of the matrix are nonzero then in order to process pivot -say- j, results of processing pivots 1 to j - 1 are required. Hence in that case all pivots must be processed strictly sequentially.

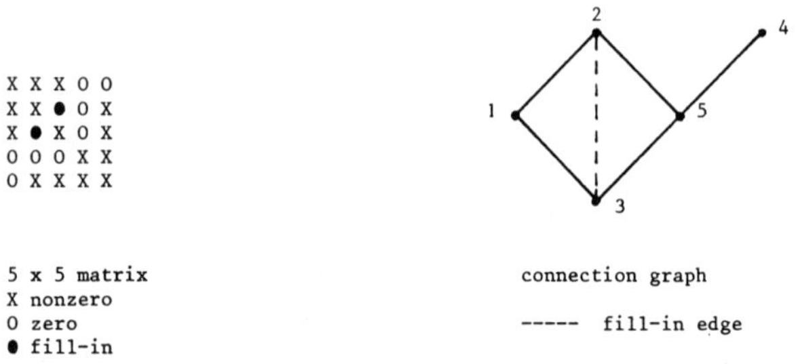

Figure 1.

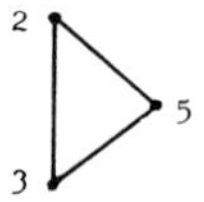

Figure 2: elimination graph after processing pivots 1 and 4 from Figure 1.

2. Processing pivots simultaneously

In case of sequential processing (that is if only one processor is available), the pivots are processed one at a time in the order given by their numbers. With parallel processing however a pivot may be processed if it is not connected with a pivot of a lower number. Recall Figure 1, where neither pivot 1 nor pivot 4 is connected with pivots of a lower number. Hence, when pivot 1 is being processed, pivot 4 may be processed as well. After processing pivot 1 and 4 the elimination graph as shown in Figure 2 remains. It is a complete graph; the associated matrix is dense and hence the remaining pivots have to be processed the one after the other.

Another example is shown in Figure 3. Only the four corner pivots 1, 2, 4 and 5 are not connected to nodes with a lower number; to start with, those four pivots may be processed. (Notice that the lower diagonal parts of the rows associated with those pivots consist of zeros only.) After processing the corner pivots the elimination graph and reduced matrix as shown in Figure 4 result. Next only pivots 3 and 6 may be pro-

X	0	X	0	0	0	X	0	0
0	X	X	0	0	0	0	0	X
X	X	X	0	0	0	●	X	●
0	0	0	X	0	X	X	0	0
0	0	0	0	X	X	0	0	X
0	0	0	X	X	X	●	X	●
X	0	●	X	0	●	X	X	●
0	0	X	0	0	X	X	X	X
0	X	●	0	X	●	●	X	X

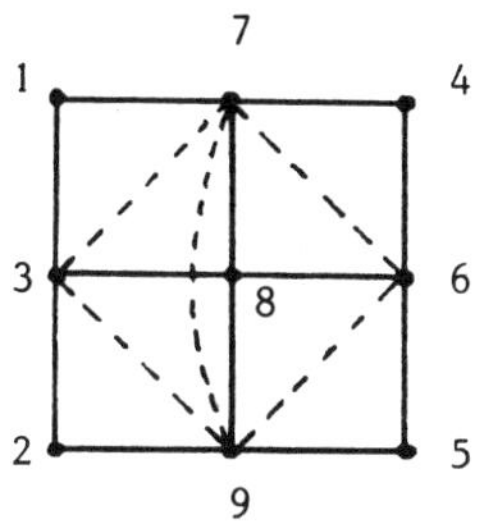

Figure 3.

3	X	0	X	X	X
6	0	X	X	X	X
7	X	X	X	X	0
8	X	X	X	X	X
9	X	X	0	X	X
	3	6	7	8	9

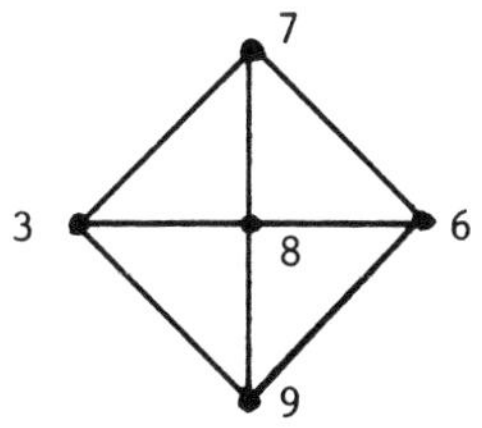

Figure 4: elimination graph and connection matrix after processing pivots 1, 2, 4 and 5 from Figure 3.

cessed. (The processing of pivot 3 may start as soon as the processing of pivots 1 and 2 are finished; in the same way the processing of pivot 6 may start as soon as the processing of pivots 4 and 5 are finished.) When 3 and 6 are finished a complete graph and dense matrix result.

With parallel pivoting a pivot may be processed as soon as the following two requirements are met:

i) a processor is available;

ii) the pivot is not connected (in the elimination graph) with pivots of a lower number.

If parallel pivoting is arranged in such a way that requirement ii is met then parallel processing yields exactly the same fill-in and the same numerical stability as sequential processing in the prescribed order. A rigorous proof may be found in (Peters 1983).

Now consider the matrix A, such that

$$A = \begin{pmatrix} A_{11} & & & \emptyset & A_{15} \\ & A_{22} & & & A_{25} \\ & \emptyset & A_{33} & & A_{35} \\ & & & A_{44} & A_{45} \\ A_{51} & A_{52} & A_{53} & A_{54} & A_{55} \end{pmatrix}$$

where A_{ii} (i=1,...,5) are square submatrices. If all submatrices are considered to be elemental then the associated connection graph is shown in Figure 5. Pivots 1, 2, 3 and 4 may all be processed simultaneously; processing of pivot 5 however must wait untill the processing of all other pivots is finished.

More generally, if the submatrices A_{ii} consist of more than one element, then all the pivots of A_{ii} (i=1,2,3,4) may be processed simultaneously with the pivots of A_{jj} ($j \neq i, j \neq 5$). Whether two pivots from A_{ii} may be processed simultaneously depends upon the sparseness structure of A_{ii}.

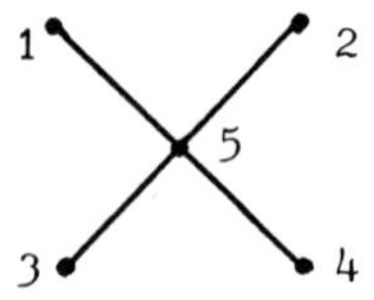

Figure 5.

In the same way whether processing of pivots from A_{55} must await processing of pivots from A_{ii} (i=1,...,4) depends upon the sparseness structure of A_{5i} and A_{i5}. If all submatrices are dense then we may employ at most four processors concurrently and the LU-decompositon consists of two stages. In the first stage pivots A_{ii} (i=1,...,4) are processed and in the second stage those from A_{55}. In the first stage at most four pivots may be processed in parallel, whereas in the second stage all pivots are processed sequentially.

Up till now we assumed the equations had to be solved in a given order. It may however be the case that the pivot order is not prescribed a priori. In that case a suitable pivot-order has to be determined. If fill-in and stability considerations are not taken into account, which pivots may then be processed in parallel? Calahan (1973) was the first to answer this question: a pivot may be processed if and only if none of its neighbours in the graph is being processed. Hence the maximum number of pivots to start with is the maximum number of non-adjacent nodes in the connection graph. See for instance Figure 3; if the pivot order is free, then processing may start with five pivots simultaneously, namely 1, 2, 4, 5 and 8. After processing those pivots, the remaining four pivots must be processed sequentially.

It is a nontrivial matter to find good orderings of pivots. For parallel processing of pivots counterexamples exist which show that fewer total parallel steps can be achieved by choosing less than the maximum number of parallel pivots. We will not go further into the investigation which ordering to choose and how to implement the algorithms. Readers interested in these subjects are referred to (Peters 1983).

We will now deal with the -as it will turn out more important- question which speed-ups are to be expected when parallel processing is applied and how many processors are needed or may be useful.

3. Speed-up

The speed-up obtained by applying more than one processor is defined as follows: let T1 be the time needed to solve a problem with one processor and Tp the time required to solve the same problem with p processors, then the speed-up is T1/Tp. If despite many processors only a small speed-up is obtained, then the processors are used rather inefficiently. An efficiency measure for applying more than one processors is the ratio of speed-up and the number of processors applied; in formula: T1 / (Tp * p) with p the number of processors. A small efficiency ratio may be caused by long idle times of the processors or by the fact that processors are duplicating each other's work. Both causes may be due to the circumstance that there is not enough useful work for that many processors.

Let us return to LU-decomposition. If all the processing of a specific pivot is done by one processor only, then the maximum number of processors to be employed equals the maximum number of pivots, that may be processed simultaneously. That number however depends -as we have seen above- upon the sparsity pattern of the matrix.

Kees (1982) and Calahan (1973) report the results of some experiments they did with parallel pivoting. Kees investigated the connection graphs of matrices associated with electronic circuits and power networks. Two different ordering criteria were used. The number of pivots ranged from 118 to 1637. Even if communication and synchronisation overhead are not taken into account, the maximum speed-up is only 5.14. To achieve this speed-up for that particular problem (with 118 pivots) 49 processors are needed.

Calahan investigated a matrix resulting from a simulation of a three-dimensional model of an automotive suspension system. The matrix has 83 pivots. From the results presented it may be concluded that by parallel pivoting at most a speed-up of 625/163 = 3.8 may be obtained. To obtain that speed-up 40 processors are needed (again communication and synchronisation overhead are not taken into account). It is worthwhile to remark that by applying only 2 processors a speed-up of 625/334 = 1.9 may be achieved. With 20 processors the speed-up is 625/166 nearly the same speed-up as with 40 processors. The efficiency ratios for 2, 20 and 40 processors respectively are 0.94, 0.19 and 0.10.

From the experiments we are led to the conclusion that parallel pivoting yields only modest speed-ups. This can also be seen in a more theoretical way.

Consider the graph G_n, the nodes of which may be thought of as being arranged in a regular grid of n rows and n columns, each node being connected to its four neighbours in the grid. See Figure 6 which shows G_5. It is well-known that to process all n^2 pivots of G_n at least $O(n^3)$ multiplications and additions are required (George 1973). Nested dissection is an ordering which achieves that bound (George 1973). Hence for sequential processing the nested dissection ordering of G_n is asymptotically optimal. Nested dissection ordering algorithms are essentially divide and conquer algorithms.

One such algorithm is as follows. Consider S, the 2n - 1 nodes which together form a cross such that if they together with their incoming edges are removed, four (nearly) equal subgraphs C_1, C_2, C_3 and C_4 remain. Then suppose we number the nodes of G_n so that those in C_i are numbered consecutively (i = 1,2,3,4), followed by those in S. Then the associated matrix will have the following structure.

$$Q = \begin{pmatrix} Q_{11} & & & \emptyset & Q_{15} \\ & Q_{22} & & & Q_{25} \\ & & Q_{33} & & Q_{35} \\ & \emptyset & & Q_{44} & Q_{45} \\ Q_{51} & Q_{52} & Q_{53} & Q_{54} & Q_{55} \end{pmatrix} .$$

Obviously we can apply the algorithm recursively to the subgraphs C_1, C_2, C_3 and C_4. The submatrices Q_{ii} (i = 1,2,3,4) then obtain the same structure as Q. The resulting ordering is a nested dissection ordering. The ideas of nested dissection ordering may be applied to other kinds of graphs as well; see Lipton et al. (1979).

Assume that G_n is provided with such an ordering. See Figure 7 which shows G_7 with a nested dissection ordering. Parallel processing of the pivots proceeds in the following recursive way. The four subgraphs C_i (i = 1,2,3,4) are processed simultaneously. After processing those subgraphs a dense (2n - 1) × (2n - 1) matrix remains, the pivots of which must be processed sequentially. This algorithm is applied recursively to each of the four subgraphs C_i (i = 1,2,3,4). This yields the simultaneously

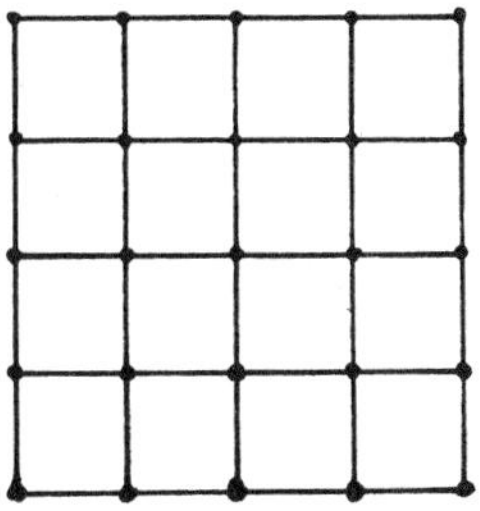

Figure 6: G_5 a regular 5 × 5 grid.

processing of 16 or more subgraphs if n is sufficiently large.

To determine the speed-up we will use the number of arithmetic operations as a measure of time. Any number of operations performed simultaneously will be counted only once. The number of operations to process the pivots of a dense $(2n-1) \times (2n-1)$ matrix is $a\,n^3 + \mathcal{O}(n^2)$ for a certain value a.

Let the number of arithmetic operations for the sequential and parallel processing of G_n be denoted by gs(n) and gp(n) respectively. For the ease of presentation we assume that $n+1$ is a power of 2, i.e. $n = 2^k - 1$ for some nonnegative integer value k. We then have the following relations:

$$gs(n) = 4\,gs(n/2) + a\,n^3 + \mathcal{O}(n^2)$$

$$gp(n) = gp(n/2) + a\,n^3 + \mathcal{O}(n^2) \ .$$

The solutions of those recurrence relations are

$$gs(n) = 2\,a\,n^3 + \mathcal{O}(n^2 \log n)$$

$$gp(n) = \frac{8}{7} a\,n^3 + \mathcal{O}(n^2)$$

(see Peters 1980, p. 63).

From these we conclude that parallel pivoting yields a speed-up of only 1.75. To achieve this speed-up $\mathcal{O}(n)$ processors are needed.

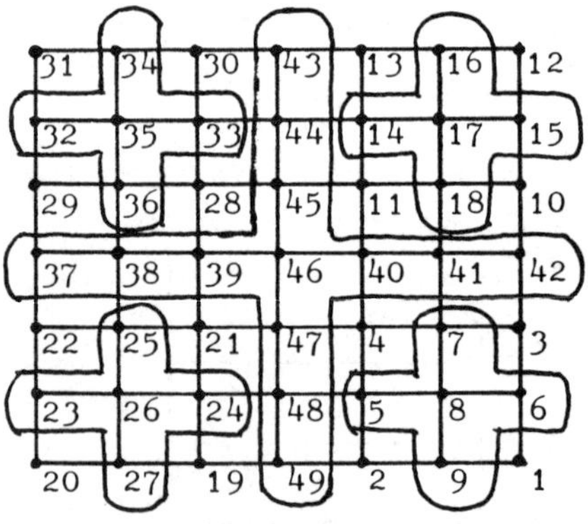

Figure 7: G_7 a regular 7×7 grid with a nested dissection ordering.

Remark: The above analysis is rather crude; moreover less arithmetic operations are needed if the graph G_n is not torn in four subgraphs, but rather into two isomorfic subgraphs of $(n^2 - n)/2$ nodes each. A precise analysis along the lines of (Peters 1980) then gives:

$$gs(n) = \frac{2487}{252} n^3 + \mathcal{O}(n^2 \log n)$$

$$gp(n) = \frac{377}{192} n^3 + \mathcal{O}(n^2) .$$

Hence the speed-up is 5.03.

4. Systolic networks

As we have seen in the foregoing, both practical experiences and theoretical investigations suggest that exploitation of the parallelism due to the sparseness of the equations yields only a modest speed-up; usually at most a factor 5. The processing of a pivot however consists of the computation of a vector-innerproduct. Therefore a larger speed-up may be obtained by employing parallelism to compute vector-innerproducts or by applying algorithms to solve dense sets of equations in a parallel way. A survey of parallel algorithms for dense sets is given by Heller (1978). Kung & Leiserson (1980) present efficient and surprising algorithms for banded sets of equations. Parallel versions of Givens' algorithms are described by Heller & Ipsen (1983). Recent results for highly parallel computers are also to be found in (Ahmed et al. 1982). These algorithms are based upon systolic networks.

Systolic networks are an architectural paradigm first proposed by Kung & Leiserson (1980) for the implementation of matrix operations in VLSI (very large scale integrated circuits). Systolic networks are special purpose devices, the design of which is made possible by new and emerging technologies. A general overview of that design activity is given by Kung (1982).

Systolic networks consist of simple processing elements laid out in a regular fashion. Interconnections (datapaths) exist between neighbouring elements only. Elements lying on the boundary of the network may have external connections to the host memory; a boundary element may receive input from and send output to external memory. Each processing element may receive input from and produce output for its neighbouring elements. The processing elements are synchronised; they act rhythmically on one or more streams of data that smoothly move across the network.

Results emerge from the network along the boundary, usually at the same rate as data are entered. The term systolic has been coined by Kung because of the resemblance with rhythmic pulsing of the blood through the arteries. Practical considerations for VLSI design require the processors to be simple and identical; those on the boundary however may differ from those in the interior.

The processing elements of a systolic network for LU-decomposition as designed by Kung & Leiserson (1980) are mutually connected so as to form a hexagonal array, see Figure 8.
The coefficients of the matrix to be decomposed are -in a certain order- fed in the lower boundary processing elements and after some time the coefficients of the triangular factor appear from the upper boundary elements. See Figure 9. The network exhibits an upward movement of the data. The details of this network are not relevant here. For the sequel of this paper we need only the following results. A systolic network with no more than n^2 processing elements can compute the LU-decomposition of an $n \times n$ matrix in 4n units of time.

Such a systolic network is equally well suited for the LU-decomposition of a sparse matrix. However sparse sets of equations are usually very large and it is impractical -because of the many processors needed- to build networks suited for very large sets of equations. A more fruitful approach is as follows. Tear the connection graph into two or more subgraphs; process the internal pivots of the subgraphs (the subgraphs may be processed simultaneously) and next process the remaining pivots. With these remaining pivots a relatively small dense matrix is associated; hence a systolic network is well suited for the processing of those pivots. This scheme may be applied recursively even if only one systolic array is available; the small dense matrices are then processed one at a time by

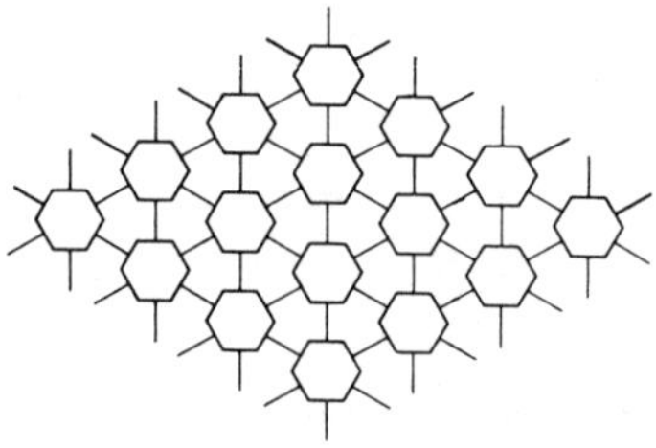

Figure 8: hexagonally connected systolic network.

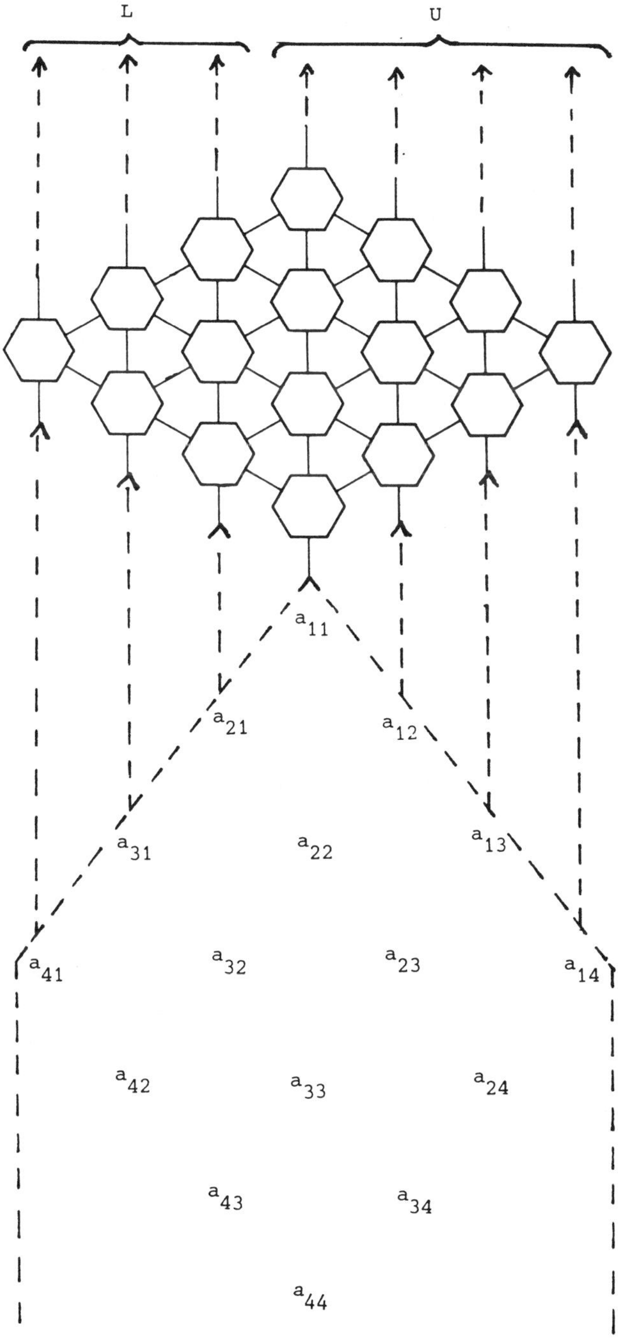

Figure 9: dataflow in a hexagonally connected systolic network for the LU-decomposition of a 4×4 matrix.

that one systolic network. To process however the pivots of the subgraphs still more networks may be employed to process the sub-subgraphs and so on simultaneously. Hence to implement the scheme we may employ a number of systolic arrays, which may be thought of as being arranged in a tree; see Peters (1981).

The following analysis is intended to get an impression of the speed-up. Let hs(n) and hp(n) denote the time to process all the pivots of G_n with only one systolic network and a tree of systolic networks respectively; G_n is the regular grid introduced earlier. Processing the $2n-1$ pivots of the dividing cross with a systolic network requires $\mathcal{O}(n)$ time, say $pn+q$. We have the following recurrence formula:

$$hs(n) = 4\,hs(n/2) + pn + q$$

$$hp(n) = hp(n/2) + pn + q$$

with as solutions

$$hs(n) = cn^2 + \mathcal{O}(n^2 \log n)$$

$$hp(n) = 2pn + \mathcal{O}(1)$$

where c is a constant determined by the "boundary conditions" hs(1).

Comparing hs(n) and hp(n) with gs(n), we see that one systolic network gives a speed-up of $\mathcal{O}(n)$, whereas a tree of systolic networks gives a speed-up of $\mathcal{O}(n^2)$. Notice that the matrix associated with G_n is a $n^2 \times n^2$ matrix. Hence $N = n^2$ is the number of rows and columns. The LU-decomposition of that $N \times N$ matrix can be computed in $\mathcal{O}(\sqrt{N})$ time. That bound can be achieved with $\mathcal{O}(\sqrt{N})$ systolic networks able to decompose dense matrices of up to $\sqrt{N}$ rows and columns. Applying conventional processes would require $\mathcal{O}(N\sqrt{N})$ time.

References

Ahmed, H.M., Delosme, J-M. & Morf, M. (1982). Highly concurrent computer structures for matrix artihmetic and signal processing. Computer, 15, 65-82.

Bunch, J.R. & Rose, D.J. (1974). Partitioning, tearing and modification of sparse linear systems. J. Math. Anal. Appl., 48, 574-593.

Calahan, D.A. (1973). untitled. Proc. 11th Allerton Conf. on Circuit and System theorey, 729-738.

George, J.A. (1973). Nested dissection of a regular finite element mesh. SIAM J. Numer. Anal., 10, 345-363.

Heller, D. (1978). A survey of parallel algorithms in numerical linear algebra. SIAM Review, 20, 740-777.

Heller, D. & Ipsen, I.C.F. (1983). Systolic networks for orthogonal decomposition. SIAM J. Sci. Stat. Comput., 4, 261-269.

Kees, H.G.M. (1982). The organization of circuit analysis on array architectures. Ph. D. thesis, Eindhoven University of Technology, Dept. of Electrical Engineering.

Kung, H.T. (1982). Why systolic architectures? Computer, 15, 37-46.

Kung, H.T. & Leiserson C.E. (1980). Algorithms for VLSI arrays. in: Introduction to VLSI systems. C. Mead and L. Conway. Addison-Wesley, Reading, 271-292.

Lipton, R.J., Rose, D.J. & Tarjan, R.E. (1979). Generalized nested dissection. SIAM J. Numer. Anal., 16, 281-285.

Parter, S.V. (1961). The use of linear graphs in Gauss elimination. SIAM Review, 3, 119-130.

Peters, F.J. (1980). Sparse matrices and substructures. MC Tract 119. Mathematical Centre, Amsterdam.

Peters, F.J. (1981). Tree machines and divide-and-conquer algorithms. in: Proc. CONPAR81 (W. Handler ed.), Lecture Notes in Computer Science 111, Springer, 25-36.

Peters, F.J. (1983). Parallel pivoting algorithms for sparse symmetric matrices. Internal report, Eindhoven University of Technology, Dept. of Mathematics and Computing Science.

Rose, D.J. (1971). A graph-theoretic study of the numerical solution of sparse positive definite systems of linear equations. in: Graph theory and computing (R. Read ed.), New York, Academic Press, 183-217.

Rose, D.J. & Tarjan, R.E. (1975). Algorithmic aspects of vertex elimination. in: Proc. Seventh Annual Symposium on Theory of Computing, 245-254.

Wilkinson, J.H. (1965). The algebraic eigenvalue problem. Oxford University Press, Oxford.

12 SPARSITY APPLICATIONS IN GEODESY AND PHOTOGRAMMETRY

Franz Steidler
WILD HEERBRUGG LTD.
CH 9442 Heerbrugg, Switzerland.

Abstract. The method of least squares adjustment is one of the most important numeric tasks to be solved in geodesy and photogrammetry. In photogrammetry the problem of simultaneously adjusting several hundreds or thousands of photos must be solved. The matrices of the normal equations are large, mostly banded and bordered with an irregular bandwidth. In contrast to this, in geodetic network adjustments, also having sparse normal equations, but the coefficients are very irregularly distributed. The generation of a digital elevation model, which is a subject of both geodesy and photogrammetry, is often accomplished in using the methods of finite elements. In this case the normal equations are a rigorously band shaped. The investigations, which are discussed here, are performed by several algorithms exploiting the sparse structure of the matrix of the normal equations. These include the method of conjugate gradients with pre-conditioning, the Gaussian algorithm modified for sparse matrices, band algorithms and dissection methods have been used. The results show impressively short computation times for sparse methods in geodetic network adjustment. In photogrammetric bundle adjustment, however the band algorithm was best and for generating digital elevation models with finite elements dissection methods seem to be most efficient. The following contribution is divided into two chapters. Chapter I explains the nature of three problems which have to be solved in geodesy and photogrammetry: geodetic network adjustment, photogrammetric aerotriangulation and generation of digital terrain models. Chapter II then shows the applications of the sparse techniques to the problems discussed in chapter I.

GEODETIC AND PHOTOGRAMMETRIC PROBLEMS FOR LARGE SYSTEMS TO BE SOLVED

The main task of Geodesy and Photogrammetry is to determine the shape and size of the earth and to depict it on maps or in a digital form for a variety of applications. The task is usually solved by defining the vertical and/or horizontal coordinates of significant points in the landscape.

The principle methods used in Geodesy and Photogrammetry are different. In Geodesy the observations are performed directly in the field while in Photogrammetry photographs or other sensored data are acquired and processed at home. Nowadays the accuracy, which can be reached in photogrammetry for point determination is comparable with the accuracy reached with geodetic methods. So the geodetic and photogrammetric methods can be competitive, but they ought to be complementary.

I. GEODETIC NETWORK ADJUSTMENTS

1. GEODETIC NETWORKS

A geodetic network consists of a certain number of points whose coordinates are to be defined. The points may be of different accuracy and they may form a framework for topographical and cadastral maps. The points are interconnected by observations, which may be distances, directions, astronomical observations (longitude, latitude , azimuths),heights or even coordinates.

Historically a geodetic network was established by creating a main framework of high accuracy, described as primary or first order. This net was subdivided using triangulation or traverses to fix the numerous points required for detailed survey. This denser network is called the tertiary, third order topographical frame. A secondary network may have intervened. Some modern methods such as special traverses or satellite triangulation, more accurate than classical triangulation, may in future be described as zero order and will control the first order framework. It is also possible that zero order control between distant points may be provided by astronomical observations of latitude and longitude combined with gravimetric determinations of the deviation of the vertical. The zero order and primary framework, together with such secondary levels as may be suitable for extending the main framework or contributing to its strength, is known as 'geodetic'. More than half the world's habitable land areas are now covered by a good geodetic framework. The layout of a triangulation system may either be a continuous net or a system of chains. The advantages of a net are that it makes less demand on the topographical triangulators and provided that the adjustment can be satisfactorily carried out, it gives the more accurate result.

Fig. 1: Part of a Triangular Network

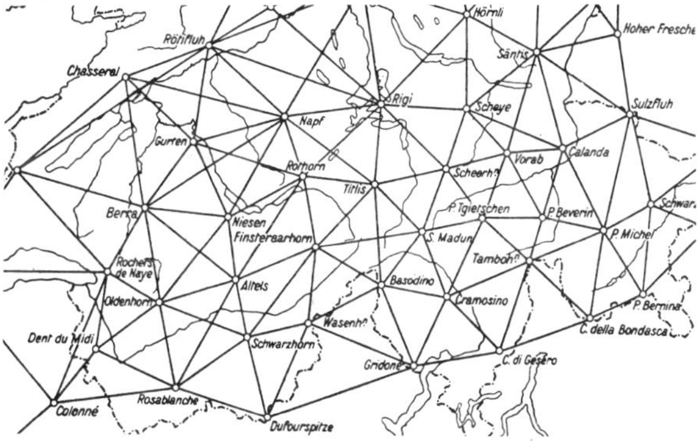

Fig. 2: A Network Formed by Chains

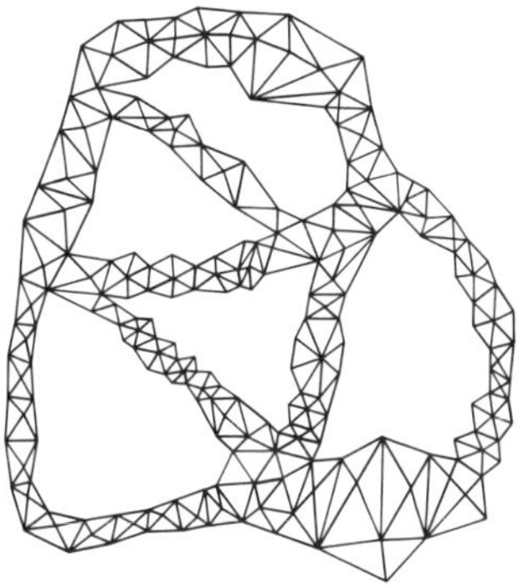

The advantages of a system of chains are that it costs less except in very small areas, is easier to adjust and is also quicker to observe and to compute.

2. THE ADJUSTMENT OF A LARGE GEODETIC NETWORK

The object of adjustment is to secure selfconsistency together with the improvement in accuracy both over-all and between points. Since a geodetic network has many more observations performed than are needed for simple determination of points, the actual system will have many redundancies. This redundancy gives the possibility of improving the reliability and the accuracy of the measured system. This is achieved by performing an adjustment. Ideally, the geodetic framework should be adjusted rigorously as a whole.

Least Squares

The most common procedure for the adjustment of a network is the method of least squares. It is preferred because the principle imposes that the sum of the squares of the residuals of the observations is minimised, the variances of the unknowns will be a minimum. Assuming that the observations are unbiased, the vector of unknowns will also be unbiased.

The observations are normally weighted but they are commonly taken as uncorrelated, which means they are independent. For large systems it is nearly impossible to correlate observations and in most cases the covariances are not known before the adjustment.

Assuming a reasonable design of the network, and the characteristically high redundancy, the networks are well conditioned in most cases. To overcome singularities (up to now the network did not have any datum, which means the position is not defined), constraints have to be introduced. It is possible for example, to use control points whose coordinates can be fixed, (in case of cadastral networks, control point coordinates are not allowed to be changed because of administration purposes) or weighted, if their standard deviation is known.

Hence in normal cases the adjustment has to be performed on non singular and well conditioned systems.

The weighted least squares procedure (referring for example to BJOERCK (1981)) reads:

$$\min_x \| P (b - A x) \|_2 \qquad (1)$$

where $\|.\|_2$ denotes the Euclidian vector norm. P is the weight matrix, which is symmetric and mostly diagonal.
We minimise

$$v'P v \longrightarrow \min \qquad (2)$$

where $$v = A x - l \qquad (3)$$

Thus:

$$\frac{d(v'P v)}{dx} = 2 v' P \frac{dv}{dx} \qquad (4)$$

Minimum condition :

$$2 v'P \frac{dv}{dx} = 0 \quad ; \qquad \left(\frac{dv}{dx} \right)' Pv = 0 \qquad (5)$$

because of (3) $$\frac{dv}{dx} = A$$

this leads to

$$A' P v = 0 \qquad \text{or} \quad A'P(A x - l) = 0 \qquad (6)$$

and so

$$(A'P A) x = A'P l \qquad (7)$$

where

$$\det(A'PA) \neq 0 \qquad \text{because } R(A) = n < m \text{ and } P \text{ positiv definit.}$$

Equation (7) is a system of n non algebraic equations for the determination of the unknowns in case that the observation equations are nonlinear. A useful procedure for the numerical solution is the GAUSS-NEWTON method (ORTEGA,RHEINBOLDT (1970):

$$A'(x^{(i)})PA(x^{(i)})(x^{(i+1)} - x^{(i)}) = A'(x^{(i)})P(l-f(x^{(i)}) \qquad (8)$$

Starting with $x^{(o)}$ for the unknowns the Jacobian is computed with the solution $x^{(i)}$.

The iterations are finished when the nonlinear observation equations are fulfilled.

The adjustment of a geodetic network needs a very high computational effort because the number of unknowns is fairly large. Very often systems of normal equations with several thousands of unknowns have to be solved . The largest project at the present time is the redefinition of the North American Geodetic Network. It spans the United States, Canada, Greenland, Mexico, the republics of Central America and parts of the Caribean area. This enormous task requires 1,500,000 observations. Most of them are horizontal directions and distances with approximately 145,000 stations, which means approx. 350,000 unknown coordinates and orientations.

Setting up the Jacobian Matrix A

Most cases of geodetic networks distinguish between horizontal and vertical adjustment. In areas where the horizontal distances are much larger than the height difference, then this approach is justified. For the examples discussed here, only the major problem of the horizontal adjustment will be considered. Also only the most common types of observations i.e. the directions and distances will be discussed. For simplification the network is assumed to be plane, which means that the reference surface is a plane. These simplifications do not affect the structure of our systems of equations.

The observations are taken as independent (uncorrelated), because their correlations are firstly unknown and secondly they are very difficult to handle.

Under the assumptions mentioned above, the observation equations for a general planimetric distance-directions network read:

$$l_j + v_j = f_j(x_i, y_i, o_i, x_k, y_k) \qquad j=1,m \qquad (9)$$

for each of the m observations between station i and k, where f is a non linear function of

$$\text{type a:}\quad ((x_i - x_k)^2 + (y_i - y_k)^2)^{1/2} \quad \text{for distances}$$

and (10)

$$\text{type b:}\quad \arctan(y_i - y_k)/(x_i - x_k)) - o_i \quad \text{for directions}$$

between two stations (x_i, y_i) and (x_k, y_k).

Because in each original observation equation there are maximal 5 unknowns, the Jacobian will be sparse. The equations are linearised and applied to the algorithm (8). 'o' is the orientation unknown of a station.

Normals and Solution

We assume a diagonal weight matrix P, which can be estimated by means of the variances of the observations, which are normally known. This means that the matrix of the normal equations A'PA has the same structure as A'A. One can also see that A'A has a zero element at places where unknown points are not connected by an observation.

In the past one tried to reduce the number of unknowns of the system to be solved in order to reduce computation time and storage capacity. The reduction was performed by pre-elimination of the orientation unknowns for instance. For small systems or if working with full matrices then this procedure was justified. More about this will be outlined in chapter II.

The normal equations are set up using the second GAUSSIAN transformation (A'PA). In most cases, this procedure was satisfactory because the normals were well conditioned and orthogonalization seemed not to be required.

The solution of the linear system of equations is usually performed by the classical GAUSS or CHOLESKY method. After the backsubstitution of the unknowns into the nonlinear observation equations it is decided whether a further iteration step is necessary (SCHEK,SCHAUER,STEIDLER (1976)).

Having finished the iterations, one optionally computes the inverse matrix of the normals in order to get the basis for analyzing the results. One can then compute the error ellipses and the errors of the adjusted coordinates. The inclusion of 'data snooping' for gross error detection is also possible.

II. PHOTOGRAMMETRIC AEROTRIANGULATION AND BLOCK ADJUSTMENT

1. WHAT IS PHOTOGRAMMETRY ?

The word photogrammetry is derived from three greek words, 'photos' meaning 'light', 'gramma' meaning something 'drawn' or 'written' and 'metron' meaning 'to measure'. of light.

A definition by the American Society of Photogrammetry 1979 reads (SLAMA (1980)): 'Photogrammetry is the art, science and technology of obtaining reliable information about physical objects and environment through processes of recording, measuring, and interpreting photographic images and patterns of electromagnetic radiant energy and other phenomena'.

The best known application of photogrammetry is the compilation of topographic maps and, complete with contour lines, based on measurements and information obtained from aerial and space photography. The compilation is usually performed by means of optical mechanical analogue instruments and/or analytic computations.

Photogrammetry is frequently divided into categories according to the types of photographs or sensing systems used. Hence when the photographs are taken from points on the ground surface , one speaks of 'terrestrial photogrammetry'. On the other hand 'aerial photogrammetry' denotes the use of photographs or other remotely sensed data which have been obtained from an airborne vehicle.

Photogrammetry also may be divided into 'stereo photogrammetry' (SP) and 'analytical photogrammetry' (AP). SP means overlapping pairs of photographs are observed or measured or interpreted in a stereoscopic viewing device, which gives a three dimensional view and creates the illusion that the observer is viewing a relief model of the object.

Fig. 3: Stereoscopic Viewing Device (WILD AMU)

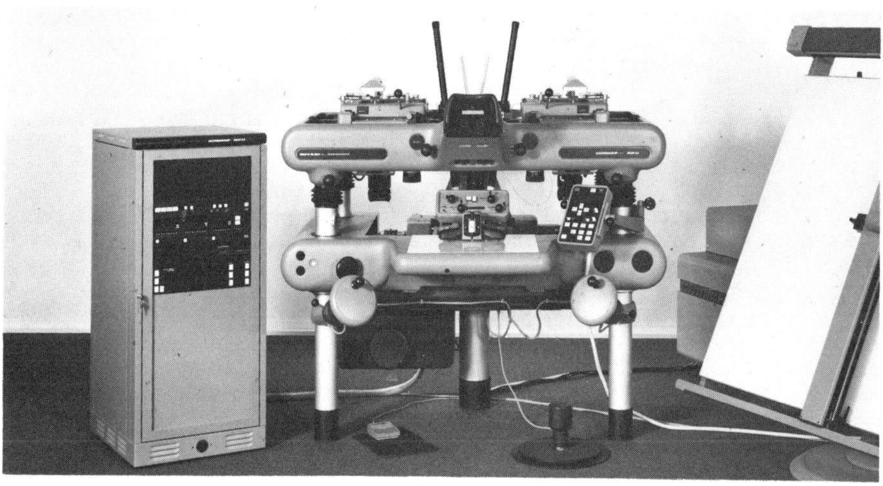

In AP problems are solved by mathematical computation, using either digital data or measurements made on photography or other remotely sensed images as input. The measurements are obtained by using comparators or analytical plotters. An analytical plotter is controlled by a computer and the orientations as well as the compilations can be performed very quickly. It is also possible to carry out other computations (for example aerial triangulation) with an analytical plotter.

Fig. 4: Analytical Plotter (WILD AVIOLYT AC1)

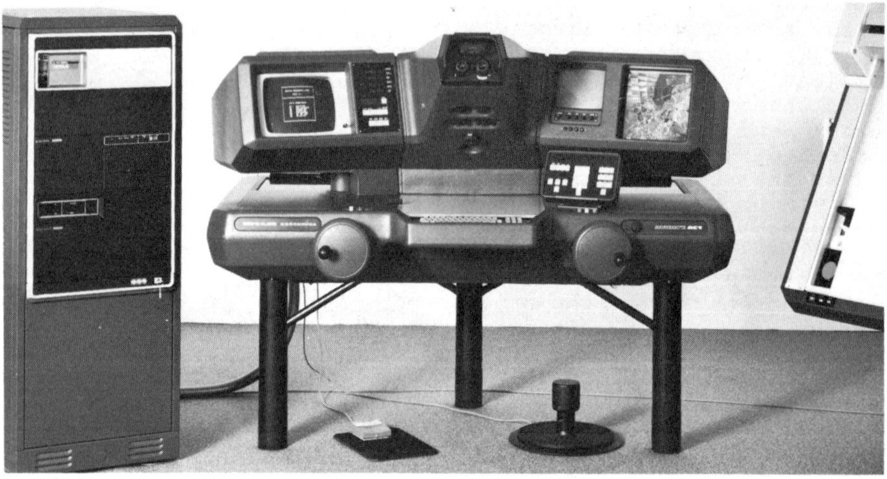

2. PHOTOGRAMMETRIC AEROTRIANGULATION

The definition of photogrammetric aerotriangulation is given by the ASP as follows (SLAMA (1980)): 'The process for the extension of horizontal and/or vertical control whereby the measurements of angles and/or distances on overlapping photographs are related into a spatial solution using the perspective principles of the photographs.'

The Aerotriangulation is divided into 'Mechanical Methods of Phototriangulation' and the 'Analytical Triangulation'. The first covers templet methods and phototriangulation using analog instruments, whereas the second covers numerical methods of phototriangulation. Mechanical methods are used very seldom nowadays.

Basic Principles of Analytical Aerotriangulation

Analytical photogrammetry can be considered as the mathematical transformation between an image point in one rectangular coordinate system (image space) and an object point in another rectangular coordinate system (object space). The difference between various methods of analytical aerotriangulation are found mainly in the procedures and algorithms used in the adjustment. In general the algorithms used depend either on collinearity or coplanarity.

The adjustment procedure, on the other hand, varies with the different users and is most likely chosen to reflect the particular requirements, configuration of camera, computers and measuring equipment used by the organisation.

Analytical aerotriangulation uses mainly one of two different methods:

(1) independent models
(2) bundle block adjustment

Fig. 5: Photogrammetric Aerotriangulation Block

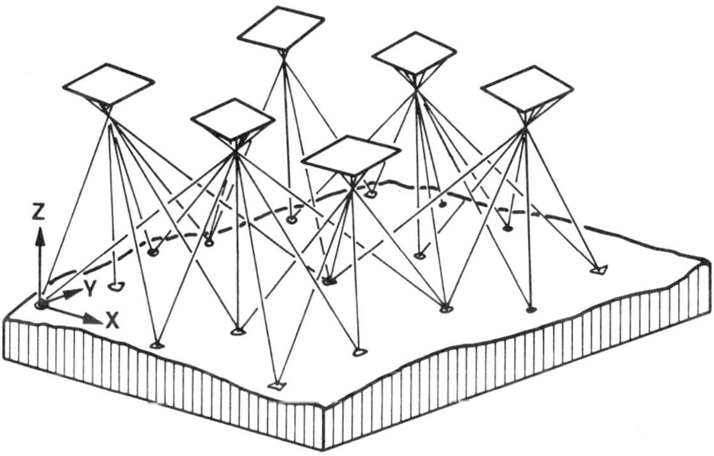

Independent Models (SLAMA (1980))

The basic unit of the independent model method is the relatively oriented photogrammetric model. The position of the model in space and its scale are arbitrary. The object points from each model are given in a three dimensional Cartesian coordinate system of arbitrary origin.

In the block adjustment procedure, all models of the block are simultaneously transformed to the terrain, using all available control points and tie points. Control points are already known in the object space and tie points are points measured in image space, that connect the photos respectively the models together. The distribution of those points are responsible for the condition number of the normals.

The mathematical formulation of the independent models method is a spatial similarity transformation, applied to every model.

If a terrain point is measured in model j (in an arbitrary coordinate system) the relationship between the model coordinates x_{ij}, y_{ij}, z_{ij} and the terrain coordinates x_i, y_i, z_i is given by:

$$\begin{bmatrix} x_i \\ y_i \\ z_i \end{bmatrix} = s_j R_j \begin{bmatrix} x_{ij} \\ y_{ij} \\ z_{ij} \end{bmatrix} + \begin{bmatrix} dx_j \\ dy_j \\ dz_j \end{bmatrix} \tag{11}$$

where s is a scale factor, R is a three-imensional orthogonal matrix and dx , dy and dz are the translations in the 'omega','phi', 'kappa' axes of model j.

For the adjustment the least squares variant of observation equations is used. Hence for each measured point in every model, three observation equations can be formed as:

$$\begin{bmatrix} v_{x_{ij}} \\ v_{y_{ij}} \\ v_{z_{ij}} \end{bmatrix} = \begin{bmatrix} x_i \\ y_i \\ z_i \end{bmatrix} - s_j R_j \begin{bmatrix} x_{ij} \\ y_{ij} \\ z_{ij} \end{bmatrix} + \begin{bmatrix} dx_i \\ dy_i \\ dz_i \end{bmatrix} \qquad (12)$$

where vx, vy and vz are the residuals of the transformed model coordinates. These nonlinear observation equations are linearised and to obtain initial values the models are roughly pre-transformed with respect to the terrain coordinate system. Usually all model coordinates are taken as uncorrelated , but with different weights.

Considering a block of m models and n terrain points, the block adjustment will contain 7 m unknown transformation parameters plus 3 n unknown terrain coordinates. Additional observation equations are written for the controls because they are in general not error free. These equations are weighted in relation to the other observation equations.

The observation equations then yield:

$$v = J x - f \qquad (13)$$

We subdivide J in two submatrices A and B for the unknown terrain coordinates and the unknown transformation parameters ; the vector x is correspondingly divided into two vectors c and p.

. Now the system of observation equations can be written:

$$v = A c + B p - f \qquad (14)$$

f vector of observations
v residuals
c unknown coordinates
p unknown parameters
A,B coefficient matrices

System (14) is solved by least squares, minimizing v'Pv (P = weight matrix). The resulting normals are:

$$\begin{bmatrix} A'PA & A'PB \\ B'PA & B'PB \end{bmatrix} \begin{bmatrix} c \\ p \end{bmatrix} = \begin{bmatrix} A'Pf \\ B'Pf \end{bmatrix} \qquad (15)$$

Normally the unknown terrain coordinates are eliminated and we get the following reduced normal equation system:

$$(B'PB - B'PA(A'PA)^{-1} A'PB)\, p = (B'Pf - B'PA(A'PA)^{-1} A'Pf \quad (16)$$

Usually direct methods are utilised for the solution of the reduced system. The most advanced procedure, used in practice is called recursive partitioning which may be based on GAUSS or CHOLESKY. The banded structure should be considered. Also band-width minimization is desirable.

The iteration process consists of a number of successive individual adjustments. Within each iteration step, reduced normal equations are directly formed and solved. With the aid of the transformation parameters the rigorously transformed model coordinates are then obtained, which are thereafter input to the next iteration step.

The unknown terrain coordinates c are computed by substituting the transformed parameters p back into the normal equation system associated with the last iteration step.

Bundle Block Adjustment

In the bundle block or simultaneous block adjustment the basic unit is the pair of coordinates x and y of an image on a photograph. Using these coordinates (along with some camera constants and the ground position of several of the images), ground coordinates of intermediate points and estimates of the camera's orientation are derived from a simultaneous bundle adjustment.(GRUEN (1976),GRAINSHAW (1980), SCHENK (1972)).

This method differs from the independent model method in that the solution leads directly to the final coordinates and does not treat the 'absolute' and 'relative' orientation seperately.

As a result, the solution and associated error propagation are more rigorous because certain correlations are not ignored. The procedure is also iterative and requires estimates of the initially unknown coordinates of ground points.

Mathematical Model

The general photogrammetric problem can be expressed mathematically as the perspective projection of a 3-dimensional object space onto a two dimensional image space. This is shown in figure 6.

The collinearity condition (Oi, Pij and Pj lying on a straight line) and the perspective projection condition lead to following equations:

$$x = z \frac{(X-Xo)r_{11} + (Y-Yo)r_{12} + (Z-Zo)r_{13}}{(X-Xo)r_{31} + (Y-Yo)r_{32} + (Z-Zo)r_{33}} \quad (17a)$$

$$y = z \frac{(X-Xo)r_{21} + (Y-Yo)r_{22} + (Z-Zo)r_{23}}{(X-Xo)r_{31} + (Y-Yo)r_{32} + (Z-Zo)r_{33}} \qquad (17b)$$

where

x,y	image coordinates
z	principal distance
X,Y,Z	coordinates of a pround point P
Xo,Yo,Zo	coordinates of a perspective center 0
r_{ii}	coefficients of an orthogonal rotation matrix R

For each image point measured on a photograph, one pair of equations of type (17) can be written. The typical photogrammetric triangulation problem is normally made up of a series of overlapping photographs, each having a certain number of measured coordinates from which a consistent solution is sought which best satisfies the basic set of condition equations. Since the number of equations are normally in excess of the number required for a unique solution, the method of least squares is used for the adjustment.

Applying Newton's method to equations (17), the linearised observation equations have a simple form. From these linearised observation equations we can form normal equations.

Fig. 6: Photogrammetric Bundles

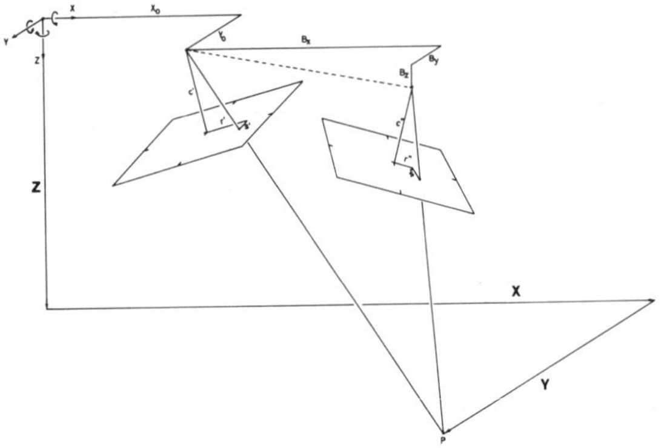

An example of the structure of the symmetric coefficient matrix of the normal equations for 70 object points and 8 pictures (GRAINSHAW (1980)) is given in Fig.7(a).

Let the partitioned normal equations be given by

$$\begin{bmatrix} N_p & N'_{ps} \\ N_{ps} & N_s \end{bmatrix} \begin{bmatrix} dx_p \\ dx_s \end{bmatrix} = \begin{bmatrix} t_p \\ t_s \end{bmatrix} \qquad (18)$$

where the suffix p indicates object point coordinates and the suffix s indicates the camera parameters.

Thus dxp and dxs represent corrections to the object coordinates and camera parameters respectively, and tp and ts are the corresponding right hand sides of the normal equations. If there are np object points and ns camera stations, the entire coefficient matrix illustrated in Fig. 7a is of order ((3np+6ns),(3np+6ns)). However submatrix Np is easily inverted and this allows the formation of the system of reduced normal equations (19).

Fig. 7: Structure of the Normals without Reduction

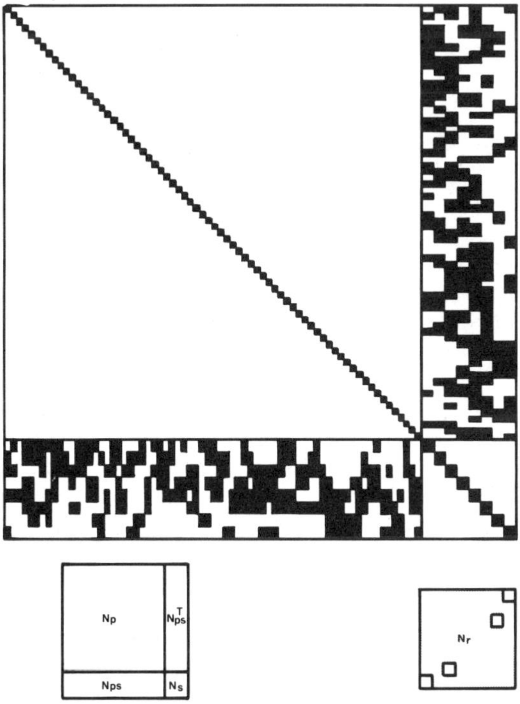

$$N_{red} - dx_s = t_r \qquad (19)$$

where

$$N_{red} = N_s - N_{ps} N_p^{-1} N'_{ps} \qquad (20a)$$

and

$$t_{red} = t_s - N_{ps} N_p^{-1} t_p \qquad (20b)$$

N_{red} is of order (6ns,6ns) (Fig. 7c)

Equation (19) can be solved directly for dx_s. Then dxp can be determined from:

$$dx_p = N_p^{-1} (t_p - N'_{ps} dx_s) \qquad (21)$$

The structure of the reduced matrix of the normal equations is similar to that of the method of independent models. This means we have non zero submatrices for all images which are connected by tie points. Very often at bundle adjustment so-called selfcalibration (EBNER (1976)) is used. Selfcalibration is a method that compensates systematic errors arising from a variety of physical sources. The aim should be to reduce the magnitude of the uncompensated systematic errors to a level, where they are not significant in relation to random errors. Parts of systematic errors are known in advance and they are reduced a priori. For the others, a set of additional parameters can be introduced to model the systematic errors. It is possible to formulate physical parameters, but it is not easy to find all descriptions of the parameters influencing the systematic errors. Hence one selects orthogonal sets of additional parameters which are uncorrelated. This additional set of parameters can be added to the equations (15). The partitioned normal equations given by equation (16) will then be augmented to the form:

$$\begin{bmatrix} N_p & N'_{ps} & N'_{pa} \\ N_{ps} & N_s & N'_{sa} \\ N_{pa} & N_{sa} & N_a \end{bmatrix} \begin{bmatrix} dx_p \\ dx_s \\ dx_a \end{bmatrix} = \begin{bmatrix} t_p \\ t_s \\ t_a \end{bmatrix} \qquad (22)$$

where the suffix 'a' denotes the additional parameters. If there are na additional parameters, the coefficient matrix of the reduced normal equations will be of order ((6ns+na), (6ns+na)). The reduced matrix of normal equations has now a banded bordered structure.

III. GENERATION OF DIGITAL ELEVATION MODELS (DEM) USING FINITE ELEMENTS.

Digital Terrain Model (DTM) is a tool for the describing or modelling features and properties that have a position on the earth's surface. Digital Elevation or Height Model (DEM or DHM) is the part of the DTM which covers the height information. The measurements for a DEM can be taken from geodetic measurements or from photogrammetric compilation.

The main applications of DTMs are automatic contour plotting, digital orthophoto production, calculation on earth-work volumes , perspective views for e.g. road plannings, flight simulations and civil engineering design.

The sampling of the terrain surface is usually achieved in two ways: elevations are measured in regular grids or only the most relevant points are selected to describe the shape of the terrain surface. The mathematical generation of the digital elevation model can result in gridded elevation data or a set of parameters that describe the shape of the surface.

One of the many possible ways to generate a digital elevation model is the use of the FINITE ELEMENT method. In geodetic practice a program package using this method was developed by EBNER,HOFMANN-WELLENHOF,REISS,STEIDLER (1980). This package has two alternative variants for the interpolation : a bilinear and a bicubic one. The bilinear variant is more general because break-lines can be considered. The following explanations are limited to this variant.

<u>DEM- Generation by Finite Elements</u>

In this case a seperate bilinear polynomial is used for each grid mesh, and the interpolation surface formed by finite elements is assumed to be continuous. The mathematical formulation is based on the use of bilinear splines Sij which are defined for all grid points Pij. (For more details see, for instance, SUENKEL (1977) . The interpolation surface is the result of a superposition of local surfaces aij, Sij. This means that each base spline Sij makes a contribution to the interpolated surface. In the case of bilinear finite elements the values aij are identical with the functional values or heights of the interpolation surface at the grid points Pij.

If a point Pk with coordinates (xk, yk) is located inside a mesh, as shown in Fig. 8, the corresponding height hk of the interpolation surface follows as linear Pi+1,j+1 and the respective coefficients are bilinear functions of the local coordinates (dxk,dyk).

The requested grid heights of the DEM are defined as unknowns and estimated from the heights hk of the available reference points and a general curvature minimization of the interpolation surface with filtering at the reference points. This problem can be interpreted as an adjustment of indirect observations and can be solved by using the least squares method.

The height hk of the respective reference point Pk is treated as an observation of the unknown height of the interpolation surface at position (xk, yk). This unknown height however , is a linear function of the four surrounding grid heights as shown above. So reference point Pk leads to an observation equation containing four unknowns; the coefficients of which are linear functions of dxk and dyk. Equation (23) represents this relation in detail, vk is the least squares correction of hk.

$$v_k = (1 - dx_k)(1 - dy_k)h_{i,j} + dx_k(1 - dy_k)h_{i+1,j} + (1 - dx_k)dy_k h_{i,j+1} + dx_k h_{i+1,j+1} - h_k \qquad (23)$$

All equations of type (23) can be weighted individually, according to the accuracy of the observations hk and to the amount of filtering desired. With the bilinear variant curvature minimisation cannot be realised rigorously, because all second derivatives of the interpolation surface are zero. Therefore second derivatives of the heights are used instead. For each grid point such second derivatives are formulated in the x- and y- directions and are interpreted as fictitious observations of amount zero and certain stochastic properties. The corresponding observation equations for grid Pij, shown in Fig. 8, then read (24).

Fig. 8: Section of a DEM

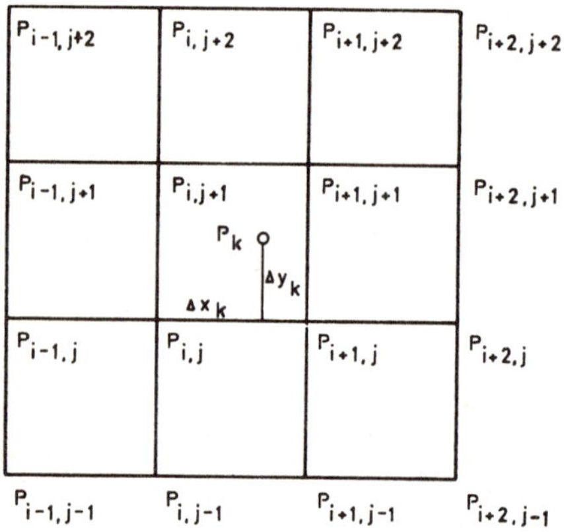

$$\Delta x_k = (x_k - x_i)/d \quad , \quad 0 \le \Delta x_k < 1$$

$$\Delta y_k = (y_k - y_j)/d \quad , \quad 0 \le \Delta y_k < 1$$

$$d = x_{i+1} - x_i = y_{j+1} - y_j$$

$$v_{1_{i,j}} = h_{i-1,j} - 2\,h_{i,j} + h_{i+1,j} - 0 \qquad (24a)$$

$$v_{2_{i,j}} = h_{i,j-1} - 2\,h_{i,j} + h_{i,j+1} - 0 \qquad (24b)$$

The simplest method is to treat all fictitious observations as uncorrelated and of equal accuracy, which is adequate for homogeneous and isotrop terrain.

Breaklines can be considered in (24), but they will not influence the zero/nonzero structure of the normal equations. Because of this their derivation is neglected here.

The equations of type (23) (set up for all reference points) and the total of equations (24) (altered accordingly to the neighborhood of breaklines) form a linear system of observation equations from which, in the case of a diagonal weight matrix, a banded structured system of normal equations is obtained (row-wise numbering assumed). Assuming a rectangular area of n grid lines and m grid points per line, the number of unknowns is m*n and the band-width is proportional to m. A direct solution of the normal equations then leads to a total computing time of const*m**3*n, whilst the corresponding computing time per grid point is proportional to m**2.

REFERENCES AND LITERATURE

Bjoerck, A. (1981): Least Squares Methods in Physics and Engineering, Cern, Geneva.

Ebner, H. (1976): Self Calibrating Block Adjustment. Bildmessung und Luftbildwesen 44(4), 128-139.

Ebner, H., Hofmann-Wellenhof, B., Reiss, P., Steidler, F. (1980): HIFI- A Minicomputer Program Package for Height Interpolation by Finite Elements, 14. Congress of ISP, Hamburg.

Grainshaw, S.I. (1980): Bundle Adjustment Methods in Engineering Photogrammetry, Phot. Record 10(56), 181-207.

Gruen, A. (1976): Die simultane Kompensation systematischer Fehler mit dem Muenchener Buendelprogramm MBOP. Presented Paper. XIII. Congress of ISP, Comm. III, Helsinki.

Ortega, J.M., Rheinboldt, W.C. (1970): Iterative Solution of Nonlinear Equations in Several Variables. Acad. Press New York - London.

Schek, H.J., Schauer, U., Steidler, F. (1976): The Adjustment of Large Geodetic Networks by Sparse Matrix Techniques. Techn. Report TR 76.07.004, Scientific Center, IBM Germany, Heidelberg.

Schenk, A. (1972): Numerische Untersuchungen zur photogrammetrischen Blocktriangulation nach der Buendelmethode. Mitt. aus dem Institut fuer Geodaesie und Photogrammetrie der ETH Zuerich, No. 16.

Slama, C.C. (1980): Manual of Photogrammetry, 4th Edition, American Society of Photogrammetry (ASP), Falls Church, Virginia.

Suenkel, H. (1977): Lokale Interpretation von Schweredaten durch Spline-Funktionen. Alpengravimetrie-Kolloqium, Wien

13 THE USE OF SPARSE MATRIX TECHNIQUES IN GEODETIC NETWORK ADJUSTMENT, PHOTOGRAMMETRIC AEROTRIANGULATION AND GENERATION OF DIGITAL ELEVATION MODELS USING FINITE ELEMENTS

Franz Steidler
WILD HEERBRUGG LTD.
CH 9442 Heerbrugg, Switzerland

This chapter describes in detail the problems that were outlined in chapter I. It is divided into three parts, namely the applied algorithms, the setting up of the sparse matrix of normal equations and the summary of different investigations which have been performed during the last seven years.

I. ALGORITHMS

Principally there are two possibilities for solving a sparse set of linear equations: The use of iterative or direct solution algorithms.

ITERATIVE METHODS:

Among the iterative methods for large sets of linear equations, the successive overrelaxation method (SOR) and the method of conjugate gradients (CG) have been investigated most fully by SCHEK, SCHAUER, STEIDLER (1976), ASHKENAZI, WOOD (1973), LINKWITZ, GRUENDIG (1975). From a programming point of view, one of the simplest iterative schemes is the well known GAUSS-SEIDEL method which can also be modified to an overrelaxation method. The overrelaxation parameter "OMEGA" has been investigated thoroughly by YOUNG (1975). For the application to geodetic nets ASHKENAZI,WOOD, HAMILTON (1972) gave an heuristic procedure for the computation of this parameter. It should be mentioned here that this procedure may be problematic because the overrelaxation parameter may differ from project to project and it can hardly be defined automatically. This means that an experienced operator is required for these programs.

On the other hand the conjugate gradient method is easier to use in practice. Theoretically, it is not a real iterative method because the solution is - as in direct factorisation - obtained after n steps (n is the number of unknowns). However, the method has one feature which makes it most attractive for the solution of linear equations; the solution of the system is reached in most practical cases after a small number of iteration steps. In combination with nonlinear problems this attribute can be very uesfully exploited. The method of CG is also easy to program. For symmetric examples, the normal occurance in least squares adjustment problems, only the upper triangular part of the normal equations is used in a compressed storage mode. This storage mode is a very important condition for effective programming. The CG method is published in GINSBURG (1971) as an ALGOL procedure. It should be noted here that within the CG program, the necessary matrix-vector multiplication may be performed efficiently without the lower triangular part. In this case each element of the upper triangle contributes twice to the solution vector. A similar coding technique can also be applied in SOR as well as in SCALING or PRECONDITIONING. SCALING and PRE- CONDITIONING are methods for improving the condition of a matrix. SCALING can be interpreted as a simple case of PRECONDITIONING. These methods are discussed in detail in BAUER (1968) and in EVANS (1967). When using iterative procedures one of the methods should be used in order to avoid convergence problems. The explanation is omitted here since the methods are published in many fundamental books.

DIRECT METHODS

The second group of algorithms for solving linear equations is DECOMPOSITION of the matrices. Efficient solution of linear equations with a sparsely populated coefficient matrix is fundamental for numerous applications, especially for network problems such as the adjustment of geodetic or photogrammetric problems. The algorithms which exploit the sparsity and are used here have been published resp. developed firstly by GUSTAVSON (1972), REID (1971), WILLOUGHBY (1969). Principally they can be divided into two groups: one where it is tried to reduce computation time by reordering the unknowns before factorisation, the other where the reordering takes place during factorisation. The first group contains the bandwidth (profile) minimisation or the method of NESTED DISSECTION. The second, the SPARSE techniques, includes MINIMUM DEGREE ordering also called MARKOWITZ criterion. The choice of the method depends on the problem involved. A detailed description of the methods may be found in the mentioned literature. It should only be mentioned that for solving the nonlinear problems, the factorisation is subdivided into a symbolical and a numerical part with a following forward-backward substitution. The symbolical part is performed only once, while the numerical is applied after each linearisation.(The non linear system is solved by substituting several linear systems, which are derived by linearisation.)

II. SETTING UP THE SPARSE MATRIX OF NORMAL EQUATIONS

An efficient algorithm for solving the systems of equations only makes sense if the corresponding program is itself efficient. The most time consuming part of a geodetic network adjustment or a photogrammetric aerotriangulation, besides the solution of the equations is the setting up of the normals. Corresponding to the factorisation, the setting up can also be divided into a symbolical and numerical part. The symbolical part is the setting up of the branch-node matrix and is performed only once. The numerical part uses the information from the symbolic one and can be applied efficiently in each iteration step. The symbolic part should not be performed for each unknown, but for unknown groups. In geodetic network adjustment this is a group of X,Y and "omega" (orientation unknown). In photogrammetric aerotriangulation it is a group of transformation parameters (6 or 7). Such unknown groups form a fully populated submatrix of the total system and one can save a lot of computation time because of the smaller bookkeeping effort. The same procedure can of course be used for symbolical factorisation.

For setting up the sparse matrix of the normal equations, a similar procedure can be used for geodesy, photogrammetry and DEMs. For that reason, only the example of geodetic networks will be used to explain the procedure of setting up. It can be recognised when studying the observation equations, that the topology of the network may be used to guide the numerical values. Only the topology is responsible for the sparse matrix of the equations. When using the method of SOR or CG the solution process can also be guided by the topological data.

Normally one avoids setting up the observation equations explicitly. Hence one has to derive directly the portions of the normal equations that arise from one observation.

GENERATION OF THE SPARSE MATRIX OF THE EQUATIONS:

a) symbolic generation

FENVES, BRANIN (1963) proposed a branch-node matrix to describe the topological situation for statical (used in civil engineering to analyse stress within structures) and electrical networks. Because of the similar net structure it can also be used for our problems. The following explanation is taken from GRUENDIG,STEIDLER (1977). The branch node matrix C is defined by

$$C(j,k) = \begin{cases} +1 & \text{if the observation } j \text{ starts from station point } k, \\ -1 & \text{if the observation } j \text{ ends at point } k \\ 0 & \text{in all other cases} \end{cases}$$

Performing the second GAUSSIAN TRANSFORMATION, the symbolic matrix of normal equation reads C'C. For the storage of C'C only two vectors are needed. They must include only the nonzero elements. The first vector holds the information about which point connects to another and the second shows the start of the connections to the next point. The following example will clarify the situation.

Points with a number > 7 are controls and do not participate at the normals.

The nonzero elements of the symbolical part of the normals (C'C) can now be written in two vectors. A column identification vector gives the numbers of the interconnected points while a second one, a row identification vector, points to the beginning of a new row.

index vectors observed from	to	pointer to the offdiag. elem. of sparsevector
3	8	0
3	1	2
3	2	6
3	7	10
3	5	9
6	2	7
6	7	17
6	1	4
6	10	0
6	5	15
4	7	13
4	1	3
4	5	12
4	9	0

	1	2	3	4	5	6	7
1	1	0	1	1	0	1	0
2		1	1	0	0	1	0
3			1	0	1	0	1
4				1	1	0	1
5				1	1	0	
6						1	1
7							1

MATRIX C'C

Index vectors give the information, from which point an other point is observed. The pointer vector is the address in the sparse matrix.

C'C= 1, 3, 4, 6, 2, 3, 6, 3, 5, 7, 4, 5, 7, 5, 6, 6, 7, 7
1, 5, 8, 11, 14, 16, 18,19

Now it is feasible to apply factorisation techniques to the symbolically generated matrix in order to find the fill-ins and the necessary auxiliary vectors of the factorised matrix. Note: the symbolical factorisation is performed only with C'C, where each number represents a fully populated submatrix !

b) <u>numerical generation</u>

The calculation of the coefficients of the sparse normal equations is guided by the index vectors, the sparse representation of C'C and the pointer vector for the off diagonal elements. The pointer vector avoids searching for the corresponding element number in the sparse C'C. The solution proposed requires either a factorisation algorithm, which operates only with the point identifications or the symbolic matrix has to be blown up to a version which works with all the unknowns' identifications.

III. RESULTS

a) GEODETIC NETWORK ADJUSTMENT

For the example computations, six geodetic networks of different order and structure have been used. The networks contain observations of distances and directions. The structure of the networks named NP131 and NP190 are especially interesting. NP131 contains many excentric points, which means that the normal equation matrix has a very irregular structure and a relatively poor condition, since distances of very different length are used. Since NP190 is very regular, the matrix of the normals is nearly banded and the distances are approximately of the same length, which makes the system well conditioned.

As may be seen from the figure 1, the solution processes are exchangeable. The ordering strategy for the direct solutions were NOORD, ROWORD and OPTORD. NOORD means that the normal equations will be processed in the original numbering sequence. ROWORD means that increasing numbering was performed only once before factorisation, according to the number of elements in each row. OPTORD means ordering using "MINIMUM DEGREE (MARKOWITZ) " strategy, where reordering takes place at each factorisation step.

Fig. 1 : Flowchart of the program

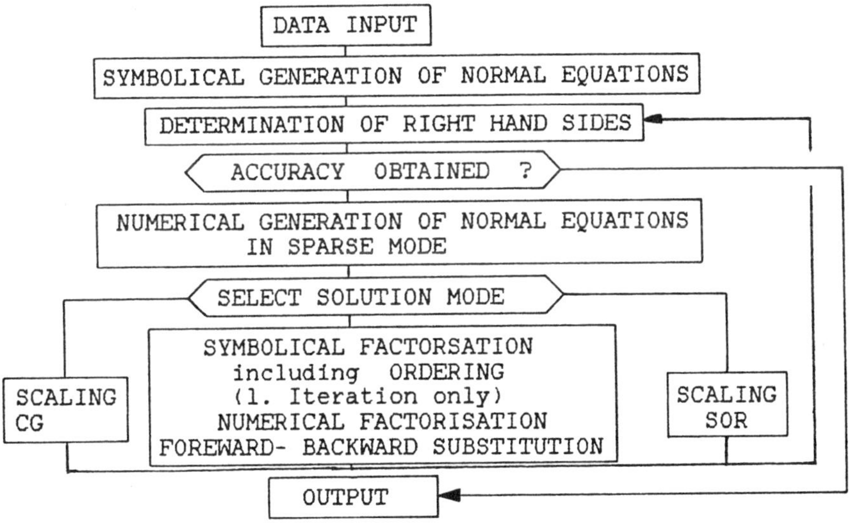

SOR or CG ?

The basic question with iterative methods is the number of iterations. The following table contains the overall number of iterations for the solution of the nonlinear problem. For both SOR and CG the same stopping criterion was used. For SOR an overrelaxation factor of 1.5 was applied. (times in cpu seconds).

NET	SOR			CG		
	Number of outer iterations	Total number of SOR-iter.	Time for 1 iter.	Number of outer iterations	Total number of CG-iter	Time for 1 CG-iter.
NP 17	4	88	0.08	2	55	0.10
NP 131	4	544	0.38	3	307	0.47
NP 190	2	25	0.48	2	30	0.59
NP 253	-	-	0.72	3	849	0.94
NP 400	-	-	1.47	6	823	1.72
NP 703	-	-	2.04	2	430	2.60

Table 1: Comparison of SOR and CG

It becomes evident that SOR fails in the inhomogeneous nets, which are poorly conditioned. On the other hand the homogeneous net NP190 shows a very fast convergence. CG shows a surprisingly stable convergence.

Influence of ordering when using direct solutions

The most critical problem in the application of the direct sparse factorisation is the number of fill-ins caused by the decomposition process. Table 2 probably contains the most interesting fact of the investigations.

One may already see that the simple ROWORD scheme reduces the number of fill-ins drastically and OPTORD again significantly improves performance. (NP17 is an exeption, because the example is too small).

The reuired storage capacity is approximatly proportional to the computation time. The times (cpu seconds) are shown at table 3.

Two important facts should be pointed out:

1. The processing of symbolical (SYFAC) and numerical (NUFAC) factorisation can be completed in a very short time.

2. The time by OPTORD is short enough to allow worthwile savings due to this ordering.

NET	NU	NE	NE L/U NOORD	NE L/U ROWORD	NE L/U OPTORD
NP 17	49	490	915	718	749
NP 131	349	2146	16217	3167	2711
NP 190	412	2678	18399	13714	5405
NP 253	741	3894	52113	9908	5453
NP 400	1075	8477	185172	20645	14176
NP 703	2012	10912	464182	48399	15791

NU = number of unknowns
NE = number of nonzeros before factorisation
NE L/U = number of nonzeros after factorisation

Table 2: Comparison of the ORDERINGS

NET	NU	NOORD		OPTORD		
		SYFAC	NUFAC	OPTORD	SYFAC	NUFAC
NP 17	49	1	1	0,2	1,3	0,7
NP 131	349	53	47	3	4	2
NP 190	412	55	41	7	8	5
NP 253	741	310	269	10	7	4
NP 400	1075	2240	1960	28	31	20
NP 703	2012	9148	8065	60	21	11

Table 3: Compution time of factorisation with NOORD and OPTORD

CG or DIRECT SPARSE factorisation ?

In the next table the comparsion of the total processing times of the complete adjustment between CG and Sparse factorisation is given. Additionally, the time for initialisation (data input, sorting, symbolic generation of normals) is shown separately.

Apart from the well conditioned case, the direct factorisation is considerably faster. The factor between the times of CG and SPARSE ranges from 2 to 14 if the initialisation time is excluded.

The investigations above are published in more detail by SCHEK, STEIDLER, SCHAUER (1978)

b) PHOTOGRAMMETRIC AEROTRIANGULATION

As previously described in chapter I, photogrammetric aerotriangulation leads in most cases to a regular (banded) structure of the matrix of the normal equations. This is similar for the bundle method and the method of independent models. When using self-calibration, the matrix will be banded bordered because the additional unknowns are ordered after the transformation parameters. Normally, an algorithm for banded bordered matrices (including bandwidth minimisation, e.g. POOLE,GIBBS,STOCKMEYER (1976) or CUTHILL- McKEE (1969)) is applied. Due to the investigation of SCHEK,STEIDLER SCHAUER (1977), where a statical network computed with a direct sparse factorisation gave a better result than a band algorithm, it seemed worthwile to apply sparse factorisation also to aerotriangulation. Because of the similarity of the structures of the normal equations of the two methods of adjustment, the following practical examples will only discuss the more rigorous method of bundle adjustment with self-calibration. Because of HALJALA's (1974) good experiences with the CG method CG is also included in the comparison.

The size of the photogrammetric aerotriangulation blocks was chosen in such a way that it was possible to calculate them in central memory, since parts of the investigations had to be computed on a CYBER 175 and TELEFUNKEN TR 440 not equipped with virtual storage.

NET	INIT-TIME	CG-TIME	SPARSE FACT-TIME
NP 17	1.3	9.2	9.2
NP 131	10	170	40
NP 190	14	43	55
NP 253	19	842	77
NP 400	48	1686	309
NP 703	110	1250	300

Table 4: Comparison CG and Sparse factorisation

Having eliminated the unknowns of the ground coordinates, the additional parameters for self-calibration were selected to be block invariant, which means that they were taken as constant for all images of the block. This causes the border to be fully populated. The number of additional parameters was 12, due to a proposal of EBNER (1976), who derived them from an orthogonal polynomial to keep them uncorrelated. The weights were chosen as zero, which means that they are taken as free unknowns. The bundle program used was developed by GRUEN (1976).

IBM 370 / 145 unknowns 828 non-zeros before(ne) factorization=40554	(1) band- algorithm	(2) techniques for sparse matrices ROWORD	(3) OPTORD
non-zeros after factorization (nn)	93943	154285	97219
$\frac{nn_i}{nn_{(1)}}$ i=(1),(2),(3)	1	1.64	1.03
number of multi-plications	5.74 Mill	19.00 Mill	7.92 Mill
$\frac{mult_i}{mult_{(1)}}$	1	3.3	1.4
time (CPU-sec.)	330	1107	470
$\frac{time_i}{time_{(1)}}$ i=(1),(2),(3)	1	3.3	1.4
storage capacity (words)	95 k	300 k	190 k

Table 5: Band algorithm and sparse factorisation with example 1

IBM 370 / 145 unknowns 570 non-zeros before(ne) factorization=38356	(1) band- algorithm	(2) techniques for sparse matrices ROWORD	(3) OPTORD
non-zeros after factorization (nn)	58100	74500	71100
$\frac{nn_i}{nn_{(1)}}$ i=(1),(2),(3)	1	1.28	1.22
number of multi-plications	3.47 Mill	6.49 Mill	6.37 Mill
$\frac{mult_i}{mult_{(1)}}$	1	1.87	1.83
time (CPU-sec.)	198	380	379
$\frac{time_i}{time_{(1)}}$ i=(1),(2),(3)	1	1.92	1.91
storage capacity (words)	60 k	150 k	145 k

Table 6: Band algorithm and sparse factorisation with example 2

Table 5 and 6 show the result of the comparisons for two examples with 828 and 570 unknowns (all symbolical calculations were performed with (6x6) submatrices to reduce the index calculations). The investigations shown below have been performed on an IBM 370/ 145.

In both cases the band algorithm needed less storage capacity and computation time. Considering only the ordering schemes, it becomes evident that the OPTORD ordering gives better results than ROWORD. The following two figures show the structure of the normal equations after factorisation with band algorithm and sparse facorisation for both examples. Each point represents a 6x6 submatrix.

Fig. 2: Example 1, structure of the normals with band and sparse factorisation algorithm.

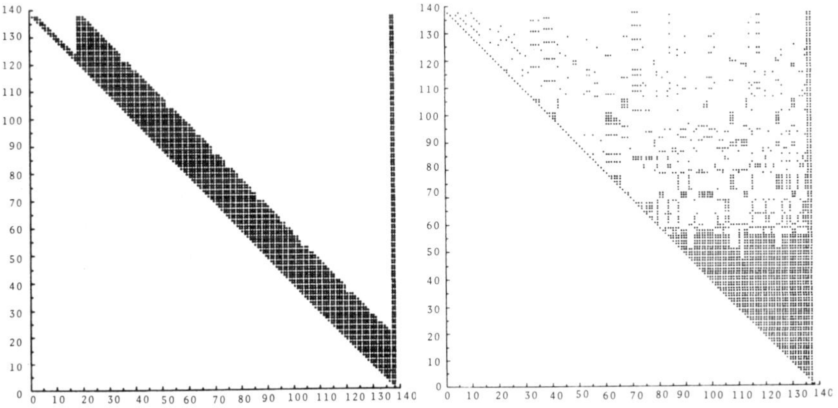

Fig. 3: Example 2, structure of the normals with sparse factorisation algorithm.

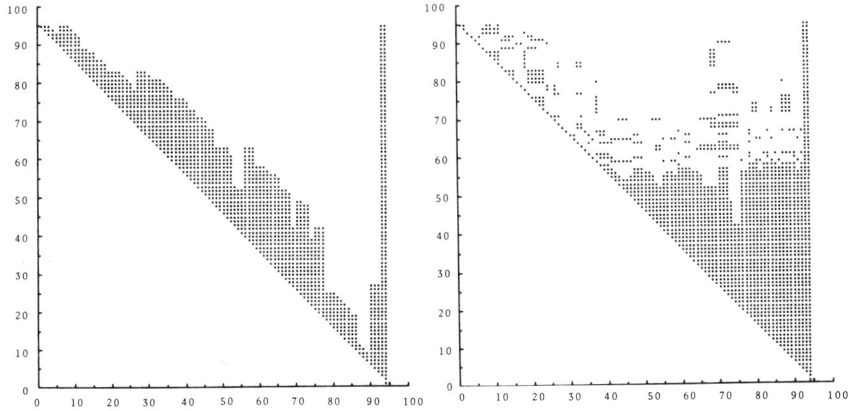

The next comparison is carried out between band algorithm and the CG method. Because of technical reasons it was performed on a TR 440 and CYBER 175. It must be mentioned here that the condition of example 2 was very poor because the distribution of control was weak. Pre-conditioning (EVANS (1967)) and SCALING (BAUER (1963)) was applied. Tables 7 and 8 show the results.

For example 1 the solution with CG was reached after 4 outer (non linear iterations) in a total of 157 CG steps. It needed 1025 cpu seconds compared to 750 cpu seconds for the bandalgorithm. Example 2 was so ill conditioned that the CG result could not be reached in a reasonable time while the result with the bandalgorithm was gained in only 82 cpu seconds.

From these investigations it can be concluded that:

- Direct techniques with sparse factorisation is not well suited for aerotriangulation. The banded structure will be destroyed and not even OPTORD ordering leads to less fill-ins compared to the conventional band algorithm. The number of multiplications for sparse factorisation is also much higher.
- CG method is inferior to band algorithm, especially for ill conditioned matrices. It should only be used if the control distribution is quite good.

The result of the comparison between band algorithm and sparse factorisation is due to the regular structure of aerotriangulation problems, where normally no geodetic observations are included. In terrestrial photogrammetry, where geodetic observations have also to be considered, sparse factorisation can be assumed to become much superior, since the structure becomes much more complicated and is similar to the structure of geodetic networks. This is a field for further investigations. In future, terrestrial photogrammetry will become more and more important because of the need of photogrammetry in industrial surveying. The main reason for this is the recent development of new analytical plotters, having many features required for industrial use. More details on the above investigation may be found in STEIDLER (1980) and (1981).

TR 440 NU = 828	band-algorithm	method of conjugate gradients (CG)
outer iterations	2	4
CG iterations	-	14, 45, 52, 46
sum of iterations	-	157
time (cpu-sec.)	750	1025

Table 7: Example 1 with band algorithm and CG

CYBER 175 NU = 570	band-algorithm	method of conjugate gradients (CG)
outer iterations	2	20
CG iterations	-	580, 580, ..., 580
sum of iterations	-	11600
time (cpu-sec.)	82	1800

Table 8: Example 2 with band algorithm and CG

c) <u>DEM generation</u>

For the generation of a DEM using a variant of FINITE ELEMENTS the band algorithm, minimum degree ordering and the method of nested dissection was used. The latter two methods were investigated by DUFF,ERISMAN,REID (1976) for the case of five and nine point finite element discretisation of a regular n by n grid. In comparison to this investigation the points in the examples discussed here are also connected with the points beyond the neighbouring points. This means it is a thirteen point finite element discretisation and leads to a larger bandwidth and to more nonzeros per row.

The pivot strategies used here are GEORGE's (1973) method of NESTED DISSECTION and the previously discussed method of MINIMUM DEGREE. The NESTED DISSECTION variant described in this contribution is one of several possibilities. It is the "complete" variant. The other variants ("incomplete" and "one way") have been investigated too in STARK,STEIDLER (1983), but "complete" showed the best result.

The numerical investigations were performed with 11 examples varying in size from 36 to 4140 unknown grid points.The data sets used are taken from square and rectangular sections of a DEM of the VERNAGT glacier, where points of intersections between contour lines and grid lines had been digitised.

The calculations were performed for data sets 1 to 5 and 8 up to 10 on the CDC CYBER 175 and for the data sets 6,7 and 11 on an AMDAHL computer. The change to the AMDAHL was necessary because storage capacity of the CYBER 175 was limited to 130 K words.

The tables (9,10,11) present the results of the computations. The comparison of the data sets shows the number of nonzeros before (neb) and after factorisation (nea), the number of multiplications and computing times. The factor neb/nea is also computed to see the increase of the number of nonzeros.

Figure 6 shows the results of the tables (9,10,11) for the square examples graphically.

It can be seen that the band algorithm needs the least computation times for all examples, but requires the largest number of multiplications and has the largest number of nonzeros after factorisation. For the factorisation processes involved in "Minimum Degree Ordering", the routine "MA17A" from the HARWELL SUBROUTINE LIBRARY, REID (1972), was used. HARWELL has a faster routine available, called "MA17E", using the same pivot strategy, but it needs more than twice the storage requirements. For this reason it was not possible to use it for these investigations. Regarding the method of "Nested Dissection", the results are always lying between "Band Algorithm" and "Minimum Degree".

	n	m	u	neb	nea	nea/neb	mult	time SYFAC (9)	NUFAC (10)	Total (9)+(10)
1	6	6	36	190	337	1.77	2 237	0.06	0.01	0.07 C
2	10	10	100	576	1 632	2.83	17 664	0.37	0.10	0.47 C
3	14	14	176	1 172	4 323	3.69	63 218	1.14	0.33	1.47 C
4	22	22	484	2 958	15 639	5.29	339 055	5.46	1.75	7.21 C
5	30	30	900	5 564	36 219	6.51	1 024 733	15.41	5.50	20.91 C
6	46	46	2 116	13 090	116 197	8.80	4 719 373	35.82	11.62	47.44 A
7	62	62	3 488	23 566	246 604	10.46	12 776 330	82.51	29.70	112.21 A
8	21	41	861	5 291	32 999	6.24	878 915	12.13	4.29	16.42 C
9	11	81	891	4 821	26 947	5.59	548 231	8.17	2.63	10.80 C
10	16	60	960	5 508	38 584	7.01	1 104 762	16.64	5.89	22.53 C
11	46	90	4 140	24 068	262 617	10.91	13 360 887	85.61	30.22	115.83 A

Table 9: Complete Nested Dissection

	n	m	u	neb	nea	nea/neb	mult	time SYFAC (9)	NUFAC (10)	Total (9)+(10)
1	6	6	36	190	316	1.66	2 032	0.09	0.02	0.11 C
2	10	10	100	576	1 434	2.49	14 432	0.94	0.11	1.05 C
3	14	14	196	1 172	3 670	3.13	47 207	3.92	0.34	4.26 C
4	22	22	484	2 958	13 463	4.55	268 710	35.06	1.79	36.85 C
5	30	30	900	5 564	32 795	5.89	924 926	166.01	6.20	172.21 C
6	46	46	2 116	13 090	100 827	7.70	4 061 732	478.05	9.47	487.52 A
7	62	62	3 844	23 566	203 373	8.63	9 594 526	1 446.54	22.55	1 469.09 A
8	21	41	861	5 291	26 963	5.10	605 325	89.63	4.17	93.80 C
9	11	81	891	4 821	19 162	3.97	274 970	28.63	1.91	30.54 C
10	16	60	960	5 508	27 018	4.91	531 497	73.60	3.71	77.31 C
11	46	90	4 140	24 068	226 993	9.43	11 414 905	1 831.75	26.16	1 857.91 A

Table 10 : Minimum Degree Ordering

Figures 4,5 and 7 show, for the 30 by 30 grid example, the structure of the normal equations matrices of the three method as well as the numbering scheme for "Nested Dissection".

Because the segmentation of the symbolic factorisation task is complex, this task is usually performed in core. Thus the application of sparse matrix algorithms like NESTED DISSECTION and MINIMUM DEGREE ORDERING is usually limited by core storage capacity.

The symbolical factorisation is relatively expensive because of the bookkeeping overhead since for DEM problems the system of normal equations is linear and the numerical factorisation has to be performed only once. Therefore band algorithms are preferred, although other strategies requiring less numerical operations for squared bordered DEMs are frequently used in practice.

In the case of non linear finite element problems with the same or a similar structure to our examples and where the numerical calculations have to be repeated because of the choice of initial values, the NESTED DISSECTION and MINIMUM DEGREE ORDERING methods are suited very well. Here the symbolic calculation of the numeric GAUSS-NEWTON iterations can be kept very small.

One possible solution to the problem of a limited storage capacity would be a combination of INCOMPLETE NESTED DISSECTION and MINIMUM DEGREE ORDERING. The dissection would be made during the early part of the process and later on the remaining blocks would be calculated using MINIMUM DEGREE ORDERING. The size of the independent blocks would be chosen with respect to the available storage capacity of the computer. In this case the desired size can often be reached after only one iteraion step.

	n	m	u	b	neb	nea	$\frac{nea}{neb}$	mult	time
1	6	6	36	13	190	390	2.0	2 866	0.02 C
2	10	10	100	21	576	1 890	3.3	21 710	0.14 C
3	14	14	196	29	1 172	5 278	4.5	82 026	0.51 C
4	22	22	484	45	2 985	20 790	7.0	490 402	2.89 C
5	30	30	900	61	5 564	53 070	9.5	1 677 530	10.07 C
6	46	46	2 116	93	13 090	192 510	14.7	9 169 226	22.01 A
7	62	62	3 844	125	23 566	472 750	20.1	30 085 562	70.10 A
8	21	41	861	43	5 291	36 120	6.8	822 416	4.84 C
9	11	81	891	23	4 821	20 240	4.2	260 326	1.51 C
10	16	60	960	33	5 508	31 152	5.7	555 824	3.39 C
11	46	90	4 140	93	24 068	380 742	15.8	18 200 314	41.32 A

Times in CP–seconds. C means CYBER 175, A means AMDAHL.

nea = number of elements after factorization

mult = number of multiplication

Table 11: Band Algorithm

Fig. 4: 30 by 30 grid, banded matrix
(before factorisation)

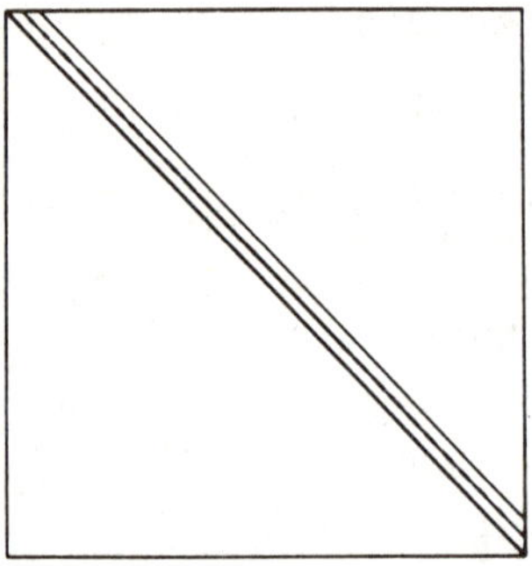

Matrix of normal equations before factorization

number of rows	30
number of columns	30
number of unknowns	900
nonzeros before factorization	5564
nonzeros after factorization	53070
number of multiplications	1677530
number of divisions	53070

Fig. 5: 30 by 30 grid, Complete Nested Dissection

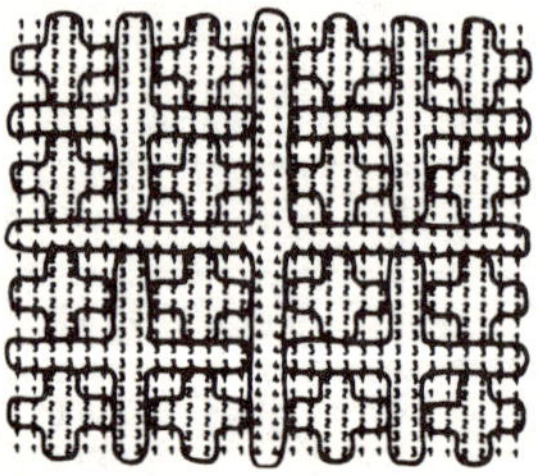

Numbering scheme of DHM

number of rows	30
number of columns	30
number of unknowns	900
nonzeros before factorization	5564
nonzeros after factorization	37217
number of multiplications	1076898
number of divisions	37217

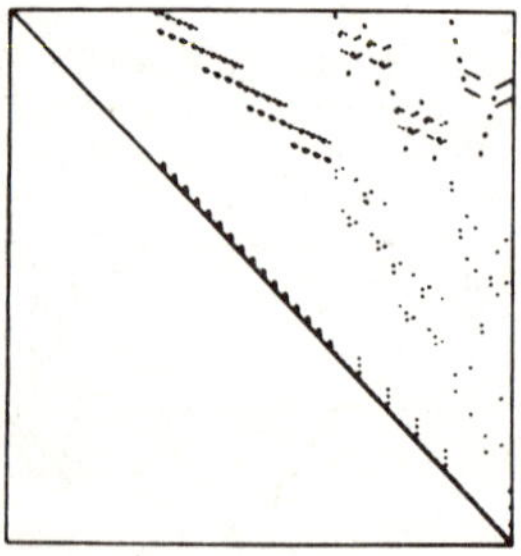

Matrix of normal equations before factorization

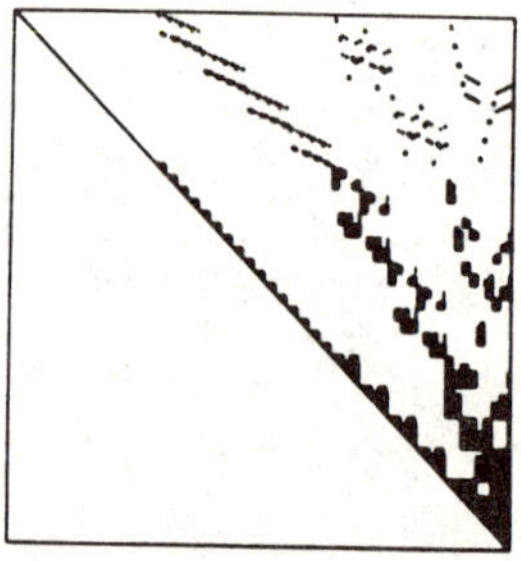

Matrix of normal equations after factorization

Fig. 6: Comparison of times, multiplications and fill-ins

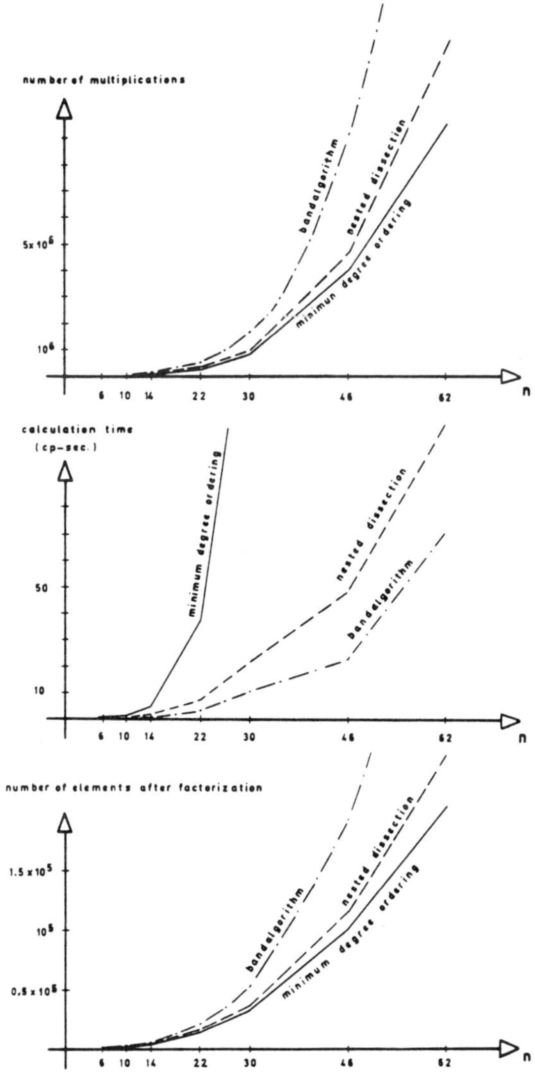

Fig. 7: 30 by 30 grid, Minimum Degree Ordering
(after factorisation)

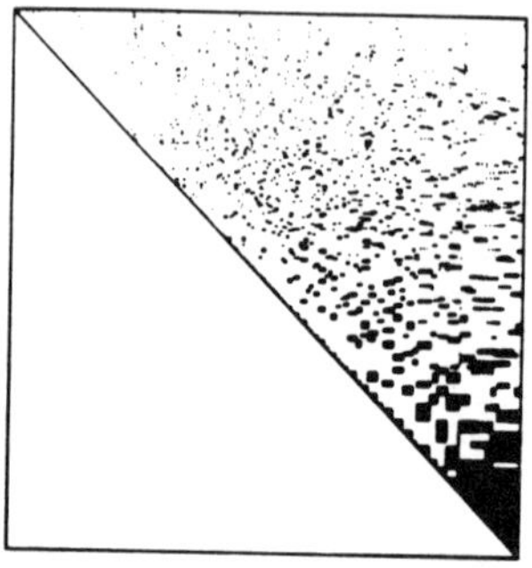

Matrix of normal equations after factorization

number of rows	30
number of columns	30
number of unknowns	900
nonzeros before factorization	5564
nonzeros after factorization	32795
number of multiplications	924926
number of divisions	32795

d) <u>Computation of Certain Elements of the Inverse</u>

A main problem in geodesy and photogrammetry is the calculation of the inverse or parts of the inverse for estimations of the achieved accuracy and the reliabilty of the defined points and the used observations. For large systems of equations the calculation of the inverse may be a problem because of the required computation time and storage capacity. In most practical cases it is sufficient to compute only those elements of the inverse, which are needed for the accuracy and reliability computations. These are mainly the elements which are non zero before factorisation and occasionally the elements which are non zero after the factorisation. The other elements are required very seldom. Hence one can use the method of ERISMAN and TINNEY (1975). A comparison between this method and conventional algorithms by UNSLEBER (1980) showed a reduction of computation time and storage requirements of a factor more than 100.

<u>CONCLUSIONS</u>

The comparisons shown above created a variety of results for the different applications. But one result is common for all; direct solution is superior and more general than iterative methods. For geodetic networks direct sparse factorisation is a quicker solution than conventional methods, but for photogrammetric aerotriangulation adjustments band algorithms are advantageous. Finite element application to digital elevation models shows that "NESTED DISSECTION" and "MINIMUM DEGREE ORDERING" are well suited, but the band algorithm is faster even although the number of multiplications and the number of elements is larger. This is due to the expensive bookkeeping overhead. Generally it can be said, that for each application a suitable method for solving the normal equations must be selected, where "MINIMUM DEGREE ORDERING" fits best for irregularly distributed coefficients while "NESTED DISSECTION" and "BAND ALGORITHM" are better suited for regular structured matrices.

REFERENCES

Ashkenazi, V., Wood, R. (1973): Merits of the Conjugate Gradient Method for Solving Geodetic Normal Equations, IAG Symposium on Computational Methods, Oxford, UK.

Ashkenazi, V., Wood, R., Hamilton, M. (1972): On the difficulties arising from the SOR iteration operating on large geodetic matrices. Bolletino de Geodetisica Teorica ed Applicata 14, No.55.

Bauer, F.L. (1968): Optimally Scaled Matrices, Num. Mathematik 5.

Cuthill, E., McKee J. (1969): Reducing the Bandwidth of Sparse Symmetric Matrices and their Applications. Proc. ACM National Conference, Assoc. for Comp. Machinery, New York.

Duff, I., Erisman, A.M., Reid, (1976): On GEORGE's Nested Dissection Method, SIAM Numer. Anal..

Erisman, A.M., Tinney, W.F. (1975): On Computing Certain Elements of the Inverse of a Sparse Matrix, in "Communications of the ACM", edited by R.A. Willoughby, vol. 18.

Ebner, H. (1976): Self Calibrating Block Adjustment, Internat. Archives of Photogrammetry, vol. 21, ISP, Comm. III, Helsinki.

Evans, D.J. (1967): The Use of Preconditioning in Iterative Methods for Solving Linear Equations with Symmetric Positive Definite Matrices. The Journal of the Institute of Mathematics and its Applications, vol.4.

Fenves, S., Branin, F. (1963): Network Topology Formulation of Structural Analysis. Proc. of the ASCE J. of the Struct. Div..

Gibbs, N., Poole jr., W., Stockmeyer, P. (1976): An Algorithm for Reducing the Bandwidth and Profile of a Sparse Matrix, SIAM J. Numer. Anal. vol.13.

Ginsburg, T. (1971): The conjugate gradient method, Handbook for Automatic Computation, (edited by F.L. Bauer), Springer Verlag, Berlin, Heidelberg, New York.

Gruen, A. (1976): Die simultane Kompensation systematischer Fehler mit dem Muenchener Buendelprogramm MBOP, ISP Cogress, Comm. III, Helsinki.

Gustavson F.G. (1972): Some basic techniques for solving sparse systems of equations, in "Sparse Matrices and their Applications", Plenum Press New York, London.

Haljala, S. (1974): Method of Conjugate Gradients for the Solution of Normal Equations in the Analytical Block Adjustment. The Photogrammetric Journal of Finland, vol. 6, No. 2, Helsinki.

Linkwitz K., Gruendig L. (1975): New Perceptions of and experiences with rigorous nonlinear least squares adjustment of extensive geodetic networks, IUGG Congress Grenoble.

Reid, J. K. (1971): Large sets of linear equations, Academic Press London New York.

Schek, H.J., Schauer, U., Steidler F. (1976): The Adjustment of Large Geodetic Networks by Sparse Matrix Techniques, IBM Report TR 76.07.004, Heidelberg Scientific Center.

Schek, H.J., Steidler, F., Schauer, U. (1978): Ausgleichung grosser geodaetischer Netze mit Verfahren fuer schwach besetzte Matrizen, Deutsche Geodaetische Kommission, Reihe A, No.87.

Steidler, F. (1981): On Solution Algorithms for Large Sparse Systems of Normal Equations in Photogrammetry, The Canadian Surveyor, vol. 35, No. 1.

Steidler, F. (1980): Darstellung und Vergleiche von Loesungsstrategien fuer grosse schwach besetzte Normalgleichungssysteme in der Geodaesie und in der Photogrammetrie. Deutsche Geodaetische Kommission, Reihe 10, No. 261, Muenchen.

Stark, W., Steidler, F. (1983): Sparse Matrix Algorithms Applied to DEM Generation, Bulletin Geodesique 57, Paris.

Willoughby, R. (1969): Sparse Matrix Proceedings, IBM Thomas Watson research Center.

Unsleber, K. (1980): Berechnung der Inversen einer schwach besetzten Matrix und deren Anwendung in der geodaetischen Netzausgleichung. Diplom thesis, Chair for Photogrammetry, Technical University of Munich.